AF167637

Health and Safety in Contemporary Britain

Paul Almond · Mike Esbester

Health and Safety in Contemporary Britain

Society, Legitimacy, and Change since 1960

Paul Almond
School of Law
University of Reading
Reading, UK

Mike Esbester
School of Area Studies, History,
Politics and Literature
University of Portsmouth
Portsmouth, UK

ISBN 978-3-030-03969-1   ISBN 978-3-030-03970-7   (eBook)
https://doi.org/10.1007/978-3-030-03970-7

Library of Congress Control Number: 2018962761

© The Editor(s) (if applicable) and The Author(s) 2019
This work is subject to copyright. All rights are solely and exclusively licensed by the Publisher, whether the whole or part of the material is concerned, specifically the rights of translation, reprinting, reuse of illustrations, recitation, broadcasting, reproduction on microfilms or in any other physical way, and transmission or information storage and retrieval, electronic adaptation, computer software, or by similar or dissimilar methodology now known or hereafter developed.
The use of general descriptive names, registered names, trademarks, service marks, etc. in this publication does not imply, even in the absence of a specific statement, that such names are exempt from the relevant protective laws and regulations and therefore free for general use. The publisher, the authors and the editors are safe to assume that the advice and information in this book are believed to be true and accurate at the date of publication. Neither the publisher nor the authors or the editors give a warranty, express or implied, with respect to the material contained herein or for any errors or omissions that may have been made. The publisher remains neutral with regard to jurisdictional claims in published maps and institutional affiliations.

Cover illustration: © iStock/Getty Images Plus

This Palgrave Macmillan imprint is published by the registered company Springer Nature Switzerland AG
The registered company address is: Gewerbestrasse 11, 6330 Cham, Switzerland

For Natalie
and
Nicki, Rosie and Thomas

Acknowledgements

We are grateful to the Institution of Occupational Safety and Health (IOSH) for funding and supporting the original research upon which this book is based. The project from which it draws, 'The Changing Legitimacy of Health and Safety at Work, 1960–2015', was a part of IOSH's 'Health and Safety in a Changing World' research programme. The five-year programme was intended to explore the landscape of occupational safety and health and its implications for developing solutions that provide effective protection for workers and their communities. This book draws on the research and materials produced as part of that programme; we are very grateful to IOSH for their agreement for us to use this material for publication purposes. In particular, we are grateful for the assistance of Robert Dingwall (Nottingham Trent University/ IOSH Research Programme Director) and Jane White (IOSH Head of Research and Information Services).

The excellent work undertaken as part of that IOSH-funded project by our research assistants, Carmen D'Cruz and Laura Mayne, was extremely important to the success of the research that underpins this book, and we gratefully recognise their multiple contributions. Thank you both. We would also like to thank the members of the project steering group for their input and support, which was valuable in shaping the direction of the research: Graham Frobisher (Marshall Aerospace and Defence Group), Arthur McIvor (University of Strathclyde), Sarah Page (Prospect), and Neal Stone (British Safety Council). In addition, at IOSH, Anne Wells provided access to archival materials, and Mary Ogungbeje provided practical assistance.

We would like to thank the focus group participants and, in particular, the 40 oral history interviewees (detailed in the Appendix), who willingly gave up their time and provided invaluable perspectives on some of the key issues of the past 60 years. We were assisted by a range of people in securing and conducting these interviews, including Jude England (British Library), Joscelyne Shaw (British Safety Council), Shahzeb Malik (IIRSM), Ryan Arthur (University of Reading), and John Wilkins (ex-UKAEA). Dave Hallam (University of Reading) provided his customarily excellent technological support to the interview transcription and analysis process. We would also like to thank Stavroula Leka and Aditya Jain, of the University of Nottingham, for their work on the quantitative data sets on public attitudes towards health and safety, which we draw on in Chapter 2.

We are grateful to all of the archives which allowed access to their holdings. However, there were a few people and organisations who went beyond what might be expected and deserve special mention. The British Safety Council, via Neal Stone, Matthew Holder, and Chris Warburton, arranged access to the BSC's archival material, at that point not publicly available. Similarly, the Royal Society for the Prevention of Accidents, via Jo Bullock, Helen Shaw and Anita Plumb, accommodated us and made RoSPA's archive available. David Walker (University of Strathclyde) provided important access to oral history testimony he had been involved in recording, notably dock and chemical workers. Those interviews, and many others of great value, are held that the Scottish Oral History Centre, archived in the University of Strathclyde Archives and Special Collections (http://atom.lib.strath.ac.uk/sohc-archive).

Guy Baxter and the team at the Museum of English Rural Life, University of Reading, helped with regards to agricultural health and safety. Lesley Whitworth (University of Brighton) and the Brighton Design Archive were extremely helpful in locating material relating to design and safety. John Stedman and the team at the Portsmouth History Centre and Bob Russell and the members of the Portsmouth Royal Dockyard Historical Trust provided great help with a variety of material related to working in and around Portsmouth. Finally, the staff at the TUC Library not only located material that we had requested, but directed us to additional material that they thought might be of interest—this was the case, and we are grateful for their initiative.

We acknowledge the support and assistance of our colleagues and respective institutions, the Universities of Reading and Portsmouth, in particular a period of sabbatical leave which Mike used to work on this book. Our editorial team at Palgrave, Molly Beck, Oliver Dyer, and Carmel Kennedy, worked with us very constructively and have been accommodating when needed: this has been appreciated.

Finally, the forbearance of our families has been given freely, and for that, we are grateful. Mike would particularly like to thank Nicki, Rosie, and Thomas Esbester for their support, patience, and humour, especially in the later stages of the writing process. Paul would particularly like to thank Natalie Almond, whose love, encouragement, and support have made all the difference, and is vastly appreciated, as always.

September 2018   Paul Almond
                                                                         Mike Esbester

CONTENTS

Abbreviations

ACoP	Approved Code of Practice
BMA	British Medical Association
BR	British Rail
BRB	British Rail Board
BSC	British Safety Council
CBI	Confederation of British Industries
COSHH	Control of Substances Hazardous to Health
DETR	Department of Environment, Transport, and the Regions
DWP	Department of Work and Pensions
EC	European Community
EEF	Current name of the former Engineering Employers' Federation
EMAS	Employment Medical Advisory Service
EU	European Union
EU-OSHA	The European Agency for Safety and Health at Work
FFI	Fee For Intervention
GCU	Glasgow Caledonian University
GMB	Current name of the former General, Municipal, Boilermakers and Allied Trade Union
HELA	HSE/Local Authority liaison committee
HSC	Health and Safety Commission
HSE	Health and Safety Executive
HSWA	Health and Safety at Work etc. Act 1974
ICI	Imperial Chemical Industries
IoD	Institute of Directors
IOSH	Institution of Occupational Health and Safety
LA	Local Authority

MERL	Museum of English Rural Life, University of Reading
MOD	Ministry of Defence
MP	Member of Parliament
MRC	Modern Records Centre, University of Warwick
NDPB	Non-Departmental Public Body
NHS	National Health Service
NPM	New Public Management
NUT	National Union of Teachers
ONR	Office for Nuclear Regulation
ONS	Office for National Statistics
ORR	Office of Rail and Road
QMV	Qualified Majority Voting
QUANGO	Quasi-Autonomous Non-Governmental Organisation
RMT	National Union of Rail, Maritime and Transport Workers
RoSPA	Royal Society for the Prevention of Accidents
SFAIRP	'So Far As Is Reasonably Practicable'
SME	Small-to-Medium sized Enterprise
SOHCA	Scottish Oral History Centre Archive
TGWU	Transport and General Workers Union
TNA	The National Archives of the UK
TUC	Trades Union Congress
VDU	Visual Display Unit

List of Cases

United Kingdom

Austin Rover v H.M. Inspector of Factories [1989] 3 WLR 520
Marshall v Gotham [1954] AC 360
R v Chargot Limited (t/a Contract Services) and Others [2008] UKHL 72

European Court of Justice

Commission of the European Communities v United Kingdom, C-127/05, 14/6/2007

LIST OF STATUTORY MATERIALS

United Kingdom

Agriculture (Safety, Health and Welfare Provisions) Act 1956
Asbestos Regulations 1969
Control of Substances Hazardous to Health Regulations 1988
Corporate Manslaughter and Corporate Homicide Act 2007
Employers' Liability (Compulsory Insurance) Act 1969
Enterprise and Regulatory Reform Act 2013
European Communities Act 1972
Factories Act 1937
Factories Act 1948
Factories Act 1961
Factory Act 1833
Factory and Workshop Act 1901
Health and Morals of Apprentices Act 1802
Health and Safety at Work etc. Act 1974
Health and Safety (Display Screen Equipment) Regulations 1992
Legislative Reform (Health and Safety Executive) Order 2008
Management of Health and Safety at Work Regulations 1992
Manual Handling Operations Regulations 1992
Mines and Quarries Act 1954
Offices, Shops, and Railway Premises Act 1963
Personal Protective Equipment at Work Regulations 1992
Provision and Use of Work Equipment Regulations 1998

LIST OF TABLES

CHAPTER 1

Introduction and Organising Ideas

INTRODUCTION

Health and safety has always been a field of policy and practice that has engaged with, and been embedded within, broader social and political contexts. But it has been noticeable that, in the years since the 2010 United Kingdom general election, health and safety events and publications have increasingly focused their attention onto questions such as 'what has gone wrong with health and safety?', or 'does health and safety have an image problem?' The public and social legitimacy of 'health and safety', as an area of practice, has become an issue of concern not just for those who work in the area, but also for others who have a stake in the issue. Everyone in Britain is a stakeholder in health and safety, in myriad—though often invisible—ways, from a working environment which doesn't kill, injure or make people ill, to public spaces which it is possible to use without coming to harm. However, these absences of ill effect go largely unnoticed. Instead, attention is focused either on moments where the normality of safe and healthy life is disrupted, or on the alleged 'nuisances' of a system claimed to be over-protective.

As a result, one of the most visible stakeholders in health and safety, the Government itself, has found it imperative to be seen to be taking action. In particular, over the past decade it has commissioned two reviews of the health and safety system which have addressed the issue of public perception in this area. The Young Review (2010) argued that:

© The Author(s) 2019
P. Almond and M. Esbester, *Health and Safety in Contemporary Britain*, https://doi.org/10.1007/978-3-030-03970-7_1

since the Health and Safety at Work etc. Act 1974 was passed we have built up an enviable record: today we have the lowest number of non-fatal accidents and the second lowest number of fatal accidents at work in Europe [...y]et at the same time the standing of health and safety in the eyes of the public has never been lower. (2010: 25)

So, even falling rates of recorded accident and injury (HSE 2012, 2013) do not translate into improved attitudes towards the system. The subsequent Löfstedt Review (2011) argued that hostile media coverage may be negatively influencing duty-holders' compliance behaviour, and recommended that steps be taken to improve public engagement with, and understanding of, risk regulation (2011: 39, 41, 92).

Both reviews reflected a more widespread recognition that a popular climate of antipathy towards health and safety regulation had taken hold in Britain. This was expressed via the Government's 'Red Tape Challenge' consultation exercise of 2011–2013,[1] which crystallised public expressions of dissatisfaction within a putative policy-making process, and via the Health and Safety Executive's (HSE, the UK's national health and safety regulator) 'Sensible Risk Management' policy (HSE 2006) and 'Myth Busters Challenge Panel',[2] both of which sought to engage with negative attitudes towards health and safety. Perhaps three main strands to this poor profile can be discerned. First, politicians have become increasingly vocal in attacking health and safety regulation during this time, with Ministers expressing their *"determin[ation] to stamp out the health and safety killjoys"*.[3] This has arguably brought health and safety more prominently into the public sphere, and also tied it to broader arguments about the desirability of state intervention in the economy and society more generally. Second, and related, has been the emergence of a vocal anti-regulatory business lobby which has called for the removal of health and safety 'burdens', for instance via a Government 'Business Taskforce', which categorised the EU's Health and Safety at Work Framework Directive as a *"barrier to starting a company and*

---

[1] The Red Tape Challenge Online Forum and the Responses to It, Are Now Archived at: http://webarchive.nationalarchives.gov.uk/20150522175321/http://www.redtapechallenge.cabinetoffice.gov.uk/home/index/.

[2] http://www.hse.gov.uk/myth/myth-busting/index.htm.

[3] Chris Grayling, Minister for Work and Pensions: "'Elf and Safety Myths' Targeted'. *Guardian*, December 18, 2011 (http://www.guardian.co.uk/uk/feedarticle/10000844).

employing people" (Business Taskforce 2013: 22). And third, media coverage of health and safety has become increasingly hostile, with press outlets seen as "*compet[ing] to write about absurdity after absurdity, all in the name of 'elf and safety*" (Young 2010: 25), resulting in a "*constant stream of stories in the press blaming health and safety [...] for preventing individuals from engaging in socially beneficial activity, overriding common sense and eroding personal responsibility*" (Löfstedt 2011: 16). This media coverage has been both widespread and influential, and has tended to focus on explicitly challenging the legitimacy of health and safety as an idea by questioning both the moral justifiability, and procedural fairness, of regulation in this area (Almond 2009: 371–374).

As a result, it has appeared as if the public profile of health and safety has become poorer now than at any time in the recent past. All of the elements identified here have contributed to a shift in popular sentiment and a dynamic of populist dissent and distrust towards health and safety that in some ways foreshadowed and mirrored the dynamics of the 2016 Brexit referendum (Smismans 2017). Like Brexit and its antecedents, this negative contemporary climate raises a series of important questions, which this book seeks to address. First, there is a need to consider whether a fundamental shift has taken place in the state of affairs surrounding health and safety in the United Kingdom, an area of activity whose origins are intimately bound up with the history of industrial relations and the modern welfare state. Health and safety was born of the moralistic, protectionist, and paternalistic reform movements of the nineteenth century, which sought to address the social costs of the industrial revolution (Bartrip and Fenn 1983; Carson 1979; Long 2011; MacDonagh 1958; Mills 2010; Thomas 1948; Ward 1962); it was also tied into the "*wave of humane feeling and high aspiration for the future*" (Mess 1926: 33) that followed World War I, and which gave rise to the nascent welfare state. Later developments have reflected the increasing importance attached to welfare-oriented social citizenship rights in the second half of the twentieth century, and the drive towards equality, widened participation, and extended state provision that this brought (Almond 2013: Ch. 4; Tucker 1995). According to this narrative, improving health and safety standards and provision has become more established and embedded as a desirable goal for modern societies to pursue via regulation and control. This broad consensus was seen to have culminated in the Robens Report of 1972, which proposed reforms to modernise the regulatory system so as to reflect the "*natural identity of interest*" (Robens 1972: 21) between parties in the area of health and safety, and in the subsequent Health and Safety at Work Act 1974, which embedded a consensual, corporatist regulatory framework into law.

As this book will show, while this framework has endured, relatively intact, to the present day, it is less clear that the social consensus around the issue, the political settlement that enacted it, and the society that it was designed for, remain the same. So while the critical policy and media climate of the last few years may appear to be a distinctive and self-contained phenomenon that constitutes a significant disjuncture with the past, there are also grounds for seeing it as reflective of a longer term shift, as a climate of neoliberal politics has become entrenched since the late 1970s (Harvey 2005; Tombs and Whyte 2010; Tucker 1995). This has arguably led to an ideological rejection of regulation and welfarism, and a reordering of regulatory systems to be less state-centred, and more focused on 'New Public Management' values of efficiency, accountability, and reduced capacity (Hood 1991). Changes in the material organisation of society—the types of work and industry undertaken, the membership and influence of representative bodies like trade unions and professional safety organisations, and the diversification of the workforce and the risks they face—have also meant that the core continuities in the area of health and safety (in terms of actors, laws, and political processes) have had to evolve and change in response to broader contextual pressures in the economic, social, and political spheres. But we should also be cautious about assuming that the degree of opposition to the idea of health and safety is necessarily greater now than in the past. This has always been a contested area of policy, and the realities of these debates and assertions must also be tested and assessed. What change has really occurred in relation to how health and safety is thought about, understood, and evaluated, and in what direction has any change moved? What are the key drivers of change; do they lie at the institutional and political levels, or in the social and cultural context against which systems of health and safety regulation operate?

This book aims to provide a focused 'modern history of health and safety in Britain', a set of historically grounded observations and assessments of contemporary trends in the profile of health and safety that speaks directly to contemporary political and policy debates about this area of law, policy, and practice. It argues that change here has been more nuanced than the simple account of decline set out above might suggest. In fact, there are core areas of continuity and stability here that suggest an enduring role for health and safety into the twenty-first century, and an underlying consensus that is harder to change than is the political and media climate. Many longer term trends and themes, such as a reliance on voluntarism, and a scepticism about the value of prescriptive rule-making, are actually recurring themes that have endured from earlier than the time span of this book. And many of the features of health and safety that are identified as most negative in terms of perceived legitimacy are actually closely interwoven with the most positive features of the same system; so a focus on flexibility of rules also gives rise to a sense of uncertainty, while the valorising of expertise also creates risk of being perceived as technocratic and aloof. One key tension of this sort is between the decentring of responsibility for health and safety from the core (the state) to the periphery of duty-holders, professionals, and individuals; this change has brought more participants into the process, but at the cost of seeming to give away some sense of control or priority. Perhaps a key unifying outcome of this investigation is a recognition that better standards of debate, discussion, and deliberation around health and safety, at all levels and among all participant groups and stakeholders, might hold the key to improving acceptance and recognition of these tensions. This lack of effective communication accounts for much of the surface-level antipathy, and some of the legitimacy challenges, that health and safety has to face. The remainder of this chapter will define and explain the scope and focus of this investigation, and the methods that have been used to conduct it.

Defining Legitimacy

This book is primarily interested in exploring the social legitimacy of health and safety as it has been manifested since 1960. Legitimacy is a multifaceted concept relating to the claims that are made about the foundation of authority which underpins the actions of a social institution. To say that an entity is 'legitimate' is to assert that there is a "*generalized perception or assumption that [its] actions…are desirable, proper, or appropriate within some socially constructed system of norms, values, beliefs, and definitions*" (Suchman 1995: 574). Legitimacy is thus a subjective quality, a "*belief in the legality of enacted rules and the right of those elevated to authority under such rules to issue commands*" held by those subject to them (Weber 1914: 215). The pursuit of legitimacy is important in regulatory circles because of the links that exist between legitimation and self-motivation; when individuals perceive regulators, rules, or norms to be legitimate, they internalise and obey them, hence this is thought to be a powerful determinant of compliance (Black 2008; Murphy et al. 2009; Tyler 1990, 2011; Wenzel and Jobling 2006). A legitimate regulator is also able to claim a mandate for action, which can translate into greater autonomy, resources, and profile (Almond 2007; Gibson and Caldeira 1995; Prosser 2010: 92–93). The legitimacy judgements of broader social audiences are arguably even more influential than those of stakeholders and policy-makers, however, as they set the wider parameters of what is possible, permissible, and acceptable in that field of activity (Almond 2015; Gunningham et al. 2004). This book does not seek to engage with the impact of legitimacy upon compliance behaviour. Instead, it asks how the legitimacy of health and safety, as a social institution, is constructed in the eyes of politicians, the media, and the public.

These questions of empirical belief, taken in isolation, do not provide a basis for assessing legitimacy, however, because they provide no justification as to why those beliefs are held, meaning that legitimacy could be manufactured via propaganda or coercion (which would not constitute a 'true' form of legitimacy: Beetham 1991: 11). Additional criteria must be found upon which assessments of legitimacy might be based, of which Suchman identifies three main types: cognitive, pragmatic, and normative (1995: 577). Cognitive legitimacy relates to the degree to which a regulatory regime is seen as inevitable or necessary. When a regulatory institution accords with the taken-for-granted expectations of society, it is accepted as legitimate regardless of whether the assessment of its work is positive, negative, or absent (Suchman 1995: 582). It is perhaps quite rare for a regulatory body to be 'taken for granted' in this way, due to the inherently political context in which they operate (Hawkins 2002), and the potential for events like disasters to upend this 'taken for grantedness' (Kica and Bowman 2012: 22). Pragmatic legitimacy refers to the self-interested calculation of the benefits (economic, political, or otherwise) that a social institution or decision made can bring for the audience. For example, a customer may perceive a utilities regulator as legitimate when it succeeds in lowering the price of that utility (Prosser 1999: 205–208; 2010: Ch. 9), or a business may see regulation as more legitimate the more it accords with their own economic self-interest (Gunningham et al. 2004; Thornton et al. 2005).

It is the third form, normative (or moral) legitimacy, which best captures issues that relate to health and safety as a social institution. This relates to the action taken by a public authority, the goals it pursues, and the values it embodies (Black 2008: 144; Suchman 1995: 579). Conceptually, it has tended to be divided into two 'families' of factors: procedural (or legal) validity and moral (or normative) justifiability (Beetham 1991: 15–25). In broad terms, this can be expressed as the distinction between recognising the legality of a decision, and the rightness of that decision. So, procedural legitimacy is generally attributed to a social institution according to how far it has acted in accordance with the laws that govern it (*'input'* legitimacy: Scharpf 1999), and whether it has observed due process in its actions and discharged its role appropriately (*'throughput'* legitimacy). If so, it can claim that it merits *"support according to recognized bench-marks"*, and so is legitimate (Baldwin 1995: 57; also Baldwin and Cave 1999: Ch. 6; Prosser 2010), irrespective of any substantive goals or outcomes (Gibson and Caldeira 1995: 460). These elements are reflected in what Black terms 'constitutional' and 'functional' legitimacy claims (2008: 146). Constitutional claims relate to the relationship between the social institution and its sources of authority: does it conform to the laws that govern it? Are there mechanisms to hold it answerable for decisions made (such as judicial review: Prosser 2010: 7–8)? Does it follow standards of fairness and evenhandedness in decision-making (Baldwin 1995: 43–44; Murphy et al. 2009)? Functional legitimacy claims relate to the ways in which the social institution achieves its outcomes. Is it efficient and effective, and does it avoid wasting money (Black 2008: 146; Hood 1991: 11)? Does it possess a competence that justifies its decisions (Baldwin 1995: 45–46; Prosser 2010: 8)? These values build public trust and official confidence in bodies like regulators (Löfstedt 2005), providing the basis on which the 'freedom to manage' is given to bodies that sit outside of direct relationships with the State (Baldwin 1995: 45; Hood 1991: 6).

The second grouping of moral legitimacy factors reflect notions of normative rightness, whether actions are justifiable by reference to a wider conception of the 'right' and the 'good' (Beetham 1991: 70). This involves content-dependent consideration of both 'input' and 'output' legitimacy factors (Scharpf 1999): whether the goals pursued and outcomes produced conform to prevailing conceptions of justice and fairness (Suchman 1995: 579; Black 2008: 145). Social institutions are evaluated according to whether they are *"congruent with a particular model of democratic governance"* (Black 2008: 146), and the prevailing values of wider society (Meyer and Rowan 1991: 50). Democratic legitimacy claims concern the processes by which decisions are made; who is represented within the mechanisms that direct the actions of social institutions? Do they reflect processes of open, rational debate in which multiple interests and a broad range of actors and constituencies can participate (Habermas 1992; Black 2000; Scharpf 1999: 10)? Finally, justice-oriented considerations of legitimacy focus on the moral character of the social institution and the justice of the outcomes produced (Black 2008: 146; Palazzo and Scherer 2006: 80). Are values pursued which reflect the interests of society as a whole? Does it act in a manner consistent with the moral values of those subject to it (Murphy et al. 2009: 3)? All of these headings of legitimacy are important, and each will help to organise the coverage of the rest of this book. Questions of cognitive and pragmatic legitimacy, as specific issues of *public belief*, are engaged with via the investigation of public attitudes in Chapter 2; while the procedural headings of constitutional (questions of legal mandate) and functional (questions of action and outcome) legitimacy are discussed in Chapters 4 and 6 respectively. The moral headings of democratic (questions of representation and coverage) and justice-based (questions of moral rightness) legitimacy are discussed in Chapters 5 and 7, respectively.

The Value of Historical Investigation

What benefit might studying the development of ideas about the legitimacy of health and safety over the past sixty years bring? One thing *not* claimed is that by studying the past, we can predict the future; the contexts are so radically different as to render this meaningless. However, studying regulatory history is a valuable means of better understanding the present, and learning for the future (Balleisen and Brake 2014; Berridge 2010; Cox 2013; Divall et al. 2015; History and Policy 2018; see also the *Journal of Policy History*). By allowing us to understand how and why decisions have been made, on the basis of what information was available at the time, it is possible to appreciate the shape and structure of the present—and potentially to use the insight gained into the *processes* involved to interrogate the basis under which decisions are made for the future. This is particularly pertinent given the apparent changes in perceptions of health and safety since 1960—and the persistence of health and safety as an area both of public interest and of significant state intervention.

Whilst the overall shape of health and safety and its regulation in Britain today is very different from that of 1960, the high-level issues remain: that health and safety is an area of concern, with challenges posed by the nature of people's occupations and workplaces. In addition, therefore, to understanding the impact of past decisions and policies on the present, critical exploration of how health and safety has been perceived may suggest new directions for meeting the challenges posed by changing occupational structures, new workplace risks, and a degree of media and political hostility which appears new. This is also a period which has received relatively little attention from historians, who have tended to focus on the nineteenth or early twentieth centuries. It is important to understand this earlier health and safety regime—which continued in place, broadly, until the 1974 Health and Safety at Work Act introduced the current regulatory regime. By exploring the period since 1960, we capture ideas about change and continuity, for a period still recent enough to be in living memory for many, and certainly to be shaping contemporary understandings and perceptions of health and safety—but also one which requires analysis in order to expose those lasting impacts and influences and to enable us to think critically about the present.

DEFINING 'HEALTH AND SAFETY' AND THE 'STATE'

The scope of this book covers the legitimacy of 'health and safety', a focus for analysis that spans both the conceptual and practical spheres, and which binds together a number of different levels of activity. Substantively, we take health and safety to connote the laws, rules, principles, perceptions, and practices that relate to the management or control of risks of injury or disease in working, public, or other organisational settings. This need not be purely occupational in nature, nor tied to any particular institutional or regulatory framework; but it does incorporate some sense of organisational activity that impinges directly upon the health, well-being, and physical integrity of individuals, whether they be workers or not. The common start-point for much of the literature looking at 'health and safety' is the formal system of legal and regulatory institutions that underpin and give shape to this concept. This includes the statutes, regulations, and case-law that set out the legal requirements that those concerned with health and safety must meet (Baldwin 1987, 1995; Gunningham and Johnstone 1999; Norrie 2001; Selznick 1985;

Sirrs 2016), but also the policies and approaches taken by regulators, and the enforcement activity that such bodies undertake (Almond 2006; Bartrip 1982; Bartrip and Fenn 1983; Dawson et al. 1988; Hawkins 2002; Hutter 1989, 1997). Much of this focuses on *occupational* matters and the role of the state, as classically understood, as for much of the period covered by the book these have been the most visible sites at which health and safety has been manifested. Accordingly, much of the evidence underlying this book considers occupational issues. However, this is certainly not exclusive. Other studies focus on the activities of embedded non-state actors, such as employers, industry bodies, and trade unions (Bronstein 2008; Genn 1993; Gunningham and Rees 1997; Long 2011; O'Rourke 2003; Walters 1996) and they remain significant in this book, along with the more public contexts in which health and safety was exposed to view.

One key feature of our study will thus be an analysis and investigation of the role of these different institutions and, perhaps most centrally, that of the 'state', broadly defined. For some time historians have been moving away from conceiving of the state as just central or local government, and have focused more on the varied agents and institutions involved in governing (Kidd 1999; Powell 2007). This includes roles for private and voluntary organisations, as well as trade unions, all of which interacted with state agencies (as classically understood) to set standards, shape expectations and contest decisions and ideas, on health and safety as much as every other aspect of interaction with the lives of citizens, consumers and employees. The contingency of these interactions is something which will be seen in relation to health and safety in particular.

But the concept of 'health and safety' has a meaning and a reach that goes beyond the state and these other institutions, encompassing a wider range of social impacts and audiences that these institutional accounts may miss. Over time, issues like health and safety have become increasingly decentralised and conducted outside of these institutional settings, as governments seek to broaden the scope of engagement and involvement, and shift away from a reliance on state-led regulatory infrastructure (Bartle and Vass 2007; Black 2007, 2008; Moran 2003), and as awareness of those issues permeates into a wider range of social settings and individuals' lives. Areas of policy and practice like health and safety are thus more complex, diffuse, and socially situated than ever before, meaning that the legitimation of the framework of values and norms that they represent is a process that can occur distinct from the

legitimation of any specific institution, relating to the broader acceptability of that organising idea (Berger et al. 1998: 380; Johnson et al.
2006: 56). We are interested in health and safety as a 'social object' or
institution, not just an organisational rule system, as we wish to understand the legitimacy of the *idea* of health and safety, not just any of the
specific things that are done in its name (though they will be relevant to
understanding that whole).

A full assessment of the legitimacy of health and safety must therefore
take into account the nature of the broader societal awareness of, and
reactions to, health and safety issues (Almond 2009); the attitudes and
opinions that are held and expressed by the general public; the political processes that shape priority-setting around the issue (Hutter and
Manning 1990); and the day-to-day experiences, decisions, and actions
of those at the front-line of health and safety, who implement the law
or are subject to it (Almond and Gray 2017; Huising and Silbey 2011).
This book seeks to understand how 'health and safety' is understood on
a *societal* level, and so will look to draw together evidence from all of
these levels in order to better understand the ways in which this concept is perceived, and the factors that shape the legitimacy it is seen to
possess. As official Government reviews have observed, much current
opposition to this area of regulation and practice is based on "*perception
rather than reality*" (Young 2010: 19), thus making these interpretations
relevant subjects of study, just as much as the formal institutional components of the regulatory system, or the 'high' political strategy of central
government.

Research Methodology

The research that underpins this book utilises a range of different methodological approaches, as befits the broad scope of this inquiry and
the multi-layered nature of health and safety as a social institution. A
mixed-methods approach of this sort allows for the relative strengths
of different research methodologies to be incorporated and to aug-
ment or verify one another, providing a greater degree of validation to
the findings generated. First, evidence was gathered as to public attitudes towards health and safety via two distinct means, which were
then integrated together. A broad literature review was undertaken
of the available empirical studies (academic and policy-oriented) into
public attitudes in this area; in addition, the findings of a wide-ranging

systematic review and analysis (Jain and Leka 2016) of existing data sets gathered by British and EU governmental sources and agencies (including the British Social Attitudes Surveys, 2001 and 2005, and the ONS Opinions and Lifestyle Survey 2004–2010 at the British level; and the Eurobarometer 'Employment and Social Policies' Survey of 2003 and 'Working Conditions in the EU' Survey of 2014, at the EU level), were also integrated into this overarching review.

Alongside this, a series of eight focus groups were conducted with 67 members of the public to discuss attitudes towards the idea of health and safety, and experiences of how it had impacted on their lives. As Kitzinger (1994) notes, focus group discussions are particularly adept at drawing out the ways in which shared meanings are negotiated and contested between participants in a discussion (particularly 'lay' knowledge and 'naturalistic' narratives: Green 1997: 160), and so are able to capture more dynamic processes of attitude-construction, as well as more personal and qualitative reasoning behind stated preferences and attitudes, than quantitative surveys are able to do. As such, drawing on findings from both allows us to construct a clearer perspective on attitudes and attitude-formation. This is particularly important in an area like health and safety, where policy discourse is increasingly shaped and driven by an understanding of surface-level public 'opinions' rather than deeper underlying 'attitudes' (Green 2006). In practical terms, these focus groups were conducted in different geographical locations across the country, with participants who were sampled to ensure broad demographic representativeness. The focus groups were semi-structured, following an agreed template or outline structure (to ensure consistent coverage), but providing flexibility to allow for the capture of spontaneous and conversational content (Kidd and Parshall 2000: 294); this structure moved from general to specific in scope and positive to negative in tenor (Krueger and Casey 2009), so that the outcomes were not unduly led. The focus groups were audio-recorded, transcribed, and analysed via the development of analytical codes (conceptual clusters of quotes and content that are generated from the bottom-up).

The second main strand of the methodology used was a series of 40 semi-structured oral history interviews, conducted during 2014–2015, with key actors and stakeholders from the recent history of health and safety, recruited purposively because of their specific contribution, reputation, and role within the field. Rather than seeking representativeness in any comprehensive sense, this process was intended to gather a range of perspectives across time, sectors, and areas of activity, each of which would provide a rich and detailed account of each interviewee's own experiences. The oral history interviews were particularly important in relation to the more recent past, where archival records, in particular, were less likely to be in the public domain. In total, 72 people were approached to take part in interviews, with 40 agreeing (56%); the range of interviewees was balanced to ensure that multiple constituencies (regulators, policy-makers and politicians, trade union actors, safety professions and professional bodies, and employers and business actors) were represented. Interviews were conducted "on-the-record," audio recorded, and transcribed to render them as oral history documents. The transcripts were then subject to analysis via an iterative process of descriptive coding. Each interview adopted a semi-structured approach, so that interviewees were given significant scope to self-direct their contributions within the context of a framework of issues for discussion. A standard template interview schedule was used as the basis of all the interviews; this was then personalised according to the experience and activity of the interviewee in question. Although this method of data collection was liable to gather highly personal and subjective accounts from interviewees, it is also the case that these subjectivities are inherently important information for this investigation to engage with; the perceptions and understandings of individuals and the organisations they represented play a key role in the legitimation processes and debates that we are interested in surveying.

In order to contextualise this data, as well as gain a more rounded view of some of the issues and events discussed by interviewees, a third main methodological approach was also utilised. A wide range of historical, archival, and written documentary sources were also surveyed and analysed. Drawing from seventeen archives or collections of material, many hundreds of sources were examined, limited only by time, survival of documents, or access restrictions—particularly affecting items under 30 years old and relating to government agencies. This was supplemented with sources available in digital databases, such as historical newspapers. Accordingly, this has produced a broad coverage of actors, including state bodies, trade unions, employers' organisations, workers, the media, and non-governmental organisations such as the British Safety Council and Royal Society for the Prevention of Accidents. Documents were initially sampled by availability to gain a broad overview, followed by more concentrated investigation of areas that emerged as being of particular importance. By their nature, the archival sources often record a more formal, and sometimes official, view of events and leave absences which may be significant; and, as with all sources, it was necessary not only to approach them critically but also to situate them in terms of their representativeness. Of necessity, only a small number of sources could be included for direct quote/reference in this book, but (unless noted as being atypical but still important to include) they have been selected to try to offer a representative impression of the parties involved at the time.

PLAN OF THE BOOK

Underlying the book, then, is a broad base of research, encompassing a variety of source types and methodologies. By combining sources and methods in this way, it is possible to explore as full a range as possible of perceptions surrounding health and safety and its legitimacy in contemporary Britain, capturing the significant moments and actors, as befits such a broad, diverse, and complex area of study. From this, it is possible to examine why health and safety appears so contested in the present, and assess how and why this has come about over the past 60 years. Chapter 2 of this book will begins by setting out and evaluating the nature of public attitudes towards health and safety—what do people really think about this issue at the moment, and in the recent past? Chapter 3 will then provide a concise and focused historical narrative of the emergence, development, and evolution of ideas of 'health and safety' within Britain. Chapter 4 will focus on analysing four strands of constitutional legitimacy challenge to health and safety, centred on the legal frameworks, and procedural and institutional settings, which govern this issue in practice. Chapter 5 will assess three fundamental contextual changes (the decline of manufacturing industries, and of trade union membership, and the growing importance of health risks) that pose democratic legitimacy challenges to health and safety. Chapter 6 will then explore four functional issues, relating to the way that health and safety operates, the outcomes it produces, and who does it, which also draw its legitimacy into question. Chapter 7 goes on to focus on three areas of justice-based legitimacy challenge, relating to the values and interests that health and safety pursues and protects. Chapter 8 provides a summary and conclusion.

REFERENCES

PRIMARY SOURCES

Business Taskforce. (2013). *Cut EU Red Tape: Report from the Business Taskforce.* London: HM Government. https://www.gov.uk/government/uploads/system/uploads/attachment_data/file/249969/TaskForce-report-15-October.pdf.

Cabinet Office. (2011). Red-Tape Challenge. http://webarchive.nationalarchives.gov.uk/20150522175321/http://www.redtapechallenge.cabinetoffice.gov.uk/home/index/.

HSE. (2006, April 13). *Sensible Risk Management* (HSE Board Paper HSE/06/46).

HSE. (2012a). *Health and Safety Executive Annual Statistics Report.* London: HSE. http://www.hse.gov.uk/statistics/overall/hssh1112.pdf).

HSE. (2012b). *Myth Busters Challenge Panel.* http://www.hse.gov.uk/myth/myth-busting/index.htm.

HSE. (2013). *Historical Picture: Trends in Work-Related Injuries and Ill Health Since the Introduction of the Health and Safety at Work Act (HSWA) 1974.* London: HSE. http://www.hse.gov.uk/statistics/history/historical-picture.pdf).

Grayling, C. (2011, December 18). 'Elf and Safety Myths' Targeted. *Guardian.* http://www.guardian.co.uk/uk/feedarticle/10000844.

Löfstedt, R. (2011). *Reclaiming Health and Safety for All: An Independent Review of Health and Safety Legislation* [The Löfstedt Review]. London: Crown.

Young, L. (2010). *Common Sense, Common Safety* [The Young Review]. London: Crown.

SECONDARY SOURCES

Almond, P. (2006). An Inspector's-Eye View: The Prospective Enforcement of Work-Related Fatality Cases. *British Journal of Criminology, 46*(5), 893–916.

Almond, P. (2007). Regulation Crisis: Evaluating the Potential Legitimizing Effects of 'Corporate Manslaughter' Cases. *Law and Policy, 29*(3), 285–310.

Almond, P. (2009). The Dangers of Hanging Baskets: Regulatory Myths' and Media Representations of Health and Safety Regulation. *Journal of Law and Society, 36*(3), 352–375.

Almond, P. (2013). *Corporate Manslaughter and Regulatory Reform.* Basingstoke: Palgrave Macmillan.

Almond, P. (2015). Revolution Blues: The Reconstruction of Health and Safety Law as 'Common-Sense' Regulation. *Journal of Law and Society, 42*(2), 202–229.

Almond, P., & Gray, G. C. (2017). Frontline Safety: Understanding the Workplace as a Site of Regulatory Engagement. *Law & Policy, 39*(1), 5–26.

Baldwin, R. (1987). Health and Safety at Work: Consensus and Self-Regulation. In R. Baldwin & C. McCrudden (Eds.), *Regulation and Public Law* (pp. 132–158). London: Weidenfeld and Nicholson.

Baldwin, R. (1995). *Rules and Government.* Oxford: Clarendon Press.

Baldwin, R., & Cave, M. (1999). *Understanding Regulation: Theory, Strategy, and Practice.* Oxford: Oxford University Press.

Balleisen, E. J., & Brake, E. K. (2014). Historical Perspective and Better Regulatory Governance: An Agenda for Institutional Reform. *Regulation & Governance, 8*(2), 222–245.

Bartle, I., & Vass, P. (2007). Self-Regulation Within the Regulatory State: Towards a New Regulatory Paradigm? *Public Administration, 85*(4), 885–905.

Bartrip, P. (1982). British Government Inspection, 1832–1875: Some Observations. *Historical Journal, 25*(3), 605–626.

Bartrip, P., & Fenn, P. T. (1983). The Evolution of Regulatory Style in the Nineteenth Century British Factory Inspectorate. *Journal of Law and Society, 10*(2), 201–222.

Beetham, D. (1991). *The Legitimacy of Power.* London: Macmillan.

Berger, J., Ridgeway, C. L., Fisek, M. H., & Norman, R. Z. (1998). The Legitimation and Delegitimation of Power and Prestige Orders. *American Sociological Review, 63*(3), 379–405.

Berridge, V. (2010). Thinking in Time: Does Health Policy Need History as Evidence? *The Lancet, 375*(9717), 798–799.

Black, J. (2000). Proceduralizing Regulation: Part I. *Oxford Journal of Legal Studies, 20*(4), 597–614.

Black, J. (2007). Tensions in the Regulatory State, *Public Law.* 58–73.

Black, J. (2008). Constructing and Contesting Legitimacy and Accountability in Polycentric Regulatory Regimes. *Regulation & Governance, 2*(2), 137–164.

Bronstein, J. (2008). *Caught in the Machinery: Workplace Accidents and Injured Workers in Nineteenth-Century Britain.* Stanford: Stanford University Press.

Carson, W. G. (1979). The Conventionalisation of Early Factory Crime. *International Journal of the Sociology of Law, 7*(1), 37–60.

Cox, P. (2013). The Future Uses of History. *History Workshop Journal, 75*(1), 125–145.

Dawson, S., Willman, P., Clinton, A., & Bamford, M. (1988). *Safety at Work: The Limits of Self-Regulation.* Cambridge: Cambridge University Press.

Divall, C., Hines, J., & Pooley, C. G. (Eds.). (2015). *Transport Policy: Learning Lessons from History.* Farnham: Routledge.

Gibson, J. L., & Caldeira, G. C. (1995). The Legitimacy of Transnational Legal Institutions: Compliance, Support, and the European Court of Justice. *American Journal of Political Science, 39*(2), 459–489.

Genn, H. (1993). Business Responses to the Regulation of Health and Safety in England. *Law and Policy, 15*(3), 219–233.

Green, J. (1997). *Risk and Misfortune: The Social Construction of Accidents.* London: UCL Press.

Green, D. (2006). Public Opinion Versus Public Judgement About Crime: Correcting the 'Comedy of Errors'. *British Journal of Criminology, 46*(1), 131–155.

Gunningham, N., & Rees, J. (1997). Industry Self-Regulation: An Institutional Perspective. *Law & Policy, 19*(4), 363–414.

Gunningham, N., & Johnstone, R. (1999). *Regulating Workplace Safety: Systems and Sanctions.* Oxford: Oxford University Press.

Gunningham, N., Kagan, R., & Thornton, D. (2004). Social License and Environmental Protection: Why Businesses Go Beyond Compliance. *Law & Social Inquiry, 29*(2), 307–341.

Habermas, J. (1992). *Faktizitat und Geltung* (Between Facts and Norms) (W. Rehg, Trans.). Cambridge: Polity Press.

Harvey, D. (2005). *A Brief History of Neoliberalism.* Oxford: Oxford University Press.

Hawkins, K. (2002). *Law as Last Resort: Prosecution Decision-Making in a Regulatory Agency.* Oxford: Oxford University Press.

History & Policy. (2018). http://www.historyandpolicy.org.

Hood, C. (1991). A Public Management for All Seasons? *Public Administration, 69*(1), 3–19.

Huising, R., & Silbey, S. S. (2011). Governing the Gap: Forging Safe Science Through Relational Regulation. *Regulation & Governance, 5*(1), 14–42.

Hutter, B. (1989). Variations in Regulatory Enforcement Styles. *Law & Policy, 11*(2), 153–174.

Hutter, B. (1997). *Compliance: Regulation and Environment.* Oxford: Clarendon Press.

Hutter, B., & Manning, P. (1990). The Contexts of Regulation: The Impacts Upon Health and Safety Inspectorates in Britain. *Law and Policy, 12*(2), 103–136.

Jain, A., & Leka, S. (2016). *Occupational Health and Safety Legitimacy in the UK: A Review of Quantitative Data.* Leicester: IOSH.

Johnson, C., Dowd, T. J., & Ridgeway, C. L. (2006). Legitimacy as a Social Process. *Annual Review of Sociology, 32*(1), 53–78.

Kica, E., & Bowman, D. (2012). Regulation by Means of Standardization: Key Legitimacy Issues of Health and Safety Nanotechnology Standards. *Jurimetrics, 53*(1), 11–56.

Kidd, A. (1999). *State, Society and the Poor in Nineteenth-Century England.* Basingstoke: Macmillan.

Kidd, P., & Parshall, M. (2000). Getting the Focus and the Group: Enhancing Analytical Rigor in Focus Group Research. *Qualitative Health Research, 10*(3), 293–308.

Kitzinger, J. (1994). The Methodology of Focus Groups: The Importance of Interaction Between Research Participants. *Sociology of Health & Illness, 16*(1), 103–121.

Krueger, R., & Casey, M. (2009). *Focus Groups* (4th ed.). London: Sage.

Löfstedt, R. (2005). *Risk Management in Post-Trust Societies*. London: Earthscan.

Long, V. (2011). *The Rise and Fall of the Healthy Factory: The Politics of Industrial Health in Britain, 1914–1960*. Basingstoke: Palgrave Macmillan.

Lord Robens. (1972). *Safety and Health at Work: Report of the Committee 1970–1972* (The Robens Report). London: HMSO.

MacDonagh, O. (1958). The Nineteenth Century Revolution in Government: A Reappraisal. *Historical Journal, 1*(1), 52–67.

Mess, H. A. (1926). *Factory Legislation and Its Administration 1891–1924*. London: P. S. King & Son.

Meyer, J. W., & Rowan, B. (1991). Institutionalized Organizations: Formal Structure and Myth and Ceremony. In W. W. Powell & P. J. DiMaggio (Eds.), *The New Institutionalism in Organizational Analysis* (pp. 41–62). Chicago: University of Chicago Press.

Mills, C. (2010). *Regulating Health and Safety in the British Mining Industries, 1800–1914*. Aldershot: Ashgate.

Moran, M. (2003). *The British Regulatory State: High Modernism and Hyper Innovation*. Oxford: Oxford University Press.

Murphy, K., Tyler, T., & Curtis, A. (2009). Nurturing Regulatory Compliance: Is Procedural Justice Effective When People Question the Legitimacy of the Law? *Regulation & Governance, 3*(1), 1–26.

Norrie, A. (2001). *Crime, Reason, and History* (2nd ed.). Cambridge: University Press.

O'Rourke, D. (2003). Outsourcing Regulation: Analyzing Nongovernmental Systems of Labor Standards and Monitoring. *Policy Studies Journal, 31*(1), 1–29.

Palazzo, G., & Scherer, A. G. (2006). Corporate Legitimacy as Deliberation: A Communicative Framework. *Journal of Business Ethics, 66*(1), 71–88.

Powell, M. A. (Ed.). (2007). *Understanding the Mixed Economy of Welfare*. Bristol: University of Bristol Press.

Prosser, T. (1999). Theorising Utility Regulation. *The Modern Law Review, 62*(2), 196–217.

Prosser, T. (2010). *The Regulatory Enterprise: Government, Regulation, and Legitimacy*. Oxford: Oxford University Press.

Scharpf, F. W. (1999). *Governing in Europe: Effective and Democratic?* Oxford: Oxford University Press.

Selznick, P. (1985). Focusing Organisational Research on Regulation. In R. Noll (Ed.), *Regulatory Policy and the Social Sciences* (pp. 363–367). Berkeley: University of California Press.

Sirrs, C. (2016). Risk, Responsibility and Robens: The Transformation of the British System of Occupational Health and Safety Regulation, 1961–1974. In T. Crook & M. Esbester (Eds.), *Governing Risks in Modern Britain: Danger, Safety and Accidents, c. 1800–2000* (pp. 249–276). London: Palgrave.

Smismans, S. (2017). Risk Regulation at Risk: Brexit, Trump It, Risk It. *European Journal of Risk Regulation, 8*(1), 33–42.

Suchman, M. (1995). Managing Legitimacy: Strategic and Institutional Approaches. *The Academy of Management Review, 20*(3), 571–610.

Thomas, M. W. (1948). *The Early Factory Legislation: A Study in Legislative and Administrative Evolution.* Leigh-on-Sea: Thames Bank.

Thornton, D., Gunningham, N., & Kagan, R. A. (2005). General Deterrence and Corporate Environmental Behavior. *Law and Policy, 27*(2), 262–288.

Tombs, S., & Whyte, D. (2010). A Deadly Consensus: Worker Safety and Regulatory Degradation Under New Labour. *British Journal of Criminology, 50*(1), 46–65.

Tucker, E. (1995). And Defeat Goes On: An Assessment of Third-Wave Health and Safety Regulation. In F. Pearce & L. Snider (Eds.), *Corporate Crime: Contemporary Debates* (pp. 245–267). Toronto: University of Toronto Press.

Tyler, T. R. (1990). *Why People Obey the Law.* Princeton: Princeton University Press.

Tyler, T. R. (2011). The Psychology of Self-Regulation: Normative Motivations for Compliance. In C. Parker & V. L. Nielsen (Eds.), *Explaining Compliance: Business Responses to Regulation* (pp. 78–102). Cheltenham: Edward Elgar.

Walters, D. (1996). Health and Safety Strategies in Europe. *Journal of Loss Prevention in the Process Industries, 9*(5), 297–308.

Ward, J. T. (1962). *The Factory Movement 1830–1855.* London: Macmillan.

Weber, M. (1914). *Economy and Society* (G. Roth & C. Wittich, Trans.). Berkeley: University of California Press.

Wenzel, M., & Jobling, P. (2006). Legitimacy of Regulatory Authorities as a Function of Inclusive Identification and Power Over Ingroups and Outgroups. *European Journal of Social Psychology, 36*(2), 239–258.

CHAPTER 2

Recent Public Attitudes Towards Health and Safety

INTRODUCTION

This book aims to provide an historical and socio-legal investigation of the legitimacy of 'health and safety', and of the institutions, laws, and actors that constitute it. The starting-point for this, as observed in Chapter 1, is a perception that the legitimacy of health and safety has declined in recent times, and certainly since 2000, to a point where the public profile of this area poses a challenge to its ongoing existence. There are many different audiences in whose eyes an undertaking such as health and safety may lose legitimacy, including regulated stakeholders (employers and employees), politicians and government actors, and the wider public (Almond 2007). While this investigation encompasses all of these dimensions, it is particularly important to consider the role that public attitudes play in creating a policy climate which is hostile to health and safety. If, as discussed in Chapter 1, legitimacy is a matter of *"generalized perceptions"* as to whether an entity or set of rules are *"desirable, proper, or appropriate"* (Suchman 1995: 574), then we can conceive of the 'public sphere' as a key site for the formation and expression of the subjective assessments that underpin this concept. Not only do public attitudes feed into, or partially reflect, the democratic processes to which public regulators are answerable, but they also create the 'social license' to operate that sets the parameters of what it is possible and acceptable for stakeholder organisations and government actors to do (Almond 2015; Gunningham et al. 2004; Lynch-Wood and Williamson 2007).

© The Author(s) 2019
P. Almond and M. Esbester, *Health and Safety in Contemporary Britain*, https://doi.org/10.1007/978-3-030-03970-7_2

This also reflects the argument that we now live in an increasingly flexible and unstable 'risk society', which presents *"new parameters of risk and danger as well as offering beneficent possibilities for humankind"* (Giddens 1991: 28; also Beck 1992), and which thus gives rise to a more prominent set of public fears and insecurities about risk (House of Lords 2000; Hood 2002; Löfstedt 2005). This view of the 'newness' of risk has not been accepted uncritically, of course, and the historiographical challenge to this view is discussed in Chapter 3. However, what is significant about the risk society model in relation to this chapter is that, on this view, regulatory projects such as 'health and safety' exist as a response to these insecurities, providing a degree of reassurance and substantive protection to citizens about the risks they face. The levels of trustworthiness, 'affect', and utility that risk regulation is seen to possess has an impact on attitudes towards the risks involved, as well as those who control them (Breakwell 2007; Slovic 1986, 1987, 1993, 2000; Slovic et al. 1979) and, importantly, are consequently regarded as forming a concrete basis for the development of public policy (House of Lords 2000; Löfstedt 2005; Vogel 2012: ch. 7). 'Public concern' has been repeatedly cited as a rationale for reforms of the law relating to health and safety, from the introduction of a 'corporate manslaughter' offence (Law Commission 1996: 1.10; Home Office 2005: 6), to the review of health and safety 'overregulation' (Young 2010), to the introduction of more punitive enforcement approaches (Baldwin 2004). Regulators invest in proactive publicity campaigns to reshape their public profile (HSE 2006; Yeung 2005), but this is not necessarily new: as far back as 1972, the Chief Inspector of Factories had seen it as *"part of the job of the Inspectorate to develop an informed public and to harness the force of its informed opinion to the improvement of industrial conditions"* (Department of Employment 1973: vi). Indeed, an awareness of the significance of public opinion extends back at least into the nineteenth century, though whilst it was something which informed policy decisions, it was not necessarily actively influenced via campaigns or deliberate effort. In addition, ideas about who constitutes the public 'worth' listening to have, of course, altered significantly over time. Nevertheless, a regulatory infrastructure that has public support can get its messages across more effectively than one which lacks this supportive basis (Almond 2007; Yeung 2005).

This chapter provides a review of evidence as to the nature of recent public perceptions of health and safety, drawing on a wide range of sources to argue that these attitudes are more complex than we might think, neither as anxiety-based and punitive, nor as sceptical and anti-regulation, as many accounts of public policy-making suggest. There is stronger evidence for the last approximately twenty years, via public surveys, than the earlier part of this book's coverage; relatedly, this also corresponds with the period in which health and safety has seemingly become an increasingly contested concept. The chapter will draw on a wide and diverse set of literatures and materials, including quantitative survey data (such as British Social Attitudes, EU-OSHA, and Labour Force surveys), much of which has been recently reanalysed by Jain and Leka (2016), qualitative studies of public attitudes towards risk and regulation, and an original focus group study undertaken as part of our research project (Almond and Esbester 2016: ch. 6).[1] Rather than presenting sources of information separately, this chapter will proceed thematically, integrating them into a single account. It will show some of the deep-seated tensions within public perceptions of health and safety, drawing a distinction between nuanced and positive public *attitudes* (which are more considered, deeper, and more reflective), and the more prominent, but perhaps less well-developed, mass of negative public *opinions* (which are more populist, shallower, and more instinctive). This duality reminds us that public attitudes are constantly being reconstructed, and are both drivers and products of their institutional, political, and historical contexts.

An Issue of Concern?

Research into public attitudes towards crime has long recognised that corporate and organisational offences, including the worst sorts of health and safety-related incident, are taken very seriously by public audiences (Almond and Colover 2010, 2012; Cullen et al. 1982; Grabosky et al. 1987; MacDonagh 1958; Rosenmerkel 2001; Wolfgang et al. 1985). In particular, the worst safety failures, such as rail accidents like the Ladbroke Grove disaster of 1999, or major hazard plants like power stations, are capable of constituting 'dread risks', prompting levels of public fear and concern which may be disproportionate to their statistical

---

[1] Eight focus groups (containing 67 participants, spread-sampled to ensure a degree of demographic representativeness) were conducted during 2014 at locations across Britain. They were structured to follow a common discussion outline, and were analysed via the development of analytical codes.

likelihoods of occurring (Chilton et al. 2007; Harrington 2003; Slovic 1987). At the same time, health and safety is an issue which is seen as socially worthwhile (Almond and Colover 2010: 336; Pidgeon et al. 2003: 38–40; Walker et al. 1998: 81–82) and as serious; Elgood et al. found (2004: 3) that 82% of the general public were concerned about it, and participants in the focus groups we conducted as part of the research underlying this book also viewed health and safety as an important issue:

> ["Do you think it's important to have health and safety laws in place?"]
> "[altogether] Yes!" (G1-7: 43.03)[2]

The normative idea of health and safety was widely endorsed, and was linked to ideas of progress and reassurance:

> "as a member of the public you want to know when you're going some-where that you're going to be safe." "Yeah, trust". (B6 and B3: 1.12.38)

'Health and safety' is an issue that attracts strong opinions, which the focus group discussions explored in depth. Participants often responded in an instinctively negative way, viewing the issue as 'boring', 'excessive', 'interfering', 'restrictive', 'rigid', and 'ridiculous', based on the 'lay' knowledge and '*naturalistic*' media-influenced narratives (Green 1997: 160) that form the 'common-sense' way of seeing this issue (Geertz 1975; also Almond 2015).

It is clear, however, that public attitudes towards 'health and safety' often reflect a lack of clarity about the term. It is applied to every level of decision-making from statute law down to the decisions of workplace supervisors, and it is used as a catch-all term to reflect a wide range of regulatory domains, from road safety and product labelling to medical negligence and waste recycling (Pidgeon et al. 2003; King et al. 2005; Almond and Colover 2010). This mirrored the findings of EU and British surveys, where attitudes towards the value of health and safety regulation (80% were satisfied with conditions in the UK) were more pronounced than levels of knowledge about it (Jain and Leka 2016: 23,

---

[2]These reference codes relate to the focus groups undertaken by Almond and Esbester (2016). The codes indicate which of the eight focus group sessions the quote comes from ('A' to 'H'), the participant code within that group ('1' to '10'), and the time point within that focus group recording where the comment was made.

39, 60). More specific and grounded forms of 'health and safety' tend to be better accepted and understood; King et al. (2005) found broad support for the notion of health and safety, but a greater acceptance of *occupational* health and safety than its other forms (2005: 4–5; also Elgood et al. 2004: 10; Pidgeon et al. 2003: 12–14). Worker surveys also tend to reflect higher levels of satisfaction and understanding when based on participants' assessments of health and safety where they work (King et al. 2005). Similarly, focus group discussions tended to become more informed, and less critical, the more they were grounded in personal experiences and concrete examples:

> you can all see the need for it, like in [participant C1's] case, but then you get these really stupid things […] everybody then looks at health and safety as a bit of a joke, and that's wrong, it should be treated properly. (C9, 06.32)

This means that we have to be cautious about concluding what the public 'think' about this complex issue. In reality, people tend to hold multiple views at once, being both resistant to health and safety in abstract, instinctive terms, but more generally accepting in relation to specific examples. The focus groups also provided many examples of attitudes being redefined through discussion with others, or in reaction to the response of the social audience (Almond and Esbester 2016: 36).[3]

As with research into public perceptions of the criminal justice system more generally (Green 2006; Hough 1996, 2002; Hutton 2005), it is possible to discern two levels of public perception, which are often held at the same time by the same person. Public *opinions* are 'top of the head', surface-level views elicited via simple methodologies such as opinion polls; these opinions tend to be reactive, instinctive, and critical, and capture the type of things people are willing to *say* about health and safety, but perhaps do not actually *wish for*. Public attitudes, on the other hand, are more considered, reflective, and nuanced judgements, which tend to be less critical than those opinions, and so are perhaps better reflective of the true preferences of the individuals

---

[3]For instance, in one example, a judgemental discussion about the fecklessness of personal injury compensation claimants was reversed into a much more measured assessment once it emerged that one of the focus group participants had made just such a claim in the past (Almond and Esbester 2016: 36).

who hold them (Green 2006). This gives some indication that the representations of public preference which shape media discourse and policy decision-making around this area may overstate the extent of any public anti-regulatory sentiment that actually exists (Almond 2009). As such, perhaps the finding of the 2001 British Social Attitudes survey, that while 49% of the population think more could be done to protect workers from health and safety risks, only 3.4% thought that too much was already being done (Jain and Leka 2016: 37), are telling. Health and safety has maintained a better public profile than is often recognised.

Changes Over Time

Across the historical span of this project, it is possible to identify areas where public attitudes have changed in response to specific events and circumstantial drivers. Despite the existence of a widespread satisfaction with Britain's health and safety framework (Jain and Leka 2016: 44–49; reporting European Working Conditions and British Social Attitudes survey findings), a significant proportion (39%) of respondents to those surveys felt that standards had declined between 2009 and 2014 (Jain and Leka 2016: 47; European Commission [EC] 2014). In particular, a key driver of these changing attitudes seems to be the increasing prevalence of psychosocial risk factors (workplace stress and ergonomic risks) within the service-oriented economy, particularly at times of economic recession. The effect of the changing workplace on health and safety was also identified within our focus group data, where participants felt that greater job insecurity and role flexibility had made the management of health and safety more challenging:

> Nowadays I don't think there are the jobs there used to be, people used to be apprentices, used to go and do things and follow rules. Now by the time you're twenty-five, chances are you could have had four or five or six jobs in different things because that's the way society is. (F7: 50.16)

Overall, the main current of change within our timeframe was a marked perception of social progress, from an era of deference and disinterest around health and safety, to one where the rights of ordinary people were respected and protected much more than in the past. Contrary to notions of the 'good old days' of support for health and safety in the mid-twentieth century, focus group participants tended to view this past as a time when health and safety was taken much less seriously and was not prioritised by either employers, workers, or the public. Such a view accords with the findings of other oral histories of work and safety, such as McIvor's study of working lives in post-war Britain, which documents the widespread and widely accepted sacrifice of worker safety and health in the pursuit of productivity (2013: ch. 5; see also Johnston and McIvor 2000, 2004; McIvor and Johnston 2007). Health and safety is broadly seen as being better now than in the past, and as important now, as a matter of individual rights-mobilisation and democratisation:

> In industry, health and safety has helped employees, because years ago you just did what your employer said, didn't you? So health and safety's probably really helped [...] you can go to your boss and say 'no, I'm not going to do this, it's not safe'. (H2: 59.38)

It is important to note, however, that our focus groups, conducted in 2015, produced many findings that were consistent with those of Pidgeon et al. (2003), conducted some fifteen years previously. Many of the frustrations and tensions identified were broadly similar, the underlying trust in the Health and Safety Executive (HSE) was consistent, and the overarching sense of progress and acceptance of health and safety as an important element of social progress, in abstract terms, was retained. An example of this consistency is provided by the following extracts from the two studies, both conducted in the town of Reading and with participants drawn primarily from the IT and service sectors:

> "[M61] We have health and safety inspecting people finding how we should sit at our desks, and how high the screens should be and stuff like that." "[F43] And using the right chairs and so on. And you do think that it is a shame that they weren't actually around in such a strong sense twenty or twenty five years ago. Because if they were then I am sure that people would have [...] perhaps had a longer life and healthier lives." (Pidgeon et al. 2003: 39)

> For your health and safety, you get your desk assessed, people come and test your technology, electronics and stuff, it's all health and safety related [...] Times have changed which I think is why the legislation has changed (D6: 47.47) So health and safety's a good thing because obviously it brought things in...to protect people. (D5: 49.14)

There is a sense, then, that while risks to health and safety have changed a lot since the middle of the last century, the attitudes of the public have not necessarily altered that much in the last fifteen years, undercutting the claims made (by e.g. Young 2010) about a hardening of attitudes towards health and safety.

Alongside the sense of progress alluded to, however, a series of more problematic changes were perceived, including a greater public tendency towards risk-aversion, an increased willingness to sue or claim damages, and a lack of common sense on the part of those to whom the law applies:

> health and safety rules, the reality of putting those rules into practice is not always common sense. This country has operated on common sense for many years previous and all of a sudden it's become politically correct [...] You read it sometimes and you think it's just craziness. (B7: 7.56)

In this sense, public attitudes mirror the tendencies that have dominated media coverage of health and safety in recent times (Almond 2009), and which the Young Review (2010) responded to, as well as the analytical concepts of the 'culture of fear' coined by Furedi (1997), which argues that society has increasingly undermined the capacity of individuals to properly determine and manage risks for themselves. More will be said on the notion of a 'compensation culture' in Chapter 6; for now, it can be observed that this contested social phenomenon (Morris 2007) is widely recognised and 'taken for granted' within public discourse, and is often linked to both the self-interest of lawyers and claimants, and to the growth of an 'Americanised' culture of '*adversarial legalism*' (Kagan 2001), which places a greater emphasis on the exercising of legal rights:

> It was the Americans, they started all this suing business [...] They sue for everything over there. (F8: 21.37)

Together, these elements of contemporary society were seen by some as having undermined the voluntary basis of social life, and led to the development of an unduly overbearing health and safety system:

> You have to have rules [...] but if you followed everything they said all the time [...] nobody would ever go anywhere, nobody would ever do anything. (F7: 14.35)

As noted above, however, this normative scepticism is offset by a more practical acceptance, grounded in the cognitive and pragmatic legitimisation (Suchman 1995) of health and safety as both necessary and useful.

A key legitimacy challenge to health and safety now derives from the sense that it is excessive and contrary to other, socially valued goals and institutions. This may be conceived of as a form of normative, justice-oriented legitimacy challenge; health and safety is not seen as reflecting the best interests of society, nor as informed by the values that many members of the public profess to hold (Black 2008). These perceptions of normative discordance reflect the influence that 'core values' have over understandings of a 'hard', technical or applied issue like workplace health and safety. Social science research suggests that individual citizens tend to rely on core principles of moral belief and normative commitment to guide their engagement with, and opinions on, specific issues of policy (Feldman 1988; Goren 2001; Pollock et al. 1993). The ability to connect these two levels of thought, to reach judgements about 'hard' policy issues, is moderated by both degrees of informedness and political engagement, and by the capacity of policy entrepreneurs to frame such issues in terms of particular 'core values' (Pollock et al. 1993: 46–47). In the case of an issue of social regulation like health and safety, core values of self-direction ('social' individualism) and self-determination ('economic' individualism) exert significant influence over policy opinions (Pitlik and Rode 2017), with economically liberal values having a particularly pronounced effect. So it appeared in the focus group discussions, where libertarian concerns about economic competitiveness and individual freedom often underpinned criticism of health and safety:

It's becoming a nonsense. It's a huge cost to the economy, we waste so much time, it's a box-ticking exercise, isn't it? (E7: 14.18)

We're used to adhering to rules [...] It's expected now that no matter what you do or where you go that there is going to be some kind of rule or law you've got to stick to. (A6: 57.58)

We might also theorise that the degree of ambivalence that emerges around public attitudes towards health and safety, particularly in terms of the shift from hostile opinions to more supportive attitudes, is itself a product of these moderating factors, not least the degree of engagement with the issue that was present. As discussions become more concrete, so they tend to become less hostile; in such situations, the 'heuristic' effect of core values (Goren 2001: 175) is set aside in favour of more grounded information. In part, this reflects the fact that public awareness is commonly constructed and mediated in a 'second-hand' form, via news coverage in the aftermath of major events (Petts et al. 2001; Pidgeon et al. 2003). This reliance allows media narratives, which are often constructed in 'core value' terms, to set the parameters of public discussion and to frame attitudes towards issues like health and safety (Almond 2015; Fyfe 2013; Pollock et al. 1993). The result, as mentioned previously, that opinions about health and safety are more widespread than knowledge of it (Jain and Leka 2016: 23, 39, 60), but that deliberative methods of engagement do offer a means of seeing beyond these narratives. In any case, the core normative values that underpin attitudes towards health and safety are more diverse than this suggests, with solidaristic, welfare-oriented concerns (such as interconnectedness, participation, trust, and justice) also playing a key role (Zwetsloot et al. 2013). While it may be that there is now a greater public acceptance of individual rather than collective responsibility for work-related risks, and beliefs in the rightness of state intervention may have declined in recent years (Mascini et al. 2013), levels of belief in the legitimacy of health and safety seem largely protected from these trends.

The HSE

One reasonably concrete component of recent public attitudes towards health and safety is the constellation of attitudes expressed in relation to the specific regulatory actors who govern the area in the United Kingdom. The HSE has been subject to several focused studies, which have generally found that it is not well known to the public and does not have a clear role or mandate in the eyes of the wider population (Elgood et al. 2004; King et al. 2005; Pidgeon et al. 2003: 9). Despite this, the British health and safety system is regarded considerably more favourably by workers than most of its European equivalents (Jain and Leka 2016: 37; based on data from the Eurobarometer survey, 45.1), and the HSE is generally recognised as the primary body responsible for health and safety. Previous studies have also provided evidence of a broad appreciation of HSE's effectiveness, expertise, and independence (King et al. 2005: 19–23; Walker et al. 1998: 85–86). Pidgeon et al., in their study of HSE's *'critical trust'* profile (2003), account for this divergence between levels of public knowledge and support via reference to the apparent normative alignment of the regulator. The sense of a good purpose or 'mission' at the heart of the regulator's activities (a broad sense that it exists to 'do good') exerts a greater influence over public perceptions than does the lack of knowledge of its role. The HSE is, like the NHS, a "*valuable public service (essentially altruistic) which, for a number of exogenous reasons, is underfunded or unable to deliver what it is set up to do*" (2003: 31), and it is trusted as a result of this public interest alignment (Walls et al. 2004: 139).

Crucially, however, this trust is capable of coexisting alongside much more critical perceptions of the regulator. The HSE's effectiveness and capacity to act were doubted by some of Walls et al.'s participants due to its reliance on central government (2004: 141–142), and it has been criticised for its tendency to introduce "*petty or restrictive*" rules, the lack of 'common sense' associated with them (Pidgeon et al. 2003: 28–30; also Almond and Colover 2012: 1008–1009), and the impact that its decisions have upon business competitiveness (King et al. 2005: 4; Pidgeon et al. 2003: 32). Within our own data, the same tendencies were visible; HSE was seen as trustworthy, competent, and well-meaning, primarily because it was perceived as acting in pursuit of altruistic and socially beneficial motives:

> I trust the HSE. When I've had meetings with them and they've come into give advice I listen to them […] I'd rather find my own information from a government source that's directly derived from the law rather than somebody just trying to capitalise. (H9: 1.12.41)

At the same time, however, it was seen as being distant from the lives of ordinary people, and as engaging in undue interference and irrational decision-making:

> In our whole lives we're not going to come across them, are we? The trouble is, you only come across the Health and Safety Executive if you have a problem […] we're just not interested, we're interested in funny stories […] you hope your whole working life you never come across them. (F7: 26.51)

HSE tends to dominate discussions of public attitudes towards the institutions of health and safety; the public regulator acts as the focal point for attitude formation, some way ahead of other bodies such as the EU (whose role is relatively little-appreciated: Jain and Leka 2016: 21–23), other government regulators (Pidgeon et al. 2003: 19), and the diffuse actors who implement health and safety regulation in workplaces. This reinforces the importance of the '*facework*' that regulators like HSE do to construct and maintain their institutional identity and public profile (Walls et al. 2004: 135; also Almond 2007; Yeung 2005), but also emphasises that, even in an era of decentred and 'polycentric' regulation (Black 2008), the apparatus of the state still exert great influence over public perceptions.

Taken together, the available evidence as to public attitudes towards the specific regulatory actors within the UK's health and safety system suggests that considerations of constitutional and functional legitimacy (accountability, effectiveness, proportionality, independence: Baldwin 1995; Black 2008) are less influential and important in shaping overall perceptions of legitimacy than more normative considerations of moral legitimacy, around the alignment of the regulator with the democratically shared values of the social sphere, and its alignment towards the pursuit of justice-oriented outcomes (Black 2008: 146). As Pidgeon et al. put it, "*HSE is perceived to be a fundamentally altruistic organisation [and...] as acting in the public interest*" (2003: 20), and this is contingent upon it not having a vested interest in the non-fulfilment of its protective duties. Because of the relative lack of understanding and knowledge of HSE's work among the general public, attitudes, trust, and acceptance of the regulator's role are primarily based upon perceptions of *why* it acts, rather than of *what* it actually does, or *how* it does it:

> Power's a privilege and in this case it's being used for good. Health and safety is a necessity, it's a good, positive thing [...] this is an instance where there can't be such a thing as too much power. (A4: 1.09.06)

Health and Safety as a 'Problem'?

Three main problems have tended to be associated with 'health and safety' in the eyes of the British public; a tendency towards disproportionate restrictiveness, a reliance upon overly legalistic rule-forms that alienate ordinary people, and the presence of a commercialised self-interest which undercuts the general perceptions of altruism identified above. Evidence as to the first of these critical evaluations was found within the data gathered by King et al., who identified health and safety's negative functional connotations of restrictiveness, interference, and 'red tape' (2005: 3–4). Pidgeon et al. (2003: 28–30) found that perceptions of 'petty' regulations and a need for 'common sense' played a significant role in undercutting trust in the regulator, with *"funny little regulations"* contributing to a degree of scepticism about the aims and value of health and safety regulators. Elsewhere, a sense of over-regulation and a media-led perception of '*health and safety gone mad*' (Almond 2009) have contributed to public resentment of health and safety (Almond and Colover 2012: 1008). Such attitudes seem to reflect a sense that the balance of interests drawn between competing goals of welfare protection and economic competitiveness might be incorrect, as well as deeper concerns over a clash between modern social conditions and the certainties of a more traditional social outlook (Almond 2009: 365). Health and safety has been the subject of an epidemic of stories of this sort (Dunlop 2014) and it is clear that they have impacted on stereotyped perceptions of health and safety as a matter of disproportionate interference in the freedoms of individuals.

The second and third criticisms relate to the proliferation of formalised legal rule-forms, and the expansion of the role and influence exerted by those who have a commercial interest in the use of those rule-forms. There is evidence of a public perception that, since the 1960s, and particularly in recent years, health and safety has become an increasingly insular, formalised, and rule-bound area of endeavour, and that this has rendered it less relevant, accessible, and legitimate. Such criticisms accord with broader critiques of regulatory systems as overly 'technocratic' (Radaelli 1999) and inward-focusing. Notions of democratic legitimacy presuppose that legal institutions are accessible and open to those they affect (Black 2008), and more procedural forms of legitimacy also place an emphasis on values of transparency, clarity, and effective oversight (Baldwin 1995). But the UK has comparatively low levels of worker participation (EU-OSHA 2012) and consultation (Jain and Leka 2016: 25–28) in health and safety, and previous public attitude data has suggested that unduly formal health and safety regulations (a *"laying down of the law"*) can result in perceptions of reduced autonomy and undermined individual responsibility (Pidgeon et al. 2003: 39–41). Our own focus group data reflected a strong sense that health and safety regulations tended to miss their mark because they were overly focused on performative process (Power 1997) rather than outcomes, and lacked sensitivity to context or the needs of users and participants:

> Some of it is just not valid though, is it? How does that measure, you attending that lecture, in what world is that ticking a real box? (H8: 10.09)

Rather than a rejection of the values that health and safety purported to protect, this criticism is instead aimed at the failure of this system to achieve what it *ought*. The pervasiveness of bureaucratic 'health and safety' was seen as a barrier, leading to a counterproductive adherence to unproductive rules:

> We've put the notice up [...] because our lawyers tell us, but have we actually advanced a healthier, safer working environment? I'm not sure we have, not as much as we'd like to think. (D9: 59.48)

This tendency constitutes an outcome of 'juridification,' the process by which legal requirements become more precise, detailed, pervasive, and numerous over time (Teubner 1987), resulting in a reduction in the perceived legitimacy of the law, which becomes more self-referential and insular, and less accessible. This juridification also results in a need for professional intermediaries (Abbott et al. 2017) to perform tasks of implementation and interpretation, and to provide expertise and assurance. Regulatory concepts and standards are constructed endogenously (between participants), and intermediary professions play a key role in creating and modelling appropriate solutions and understandings of the law (Edelman et al. 1999). The problem, however, is that these intermediary roles tend, in practice, to expand in scope and influence over time, and the drivers of this growth are often associated with the self-interest of those who perform the roles (van der Heijden 2017), including the creation of a commercialised arena of service-provision. This trend is one that has a significantly negative effect upon public attitudes in the field of health and safety, where an emergent 'safety industry' (of safety professionals and institutions) was seen to have led to an exponential growth in the requirements imposed upon businesses and individuals:

> It's almost a whole commercial enterprise [...] there's so much money being made. That's why I think we're going to struggle to [...] stop it. (H8: 1.11.00)

A crucial component of this objection was the avowedly commercial nature of health and safety. This had arguably made it a more self-serving undertaking rather than a matter of public interest:

> It's turned into an industry, it's gone from being rules and regulations that you have to enforce because it's common sense entrenched in law, to an industry that has generated surveyors, health and safety professionals, a slew of qualifications […] it's started being a means for consultancies and firms to generate probably quite a lot of cash. (D10: 10.23)

This brings us to the key legitimacy challenge posed by this perceived commercialisation; that it challenges the altruistic normative alignment identified previously in relation to ideas of 'critical trust', and calls into question the idea that those in the area have good motives for doing what they do. This change of normative positioning had a marked effect on the trust that participants were willing to invest in such actors:

> "Yeah [it affects trust], you think 'how much money are they making out of it? […] and who's telling them what they can and can't do?'" (H2: 1.11.21)

It also placed HSE and others involved in health and safety on the wrong side of the 'safety vs profit' heuristic identified by previous research into public attitudes in the area (Pidgeon et al. 2003: 16; Walker et al. 1998). A perceived lack of concern with money is a powerful means of deducing a socially beneficial alignment and a claim to normative legitimacy rooted in justice-based 'public interest' considerations. Instead, such actors were perceived in much more cynical terms, and health and safety itself, as a goal and undertaking, was increasingly seen as a matter of financial self-interest, including as a means of reducing insurance costs, of securing the employment of those who implement it, and of protecting the bottom-line of an employer from having to spend on staff welfare. From this, we might conclude that the movement towards a more commercialised health and safety system causes considerable tensions in terms of public attitudes.

Conclusion: The Paradox of Health and Safety

In conclusion, three important observations are worth making about public attitudes towards health and safety, each of which reveals something important about the ways in which the legitimacy of this area of activity is constructed, and also about the distinction and interrelationship between public opinions and attitudes (Green 2006). First, as mentioned previously, there is a disjunction between levels of knowledge about health and safety, and levels of support for it, with people who lack first-hand knowledge and awareness of health and safety often demonstrating much more critical or negative views of the subject (for instance, Pidgeon et al. 2003: 13–15). This seems to be a product of the need to base assessments on mediated, second-hand information (particularly from news media: Almond 2009), resulting in an engagement at the level of opinion—impressionistic, instinctive, and more negative than might otherwise be the case. The relatively low levels of participation in health and safety found within British workplaces, and the decline in trade union membership and activity within those same workplaces (discussed in more detail in Chapter 5) have undoubtedly had an effect on perceptions of health and safety as a whole (Jain and Leka 2016: 25–32), as they have shifted the bases for those perceptions away from experience and towards more distant sources of knowledge. Health and safety is an issue about which people have strong ideas, rooted in their own lay knowledge, but which are often reflective of fairly weak levels of formalised knowledge or systematic experience.

Second, there are inherent tensions within these public attitudes and perceptions, which help to illustrate that health and safety is often the subject of contradictory judgements. As an example, a key objection to health and safety has often been that it is dominated by professionals whose boring, rigid, and officious approach has an alienating effect on workers and the public:

> We've got a whole bunch of health and safety 'white coats' […] one guy who goes around and just talks about health and safety legislation. (H5: 34.18)

Health and safety professionals, a category including both external regulators and inspectors, and those who work internally within organisations to secure compliance, are variously seen as serious, forbidding, cautious, intimidating, and lacking in common sense. These negative stereotypes were regularly linked to the perceptions of over-restrictive intervention, and of rigid formalism, identified previously. This tendency was not so pronounced in earlier research, suggesting that it is an outcome of more modern 'regulatory myths' which have eroded public trust and deference to 'over-zealous' experts (Almond 2009: 368). At the same time, however, when we asked our focus group participants to set out what features they wished for the health and safety professionals in their workplace to have, a positive profile was evoked (stern, rigid, detached and single-minded) which was very similar to the officious negative stereotype set out above. In particular, the expertise of those who implement health and safety was central to the question of trust and acceptance. The qualifications that safety people have, their experience of real-world issues, and their capacity to communicate this expertise were all vital to this process. This is an amusing but important paradox; health and safety is accepted as legitimate *despite* being unpopular; it need not be enthusiastically accepted to be cognitively and pragmatically legitimate as a necessary and useful institution (Suchman 1995).

Finally, it is worth reflecting on the overall tenor of public attitudes towards health and safety, which appear to be much more positive, and much less committedly critical, than policy and media narratives around the issue might suggest. As well as the positive trust profile (Pidgeon et al. 2003; Walls et al. 2004), sense of social importance (Almond and Colover 2010; Walker et al. 1998), and perception of effectiveness (King et al. 2005) identified in previous studies, it is also worth recognising that there is an overarching idealistic attachment to health and safety as an idea. The 'right to be safe', or the pursuit of health and safety as a component of individuals' social citizenship rights, was widely accepted and supported within our focus groups:

> "I think it's become a right now." (D3) […] "yeah, you've got the right to work in a safe environment." (D5) "That sounds just like a poster, it's true though!" (D8: 58.47)

Members of the public were willing to commit to the idea of health and safety and to advocate it as a sign of social progress and development; they saw this value (that of protecting the right of individuals to be treated with respect) as both justifying the process of regulation, and providing a positive and admirable motive for those who work in the area. At the same time as this high-minded motivation, however, individuals also possess a relatively irreverent interest in the 'lowbrow' discussion of health and safety stories in the media, participants both internalising the content of these stories, but also questioning their veracity and accuracy. Health and safety is thus a forum for the negotiation of competing strands of discussion, reflecting both fundamental ideas, and novelty interest. Crucially, and to conclude, it is worth remembering that while the latter discussion of health and safety as a matter of critical public opinion attracts a great deal of attention and profile, the value-based public attitudes that underpin the former tend to mean that those criticisms do not necessarily translate into deep-seated opposition to the issue. Public attitudes towards health and safety are more robust, and more positive, than might be thought.

REFERENCES

PRIMARY SOURCES

Department of Employment (1973). *Annual Report, H.M. Chief Inspector of Factories, 1972.* Cmnd. 5398. London: Crown.

European Commission (EC). (2014). *Flash Eurobarometer 398: Results for the UK.* Luxembourg: Communications Office of the EU.

EU-OSHA. (2012). *Worker Representation and Consultation on Health and Safety: An Analysis of the Findings of the European Survey of Enterprises on New and Emerging Risks (ESENER).* Luxembourg: Publications Office of the EU.

Home Office. (2005). *Corporate Manslaughter: The Government's Draft Bill for Reform.* London: The Stationery Office.

House of Lords. (2000). *Science and Society: Third Report of the Select Committee on Science and Technology.* London: HMSO. http://www.publications.parliament.uk/pa/ld199900/ldselect/ldsctech/38/3801.htm.

HSE. (2006, April 13). *Sensible Risk Management,* HSE Board Paper HSE/06/46.

Law Commission. (1996). *Legislating the Criminal Code: Involuntary Manslaughter* (Report 237). London: HMSO.

Young, Lord. (2010). *Common Sense, Common Safety* (The Young Review). London: Crown.

SECONDARY SOURCES

Abbott, K., Levi-Faur, D., & Snidal, D. (2017). Theorizing Regulatory Intermediaries: The RIT Model. *The ANNALS of the American Academy of Political and Social Science, 670*(1), 14–35.

Almond, P. (2007). Regulation Crisis: Evaluating the Potential Legitimizing Effects of 'Corporate Manslaughter' Cases. *Law and Policy, 29*(3), 285–310.

Almond, P. (2009). The Dangers of Hanging Baskets: Regulatory Myths' and Media Representations of Health and Safety Regulation. *Journal of Law and Society, 36*(3), 352–375.

Almond, P. (2015). Revolution Blues: The Reconstruction of Health and Safety Law as 'Common-Sense' Regulation. *Journal of Law and Society, 42*(2), 202–229.

Almond, P., & Colover, S. (2010). Mediating Punitiveness: Understanding Public Attitudes Towards Work-Related Fatality Cases. *European Journal of Criminology, 7*(5), 323–338.

Almond, P., & Colover, S. (2012). Communication and Social Regulation: The Criminalization of Work-Related Death. *British Journal of Criminology, 52*(5), 997–1016.

Almond, P., & Esbester, M. (2016). [With D'Cruz, C., and Mayne, L.] *The Changing Legitimacy of Health and Safety, 1960–2015.* Leicester: IOSH.

Baldwin, R. (1995). *Rules and Government.* Oxford: Clarendon Press.

Baldwin, R. (2004). The New Punitive Regulation. *Modern Law Review, 67*(3), 351–383.

Beck, U. (1992). *Risk Society: Towards a New Modernity* (M. Ritter, Trans.). London: Sage.

Black, J. (2008). Constructing and Contesting Legitimacy and Accountability in Polycentric Regulatory Regimes. *Regulation & Governance, 2*(2), 137–164.

Breakwell, G. (2007). *The Psychology of Risk.* Cambridge: Cambridge University Press.

Chilton, S., Jones-Lee, M., Metcalfe, H., Loomes, G., Robinson, A., Covey, J., Spencer, A., & Spackman, M. (2007). *Valuation of Health and Safety Benefits: Dread Risks* (Research Report RR541). Sudbury: HSE Books.

Cullen, F. T., Link, B. G., & Polanzi, C. W. (1982). The Seriousness of Crime Revisited: Have Attitudes Toward White-Collar Crime Changed? *Criminology, 20*(1), 83–102.

Dunlop, C. (2014). *Health and Safety Myth-Busters Challenge Panel: Case Analysis* (University of Exeter Research Paper).

Edelman, L., Uggen, C., & Erlanger, H. (1999). The Endogeneity of Legal Regulation: Grievance Procedures as Rational Myth. *American Journal of Sociology, 105*(2), 406–454.

Elgood, J., Gilby, N., & Pearson, H. (2004). *Attitudes Towards Health and Safety: A Quantitative Survey of Stakeholder Opinion* (MORI Social Research Institute Report for the HSE). London: Crown.

Feldman, S. (1988). Structure and Consistency in Public Opinion: The Role of Core Beliefs and Values. *American Journal of Political Science, 32*(2), 416–440.

Furedi, F. (1997). *Culture of Fear: Risk-Taking and the Morality of Low Expectation.* London: Cassell.

Fyfe, P. (2013). Illustrating the Accident: Railways and the Catastrophic Picturesque in the *Illustrated London News. Victorian Periodicals Review, 46*(1), 61–91.

Geertz, C. (1975). Common Sense as a Cultural System. *The Antioch Review, 33*(1), 5–26.

Giddens, A. (1991). *Modernity and Self-Identity: Self and Society in the Late Modern Age*. Cambridge: Polity.

Goren, P. (2001). Core Principles and Policy Reasoning in Mass Publics: A Test of Two Theories. *British Journal of Political Science, 31*(1), 159–177.

Grabosky, P. N., Braithwaite, J., & Wilson, P. R. (1987). The Myth of Community Tolerance Toward White-Collar Crime. *Australia and New Zealand Journal of Criminology, 20*(1), 33–44.

Green, D. (2006). Public Opinion Versus Public Judgement About Crime: Correcting the 'Comedy of Errors'. *British Journal of Criminology, 46*(1), 131–155.

Green, J. (1997). *Risk and Misfortune: The Social Construction of Accidents*. London: UCL Press.

Gunningham, N., Kagan, R., & Thornton, D. (2004). Social License and Environmental Protection: Why Businesses Go Beyond Compliance. *Law & Social Inquiry, 29*(2), 307–341.

Harrington, R. (2003). Railway Safety and Railway Slaughter: Railway Accidents, Government and Public in Victorian Britain. *Journal of Victorian Culture, 8*(2), 187–207.

Hood, C. (2002). The Risk Game and the Blame Game. *Government and Opposition, 37*(1), 15–37.

Hough, M. (1996). People Talking About Punishment. *The Howard Journal of Criminal Justice, 35*(3), 191–213.

Hutton, N. (2005). Beyond Populist Punitiveness? *Punishment & Society, 7*(3), 243–258.

Jain, A., & Leka, S. (2016). *Occupational Health and Safety Legitimacy in the UK: A Review of Quantitative Data*. Leicester: IOSH.

Johnston, R., & McIvor, A. (2000). *Lethal Work: A History of the Asbestos Tragedy in Scotland*. East Linton: Tuckwell Press.

Johnston, R., & McIvor, A. (2004). Dangerous Work, Hard Men and Broken Bodies: Masculinity in the Clydeside Heavy Industries, c.1930–1970s. *Labour History Review, 69*(2), 135–151.

Kagan, R. A. (2001). *Adversarial Legalism: The American Way of Law*. Cambridge, MA: Harvard University Press.

King, S., Dyball, M., & Waller, L. (2005). *Public Protection Consultation Study* (Research Report RR541). Sudbury: HSE Books.

Löfstedt, R. (2005). *Risk Management in Post-Trust Societies*. London: Earthscan.

Lynch-Wood, G., & Williamson, D. (2007). The Social Licence as a Form of Regulation for Small and Medium Enterprises. *Journal of Law and Society, 34*(3), 321–341.

MacDonagh, O. (1958). The Nineteenth Century Revolution in Government: A Reappraisal. *Historical Journal, 1*(1), 52–67.

Mascini, P., Achterberg, P., & Houtman, D. (2013). Neoliberalism and Work-Related Risks: Individual or Collective Responsibilization? *Journal of Risk Research, 16*(10), 1209–1224.

McIvor, A. (2013). *Working Lives: Work in Britain Since 1945.* Basingstoke: Palgrave Macmillan.

McIvor, A., & Johnston, R. (2007). *Miners' Lung. A History of Dust Disease in British Coal Mining.* Aldershot: Ashgate.

Morris, A. (2007). Spiralling or Stabilising? The Compensation Culture and Our Propensity to Claim Damages for Personal Injury. *Modern Law Review, 70*(3), 349–378.

Petts, J., Horlick-Jones, T., & Murdock, G. (2001). *Social Amplification of Risk: The Media and the Public* (Research Report 329). Sudbury: HSE Books.

Pidgeon, N., Walls, J., Weyman, A., & Horlick-Jones, T. (2003). *Perceptions of and Trust in the Health and Safety Executive as a Risk Regulator* (Research Report 100). Sudbury: HSE Books.

Pitlik, H., & Rode, M. (2017). Individualistic Values, Institutional Trust, and Interventionist Attitudes. *Journal of Institutional Economics, 13*(3), 575–598.

Pollock, P. H., Lilie, S. A., & Vittes, M. E. (1993). Hard Issues, Core Values and Vertical Constraint: The Case of Nuclear Power. *British Journal of Political Science, 23*(1), 29–50.

Power, M. (1997). *The Audit Society: Rituals of Verification.* Oxford: Oxford University Press.

Radaelli, C. M. (1999). The Public Policy of the European Union: Whither Politics of Expertise? *Journal of European Public Policy, 6*(5), 757–774.

Rosenmerkel, S. P. (2001). Wrongfulness and Harmfulness as Components of Seriousness of White-Collar Offenses. *Journal of Contemporary Criminal Justice, 17*(4), 308–327.

Slovic, P. (1986). Informing and Educating the Public About Risk. *Risk Analysis, 6*(4), 403–415.

Slovic, P. (1987). Perception of Risk. *Science, 236,* 280–285.

Slovic, P. (1993). Perceived Risk, Trust, and Democracy. *Risk Analysis, 13*(6), 675–682.

Slovic, P. (2000). *The Perception of Risk.* London: Earthscan.

Slovic, P., Fischhoff, B., & Lichtenstein, S. (1979). Rating the Risks. *Environment, 2*(3), 14–39.

Suchman, M. (1995). Managing Legitimacy: Strategic and Institutional Approaches. *The Academy of Management Review, 20*(3), 571–610.

Teubner, G. (1987). Juridification: Concepts, Aspects, Limits, Solutions. In R. Baldwin, C. Scott, and C. Hood (Eds.), *A Reader on Regulation* (pp. 389–440). Oxford: Oxford University Press.

van der Heijden, J. (2017). Brighter and Darker Sides of Intermediation: Target-Oriented and Self-Interested Intermediaries in the Regulatory Governance of Buildings. *The ANNALS of the American Academy of Political and Social Science, 670*(1), 207–224.

Vogel, D. (2012). *The Politics of Precaution: Regulating Health, Safety and Environmental Risks in Europe and the United States.* Princeton: Princeton University Press.

Walker, G., Simmons, P., Wynne, B., & Irvine, A. (1998). *Public Perceptions of Risks Associated with Major Accident Hazards* (Contract Research Report 194/1998). Sudbury: HSE Books.

Walls, J., Pidgeon, N., Weyman, A., & Horlick-Jones, T. (2004). Critical Trust: Understanding Lay Perceptions of Health and Safety Risk Regulation. *Health, Risk, and Society, 6*(2), 133–150.

Wolfgang, M. E., Figlio, R. M., Tracy, P. E., & Singer, S. I. (1985). *The National Survey of Crime Severity.* Washington, DC: U.S. Department of Justice, Bureau of Justice Statistics.

Yeung, K. (2005). Government by Publicity Management: Sunlight or Spin? *Public Law, 2*, 360–383.

Zwetsloot, G., van Scheppingen, A. R., Bos, E. H., Dijkman, A., & Starren, A. (2013). The Core Values that Support Health, Safety, and Well-Being at Work. *Safety and Health at Work, 4*(4), 187–196.

CHAPTER 3

Shaping Health and Safety, 1800–2015

Introduction

As Chapter 2 has made clear, public attitudes towards health and safety are sophisticated, multi-layered, and encompass a range of aspects that impact on the perceived legitimacy of both health and safety in general and on regulatory action in this area. These attitudes have arisen over a long period, something which became clear when focus group members were asked to consider health and safety over time. Regulatory systems and approaches are also historically constituted phenomena, which must be placed in their proper contexts if they are to be understood (Balleisen and Brake 2014). It is important, therefore, to look at the origins of contemporary understandings of health and safety from a longer term perspective if we are to appreciate the current position.

This book's immediate concern is the period since 1960, a roughly sixty-year span during which the social, political, cultural, and economic contexts of health and safety changed dramatically. It saw the passing of the last of the Factories Acts (in 1961), the publication of the Robens Report (1972) and the passing into law of the fundamentally important Health and Safety at Work Act (HSWA) (1974), and the development of a self-regulatory framework during the 1970s and 1980s. It has witnessed downward trends in rates of workplace injury and death, but also disasters at Aberfan (1966), Flixborough (1974), and Piper Alpha (1988), the Clapham Junction (1988), Southall (1997), and Ladbroke Grove (1999) rail crashes, and the emergence of new concerns

© The Author(s) 2019
P. Almond and M. Esbester, *Health and Safety in Contemporary Britain*, https://doi.org/10.1007/978-3-030-03970-7_3

around work-related health issues. Finally, this period has seen significant changes in wider society which have also impacted upon health and safety, with traditional heavy industries and trade union membership both declining, and a growth in flexible, service sector employment and the demographic diversification of the workforce (McIvor 2013). The rising influence of new institutional structures such as the European Union (Baldwin 1996; Eichener 1997; Majone 1997; Walters 1996), the diversification of regulatory institutions (Black 2008; Gilad 2011), changes in the level of public trust towards regulators (Löfstedt 2005, 2011), and changes in the political context around issues of welfare and state intervention (Harvey 2005), have all made this a significant period for health and safety.

All of these developments will be explored in more detail in the chapters that follow. However, focusing *only* on the post-1960 period would overlook the much earlier origins of debates about the legitimacy of health and safety and about the most appropriate methods and means of regulating it. This chapter explores the development of health and safety, its legitimacy and in particular the role of the state, from the early nineteenth century through to the present. It does so in order to provide a sense of the ways in which the past has fundamentally shaped the present. Drawing extensively on existing scholarship, the chapter explores where 'legitimacy' has featured in the history of occupational health and safety and which histories of occupational health and safety have so far been written (and which have been missed), concluding with observations about the legacy of the past in the present.

Legitimacy and the Early History of Occupational Health and Safety

The concept of 'legitimacy', as discussed in Chapter 1, is largely unexplored in historical scholarship on health and safety—at least as an explicit formulation. However, much of the existing literature speaks to important legitimacy-related issues and ideas concerning the role of the state in relation to health and safety, chiefly concerning the questions of who has the right to act and the moral basis for intervening in health and safety. There is a growing field of work that relates to health and safety broadly conceived—not just in occupational settings, although this is where traditionally the strongest focus has been found, but also touching upon public health, accidents and public safety issues (Cooter and Luckin 1997: 1–16; Crook and Esbester 2016: 1–26). Many of these early interventions in the field were concerned with Victorian public health (for example, see Cox 2007; Frazer 1950; MacLeod 1967; Porter and Porter 1988; Walkowitz 1980), and demonstrate the ways in which public debates about health and the appropriate remedial actions shaped the actions which regulators were able to take. This is significant, as occupational health and safety was related to these topics and so part of a broader pattern of debate. Whilst legitimacy was not discussed in the terms theorised by more recent socio-legal scholars, it can be seen underlying these broad and congruent areas.

In terms of understanding the historical development of occupational health and safety specifically, much of the focus has been on the period before 1945. The dramatic and very visible changes in workplaces and working practices in the nineteenth century occasioned by industrialisation, urbanisation, concentration and mechanisation all impacted on worker health and safety. How these issues were to be resolved, and by whom, attracted attention from the state, trades unions and employers, all of whom feature in the historiography of the topic. This was often expressed in moral terms, though there has also been some analysis of how far regulation worked in practice. The role of the state in relation to health and safety is generally understood to have grown during the nineteenth century, albeit slowly and through a process of negotiation between interest groups (Bartrip 1983; Leka et al. 2012; MacDonagh 1958; Mills 2010; Roberts 1960: 287–293; Yarmie 1984), and as part of a wider shift from an individualist to collectivist approach (Evans 1978; Greenleaf 1983). Concerns about health and safety were initially expressed in justice-based moral terms, phrased around the employment of children and women (Bartrip 1983: 68; Humphries 1988; Kirby 2013; McIvor 1997: 126; Nardinelli 1980). Provisions were gradually extended to include adult male workers, though this was particularly controversial, as it was widely believed that the contract between worker and employer was sacrosanct (Kostal 1994: 257–279, 313–321). There was, however, increasing pressure on the state to act, notably in the final quarter of the nineteenth century coming from MPs sympathetic to the labour cause (Bartrip 1996; Clegg et al. 1964; Clegg 1985), although there has been debate about the extent to which wages and conditions were prioritised by trades unions at the expense of health and safety issues (Bartrip 2002; Bowden and Tweedale 2003; Johnston and McIvor 2007; McIvor 1997: 134–136; 2001: 130; Williams 1960: 51).

The Factory Act 1833 established the Factory Inspectorate (Bartrip and Fenn 1983), and is often regarded as *"the turning point of factory legislation"* (Hutchins and Harrison 1911: 40) because of the vast improvement in administration brought about by inspection and enforcement via the executive authority of the state. Alongside this, regulations were passed to control specific issues or processes; over time these regulations accreted into a confusing mass of detail and of overlapping and uncertain regulatory jurisdictions. There were limits to the perceived constitutional legitimacy and acceptability of state action (Bartrip 1983; Carson 1979; Ward 1962: 115); the solution often proposed was that voluntarism and self-regulation were sufficient to improve health and safety (Yarmie 1984; Esbester, forthcoming). The practical work of the inspectorates focused largely on reactive investigation and on controlling rather than eliminating dangers via cooperation with employers (McIvor 2001: 157). This has been seen as a practical response to the limited resources available to the inspectorates, as well as a means of persuading employers to go beyond the minimum standards laid down at law (Bartrip 1983: 74; Braithwaite 1985; Dawson et al. 1988: 208–210, 235; Jones 1983a: 260–262; Rhodes 1981). Treatment of health and safety issues during this period left something to be desired in a number of aspects, not least the gap between the laws as enacted and the effectiveness of their enforcement and outcomes (Bartrip 1982, 1983; Carson 1979), impacting heavily on their perceived functional legitimacy. Recalcitrant employers, as well as workers who wished to avoid state regulation, had the power to resist implementation at the point of work (McIvor 1997: 126; Mills 2010: 69–97). And sectoral coverage by inspectors was limited; until 1974, a very large number of workers remained outside the remit of the various inspectorates (Leka et al. 2012: 9). Before 1960 the focus remained largely on safety issues, as they were more easily amenable to inspection and control. Attention to health issues slowly increased, however, focused on particular 'dangerous trades', roles or substances/diseases such as lead, arsenic, anthrax and asbestos (Bartrip 2002; Harrison 1989, 1990). Finally, gender remained a significant factor, with inspectorates clustered in industries that were often male-dominated. Only in 1893 were the first Lady Inspectors appointed, tasked with examining health and safety issues specific to women (Bartrip 1996; Harrison 1989, 1990; Harrison and Nolan 2004; Livesey 2004).

The twentieth century saw the gradual creation of a welfare state, as well as further slow extension of the scope of the health and safety regulatory regime. There were moments of rather more dramatic change—notably the two World Wars (McIvor 2001: 130–144; Price 1993; Roberts 2007), but these remained exceptional and did not necessarily produce immediate lasting change. A more continuous process of change over time can be seen in relation to the Factory Inspectorate, with Acts in 1901 and 1937 attempting to consolidate existing legislation and in 1948 extending limited regulation to some new areas, including building sites. Throughout the interwar years Jones has shown how inspectors continued to prefer persuasion over punishment, with the state hoping that industry would become self-regulative (Jones 1985). This was seen, for example, in the slow and uneven formation of voluntary 'Safety First' councils, which were often cited as a reason why further state regulation was unnecessary, but which also provided something akin to an early experiment in tripartism (Jones 1983a: 265; b). Ideas of self-regulation; a focus on hardware such as machine guarding rather than processes and systems; a managerial determination to concede as little ground as possible to trades unions; and a 'victim-blaming' approach, established in the nineteenth century, remained firmly entrenched (Higgins and Tweedale 2010; Jones 1985: 223–239; McIvor 1997: 130; Nichols 1997: 12, 53–54; Rhodes 1981: 70). While the 1920s saw some safety legislation introduced, linked to a post-war *"wave of humane feeling and high aspiration for the future"* (Mess 1926: 33), the economic difficulties of the 1920s and 1930s, and the novel dangers introduced by new technologies, meant that for most workers there was little change in the standard of health and safety between 1900 and 1939 (Jones 1994: 70–71; 1985: 223–239; McIvor 1997: 125, 131).

The post-World War 2 Labour government paid more attention to health and safety, as part of a growing corporatism and wide-ranging, cross-party interest in consensual relations and improving welfare (Ackers 2014: 64; McIvor 2001: 227–229; Wrigley 1996: 38–39), enhancing the democratic legitimacy of the issue. New inspectorates were established in emergent sectors (such as the Nuclear Installations Inspectorate, 1959) or those where concern was increasing (such as Agriculture, 1956). In addition, nationalisation brought under state control many industries that had been reluctant to engage with health and safety regulation, including the railway and coal industries. In the twentieth century, as in the nineteenth, safety dominated attention, though some health issues were gradually recognised (Long 2011), including those surrounding asbestos (Bartrip 1998, 2001, 2014; Greenberg and Wikeley 1999; Tweedale and Hanson 1998; Tweedale 2000) and stress (Melling 2014; Jackson 2015). A final significant point to note for the pre-1960 period is the emergence of an idea of a safety movement and a professional cadre of health and safety officials. International in its origins (Swuste et al. 2010), drawing from the USA's 'Safety First' movement (Aldrich 1997), the railway industry introduced the first sustained workplace safety campaign in 1913 (Esbester, forthcoming). This voluntary effort spread across other industries; with it came officials later to call themselves 'safety officers'. The British Industrial 'Safety First' Association was formed in 1917, going on to be part of the National 'Safety First' Association in 1923, renamed RoSPA in 1941. From within RoSPA, industrial safety officers coalesced to form the Institution of Industrial Safety Officers, eventually becoming the Institution of Occupational Safety and Health; and the British Safety Council was formed in 1957. Such bodies all spoke to the increasing prominence of health and safety issues and practice in the twentieth century and played their part in informing debates about health and safety and its regulation.

1960–1970: Challenges to the Existing Regime

Much less analysis has been undertaken of the post-1960 period, and hence the historiographical engagement here decreases; instead it is necessary to introduce a more narrative element, supported by archival and oral history accounts. During the 1960s and early 1970s, questions about the efficacy of the health and safety regime became more pronounced. The Factories Act 1961 had consolidated previous legislation but introduced no big changes, and factories were inspected on a four-yearly cycle, regardless of the relative risks posed by their operations; the law continued to prescribe particular standards, often in minute detail; education remained focused on the 'careless worker'; and persuasion remained core to the inspectors' approach, as they still lacked strong enforcement powers. As the Chief Inspector of Factories noted in his 1969 report, "*better compliance for most of the time can be secured in most premises if one persuades the occupier of the need for compliance as a matter of good practice, rather than to avoid conflict with the law*" (Department of Employment and Productivity 1970: xii). Correspondingly, voluntary action remained a core part of the health and safety system, although results appear to have been disappointing: in agriculture the failure to establish active safety committees was repeatedly commented upon[1] and in 1967, a Factory Inspectorate survey revealed only 33% of SMEs (50–500 employees) had joint safety committees in place.[2] Similarly, in 1968, the Trades Union Congress (TUC) representatives to the Industrial Safety Advisory Council's Joint Safety Organisation noted that *"existing safety organisation reflects the piece-meal nature of its development and is not capable of meeting present day needs"*.[3] This was partly a product of employers' continued preference for a narrow conception of the regulators' constitutional and democratic legitimacy, based upon a dislike of formal bodies that gave a place to unions and which impinged upon managerial autonomy. The trade unions were relatively powerful during this period and, although industrial relations were at times tense (Wrigley 1996), for the most part the 1960s were marked by a relative degree of cooperation and understanding, though with wide variations across sectors (McIvor 2013: 201–227). There was also a growing interest in the implications of health and safety for efficient management, driven, so far as the state was concerned, by a desire to promote self-regulation (Sirrs 2016b: 319–323).

The most serious challenge to the current regime was posed by the rising numbers of workplace deaths and injuries. 1961 was the third year in a row in which casualties increased, rising from nearly 168,000 in 1958 to over 192,500 in 1961 (Ministry of Labour 1962: 7). At the same time, many millions of workers remained outside the scope of state regulation, and it was known that as many as 40–66% of all casualties went unreported.[4] By 1969, just under 323,000 incidents were reported to the Factory Inspectorate (Department of Employment and Productivity 1970: xiv). Such rises increasingly challenged the

---

[1] RoSPA, 'Sixth Annual Report of Head of Agricultural Safety Division', 12 June 1964, p. 2 (National Archives Scotland, Edinburgh (NAS), AF59.152); RoSPA, Minutes of the National Agricultural Safety Committee, 24 September 1965, p. 2 (RoSPA Archives, Birmingham, D/266/2/25).

[2] Untitled Memorandum, Chief Inspector of Factories to Managing Directors and Managers of Firms Employing More Than 50 (n.d., c. 1969), p. 1. Modern Records Centre, University of Warwick (MRC), MSS.292D/146/18/1.

[3] 'TUC Representatives' Comments, Industrial Safety Advisory Council, Joint Safety Organisation', 10 July 1968, p. 1. MRC, MSS.292B/146/17/2.

[4] TUC, 'National Joint Advisory Council, Industrial Safety Sub-Committee, Notes for Meeting 17 November 1964', p. 1. MRC, MSS.292B/146.17.1.

existing regulatory system, suggesting that it was failing to protect workers and failing to keep pace with industrial and economic changes. The old economic bases of the British economy were slowly being eroded, as heavy industries like mining, manufacturing, and shipbuilding started to decline. The pace of technological change rendered prescriptive legislation ineffective, as by the time new legislation was enacted, the ill that was being remedied had often been replaced by another. As one source noted *"The 1961 Factory Act had not been successful in regulating the use of chemicals [...] the law did not keep up with the vast range of new materials and processes being introduced onto the shop-floor"*.[5] At the same time, other occupations started to grow in importance in the economy, notably office work and the retail sector; inspectors, unions and employers had to negotiate what was appropriate for these new environments.

On the other hand, across this period, there remained a broad (if often tacit) acceptance of health and safety regulation's cognitive legitimacy as both necessary and important; indeed there was also a moral component to this acceptance, with the Chief Inspector of Factories arguing in his 1963 report that there was an imperative *"to accept responsibilities that are wider than those imposed by law [...] Legislation cannot be a complete guide to what should be done"* (Ministry of Labour 1964: 47). Although additional inspectors were appointed across this period, by the early 1970s, there were still calls for further increases from organisations such as the Dock and Harbour Authorities' Association[6] and RoSPA.[7] There were, of course, some responses to the changing conditions: most notably, in 1963, with the Offices, Shops, and Railway Premises Act, which extended regulation to an additional one million non-industrial workplaces (Allen 1966: 691) and the establishment of an Industrial Hygiene Unit in 1966 to address occupational health issues more scientifically. The Asbestos Regulations of 1969 reflected ongoing concern about a long-term, though increasingly visible, issue; in his report of the same year, the Chief Inspector of Factories observed that *"at least as much effort must be put into the control of toxic contaminants*

---

[5] R. Williams, 'A Key Issue', *Work Hazards* 21 (c. 1979), p. 5. Samuel Barr Collection, Glasgow Caledonian University, GCU DC 140/2/1/2.

[6] Written Evidence of the Dock and Harbour Authorities' Association to the Robens Committee, (n.d., c. 1971), p. 3. TNA, LAB 96/57.

[7] Written Evidence of RoSPA to the Robens Committee, (n.d., c. 1971), p. 7. TNA, LAB 96/75.

in the atmosphere as into the elimination of the physical causes of accidents" (Department of Employment and Productivity 1970: xiv).

Finally, occupational health and safety took on an occasional public dimension, with several major incidents affecting people outside the workplace: but despite events like the Brent Cross crane collapse in 1964 and the collapse of the coal tip at Aberfan in 1966, widespread public interest in occupational health and safety remained relatively limited (McLean and Johnes 2000; Pantti and Wahl-Jorgensen 2011). According to a 1971 article in the *Times*, "*Industrial safety is a dry subject which arouses the passions of a limited number of people directly concerned with preventing accidents at work*" (Spiegelberg 1971: 5). Indeed, there was a prominent impression that there was widespread apathy towards health and safety, even amongst workers and managers; in 1966 one inspector commented that there was an "*almost complete indifference on the part of the individual workers as to their health, welfare and safety during their working day*" (Allen 1966: 691). Nevertheless, from a variety of sources, there was an increasing feeling that the existing health and safety regime was no longer fit for purpose. Whilst several attempts to introduce new legislation in the latter half of the 1960s failed, they did lead to a formal investigation of the condition of the British health and safety system—the Committee on Safety and Health at Work, or Robens Committee (Sirrs 2016a).

1970–1979: ROBENS, THE 1974 ACT, AND THE NEW ERA

Contextually, the 1970s saw some significant changes occur within the British economy and workplace, including to the relative importance to the economy of heavy industry and service sector work. Numbers employed in the textile industry, for example, declined from 309,000 in 1971 to 176,000 by 1981, whereas administrators and managers increased from 923,000 to 1,343,000 in the same period (Mitchell 1988: 107). One heavy industry that saw considerable expansion was the offshore oil and gas sector. This was a period of significant exploration and production (Brotherstone and Manson 2007), with the Department of Energy responsible for health and safety as well as production, a dual role that would later be criticised (Carson 1982; Paterson 2007). There was a slight increase in the percentage of women as part of the overall working population, from 37% in 1971 to 39% in 1981 (Mitchell 1988: 107). Trade unionism was strong, with membership remaining at around 11 million for most of the 1970s and peaking at over 13 million in 1979 (BIS 2014: 21). Connected to this growth, industrial relations became more conflictual as the decade wore on, with a number of high-profile disputes including the coal miners' strike in 1974 and the 'winter of discontent' of 1978–1979. There were a number of economic difficulties, including the oil crisis of 1973 and ongoing struggles to control inflation; by the end of the decade, these pressures started to impact on health and safety, noted in the Health and Safety Executive (HSE)'s 1978 annual report as producing a functional legitimacy-based, and *"much more overt questioning of the costs and benefits of health and safety legislation"* (HSE 1980: v).

In May 1970, Barbara Castle, Labour MP and Secretary of State for Employment and Productivity, appointed a committee of inquiry to examine health and safety in the broadest possible terms; this was the first time that such an extensive overview had been undertaken. The committee was chaired by Lord Alfred Robens, formerly a Labour MP and at that point Chair of the National Coal Board; it comprised a further six members, including a Conservative MP, a trade unionist, an academic lawyer, and a radiologist. It sat for two years from June 1970 and received written evidence from 183 organisations or individuals, as well as further oral testimony and site visits in the UK and abroad. The Committee's terms of reference were huge: *"to review the provision made for the safety and health of persons in the course of their employment"* and the need for any changes in relation to the law. In a new departure, the Committee was specifically tasked with considering how the public might best be protected from dangers produced in, but extending beyond, the workplace (Robens 1972: v). There were some exclusions, including transport workers (covered by other provisions), the movement of hazardous goods, compensation and environmental pollution, although the Committee noted that these issues were raised in evidence, as *"[s]afety and health at work is not a subject that is easily delimited"* (Robens 1972: para. 3). The Committee concluded that the existing regulatory system was no longer fit for purpose and deaths, injuries, suffering and economic loss were *"unacceptably high"* (Robens 1972: para. 456). The *"haphazard mass of law which is intricate in detail, unprogressive, often difficult to comprehend and difficult to amend and keep up to date"* was a significant problem, along with *"excessively fragmented"* administrative arrangements (Robens 1972: para. 458).

A new regime was called for, covering more workers as well as the public, with greater flexibility and less prescription. Apathy, a product of the existing approach which encouraged people "*to think and behave as if safety and health at work were primarily a matter of detailed regulation by external agencies*", was the "*greatest single obstacle*" to improving health and safety (Robens 1972: paras. 456–457). Instead, the Committee called for greater self-regulation, viewing employers and employees as best placed to ameliorate them. It was recommended that existing legislation be simplified under a single Act to be administered by a new national body composed of many of the existing inspectorates (Robens 1972: paras. 91–115). This would be supplemented by additional regulations and voluntary standards and codes of practice "*developed within industry and by independent bodies*". Aware that this was likely to be controversial, especially amongst trades unionists, it was explicitly stated that "*[w]e are not advocating a slacker approach*" but one which was "*more flexibly based*" (Robens 1972: para. 148). Finally, the Committee recommended that employees be consulted about health and safety matters, but did not call for the compulsory adoption of safety committees as favoured by trades unions (Robens 1972: paras. 68–71). At the heart of the Committee's approach was the idea that there was a natural "*identity of interest*" between employers and workers (Robens 1972: para. 66), underpinning the democratic legitimacy claims that the law could make. To this end, it was proposed that key stakeholder groups were "*to play an effective part in the management of the new institution*" (Robens 1972: para. 114); these would include industrial managers, trades unions, medical practitioners and local authority representatives. In this way, the Committee tried to secure legitimacy for the new regime from stakeholders by involving them—though notably absent were the public. The approach taken by the Robens Committee and the subsequent HSWA 1974, then, is generally seen as the zenith of this consensual approach and the corporatist, tripartite decision-making systems that implemented it.

Now at over 40 years' distance from the Committee and Report, it is perhaps hard to give a sense of just how radical they were. The changes they heralded were huge, although a product of their time (Sirrs 2016a). Politically, there was broad cross-party support—the 1974 Act stemmed from a Bill introduced by the 1970–1974 Conservative government, based on a report commissioned by the 1966–1970 Labour one, that was then carried into legislation, with only minor amendments, by the 1974 minority Labour government. Commentators in the press endorsed self-regulation, as in a May 1971 article which claimed

"How far Government should consider itself responsible for safety at work is a highly debateable point [...] If there is any general agreement, it is that beyond a certain point the responsibility must lie with industry" (Spiegelberg 1971: 5). However, there were voices of dissent, typically from the political left which tended to favour a stronger state role. One Amalgamated Union of Engineering Workers representative told the TUC that *"we deplore the suggestion in the [Robens] Report that there is too much law. We reject this entirely. How ridiculous can you get, making a suggestion like this?"*[8] Lady Summerskill, a Labour peer, was reported as arguing that Robens' report *"would bring rejoicing to the hearts of every irresponsible employer of labour who used the voluntary approach as an escape from his obligations"* (*Times* 1973: 7). Caught in the middle, the Chief Inspector of Factories stated that *"impartial we must be, for no side – employer or workforce – has a monopoly of rectitude in safety and health at work. We must also exercise a strict impartiality if we are to hold – and deserve to hold – the trust of managers and workers"* (Department of Employment 1973: xiii). Academics have criticised Robens' conclusions, arguing that business interests dominated (Beck and Woolfson 2000), while public awareness of Robens' recommendations was much more limited, and the response more muted.

In brief, the 1974 Act followed the recommendations of the Robens Committee closely. It unified most of the multiple existing inspectorates under a body that became the HSE (the key exceptions to this were the railway and offshore industries). The old prescriptive law was replaced with codes of practice and general duties to reduce risks 'so far as is reasonably practicable'. This was intended to allow flexibility in creating and determining appropriate standards, and placed proportionality at the heart of the system, judging the elimination of risks against the cost and difficulties of making changes. Coverage was extended to include virtually every worker, and, via Sect. "1960–1970: Challenges to the Existing Regime", those members of the public who might be endangered by workplace activities. As well as establishing the HSE, the 1974 Act created the Health and Safety Commission (HSC). This was the management board, tripartite following Robens' recommendations, comprising representatives from industry, trade unions and local government, as well as safety experts and policy-makers.

---

[8]Verbatim Report of TUC Conference on the Robens Committee Report, 12 October 1972, p. 36. TUC Library, HD 7273.

This paved the way for some degree of accommodation between interest groups, enhancing the democratic legitimacy claims of the policy-making process while also, to some extent, depoliticising health and safety issues and downplaying issues of justice-based legitimacy. As Rex Symons, a former HSC Commissioner (1989–2002), observed:

> the politics were completely submerged in the issues. The fact that I've sat opposite a trade unionist and he sat opposite me and we probably knew how we both voted didn't make the slightest difference (Rex Symons Interview, para. 58).

It has been suggested that the functioning of tripartism in practice was less collaborative, representative, or equal than Robens intended (Baldwin 1987: 139; Dawson et al. 1988: 202; Hutter 1997: 24). For example, one letter to the journal of the National Union of Agricultural and Allied Workers claimed in 1979 that the HSWA was "*in too many cases [...] sloppy, out of date, ineffectively worded and biased to the extreme in the general direction of the farmer or employer*".[9] The presence of tripartism did not prevent accusations that Robens and the HSWA remained essentially limited by the political power of business.

In practice, inspectors had an additional five million workers to cover, which even despite some increases in resources and staffing led to criticism from the political left: in 1975 the *Morning Star* claimed that "*[t]he present 800 factory inspectors have proved ludicrously inadequate to cope with the safety and health of hundreds of thousands of workers*" (Paterson 1975: 4). Whilst not perhaps stretching to this end of the spectrum, the rhetoric of the identity of interest was noted publicly by the HSE as negotiable:

> it is too simplistic to assume that there will always be an identity of approach between management and trade unions [...] The severity or extent of a potential hazard, the costs and benefits of [...] preventative measures and the allocation of priorities are all matters of legitimate discussion between managers, safety representatives and inspectors. (HSE 1980: v)

The importance of worker action (in terms of democratic legitimacy) was noted by the HSE and the labour movement, and although the

---

[9]J. Tubby, 'Safety Act Not Good Enough', *The Land Worker*, April 1979, p. 6. Museum of English Rural Life (MERL), University of Reading.

compulsory appointment of safety representatives had proven a controversial omission from the HSWA, allowance was eventually made for their appointment in 1977. Some, like an ICI plant manager in 1970s, saw the impact of the HSWA:

> I think the unions were very active and effective in promoting health and safety until the advent of the Act [...] whereas at the beginning I hadn't thought much about it by the mid-1970s everybody's safety consciousness and health consciousness increased.[10]

As in earlier years, the extent to which trade unions prioritised health and safety issues has been debated, though careful acknowledgement of the constraints under which unions acted is vital (Bowden and Tweedale 2003; Higgison 2005; Walters 1996). Safety remained the major focus, though the Employment Medical Advisory Service (EMAS) was set up in 1972 to bring together medical practitioners working within the field of occupational health (Johnston and McIvor 2000b).

The aspirations of political decision-makers to create a broadly apolitical regulatory body led to the HSE's creation as a non-Departmental Government body, remaining within state oversight via a 'sponsor' minister, initially the Secretary of State for Employment. For Robens, this was a matter of ensuring operational independence: "*it should have autonomy in its day-to-day operations [...] it should be allowed to do it without unnecessary interference*" (Robens 1972: para. 112). This would ensure that it retained a healthy degree of constitutional legitimacy as both fair and impartial, but also appropriately accountable to government. For politicians, it arguably was a means of insulating issues of health and safety from the cut-and-thrust of wider industrial relations and politics at the time by creating a space for these issues to be settled where neither militant unionism nor deregulatory conservatism could exert undue influence (Wilson 1983: 187). This initial consensus did not extend to the creation of the HSE within Whitehall, a process Michael Foot described as "*a first-class Whitehall row*" (Hutter 1997: 27). According to Jim Hammer, first Chief Inspector of Factories under the HSE, the Home Office, MOD and Departments of Health and Education were

---

[10]Brian Watson, Interviewed by David Walker (2005), Scottish Oral History Centre. University of Strathclyde Archives and Special Collections, GB249 SOHC7 (2004–2005).

extremely concerned at the thought of the all-embracing obligations being applied to the police, prison service, fire brigades, hospitals, schools, universities, secure Establishments and the armed forces, not any longer to be protected by Crown Immunity.

Perhaps more importantly, departments hosting the existing inspectorates were reluctant to lose 'their' staff: "*every department with an Inspectorate proposed for integration, fought to retain their own*" (Hammer 2014: 1). Unification of the HSE was a long-running, and bumpy, process of reconciling internal rivalries and competing institutional interests, and is discussed in greater depth in Chapter 4.

One of the most significant features of these reforms was the determination to systematise thinking about health and safety which had previously taken place in a rather piecemeal, isolated fashion, spread between inspectors, company safety officers, trade union representatives, professional organisations such as IOSH, RoSPA and the BSC, and on the shopfloor. The HSWA did more than just create a unified regulator; it promoted the *idea* of health and safety and put in place practical measures by which those who had a professional interest in the area might coordinate efforts (Rimington 2008: 3). This included the notion of health and safety as an integral part of good management, in keeping with thinking emerging at the time (seen, for example, in the 'total loss control' system promoted by the British Safety Council in the early 1970s) (Industrial Health and Safety Bulletin 1971: 7). Another fundamental point was the expansion of the scope of health and safety to include the public, which came to be of great importance in the 1980s, and which was very much in line with the beliefs of some in the labour movement; in 1976, for example, trade unionist Patrick Kinnersly noted that in the wake of incidents like Flixborough "*[a]gain and again we find that workplace and community hazards are interlinked […] the task [is] building a united front to fight pollution inside and outside the plant*" (Kinnersly 1976: 4, 7).

1979–1990: Burdens, Risk, and Crisis

In many respects, the trends of the 1960s and 1970s intensified in the 1980s: particularly de-industrialisation and economic crisis and growth. Yet, the 1980s also saw significant new challenges and opportunities for the legitimation of health and safety. The economic context of the early 1980s remained difficult, with a continuing push to reduce government spending provoking the HSC to state as early as July 1980 that it would be unable to fulfil its legal obligations if further cuts were imposed (*Times* 1980: 3). Yet, partly, the focus of attention was political, reflecting an emerging Conservative agenda that would become known as neo-liberalism and which was focused on rolling back the frontiers of the state (Gamble 1994; Harvey 2005; Tucker 1995). This produced an political environment that was increasingly hostile to regulation; multiple reviews of state involvement in social and economic life, including the White Papers *Lifting the Burden* (1985) and *Building Businesses...Not Barriers* (1986), focused on the perceived disadvantages of regulation and 'red tape', often in largely economic terms, as well as giving business more of a voice in the creation of regulations, and potentially reducing routine inspections for some businesses. Unsurprisingly, this latter point was strongly opposed (*Building Businesses...Not Barriers* 1986: 38). What is clear is that the 1980s saw both an increasing dominance of ideas of free-market enterprise and notions of health and safety becoming more contested (Baldwin 1987; Dawson et al. 1988; Hawkins 2002; Hutter and Manning 1990; Pearce and Tombs 1990).

One of the reasons that the legitimacy of health and safety regulation was increasingly questioned was the perception that it was seen as part of an outdated, interventionist post-war welfare state, and closely connected with declining traditional industries (Beck and Woolfson 2000; Hutter and Manning 1990: 107–110; Leka et al. 2012: 32). Broadly, this period saw the gradual decline of consensus politics, including tripartism, as Roger Bibbings, formerly of the TUC, recalled:

> Trades Unions' participation in government had been fashionable in 1972, but no longer was in 1992. (Roger Bibbings Interview, para. 19)

The complex realities around this political movement are assessed in Chapter 7. The 1980s also saw a decline in trade union membership, from the high in 1979 of over 13 million to something under 10 million in 1990 (BIS 2014: 21); there was a consequent decrease in political power enjoyed by the unions, and industrial relations became increasingly fractious, especially in the older, declining industries, with a number of high-profile strikes, like the miners' strike of 1984–1985. Existing accounts have stressed the impact of this anti-regulatory environment on the HSE during the 1980s, including having to use cost–benefit analysis when creating regulations, increased calls for accountability in decision-making, and reduced resources (Baldwin 1987; Dawson et al. 1988; Hawkins 2002). Britain's relationship with Europe was also starting to factor in health and safety matters, though often it was of greater concern to politicians and policy-makers than the wider public. So, John Rimington, Director General of the HSE 1983–1995, remembered that:

> By 1986 "Europe" was producing regulations which HSE (under supervision from the Department of Employment and the Foreign Office) had to negotiate…as time went on, several strands of political phobia became engaged and entangled in the process. These were (1) dislike of "European" interference in "British" law; (2) the idea of the "nanny state"; (3) the traditional Tory dislike of "burdens on business" and susceptibility to the views of small companies. (John Rimington Interview, para. 66)

During the 1980s, the public and public opinion became more prominent in legitimacy debates around health and safety (Almond and Esbester 2016; Burgess 2011; Leka et al. 2012: 29). Debates and controversies over health and safety were increasingly played out in the media, and so reached well beyond those who might have otherwise had a 'vested' interest in the area, such as regulators, politicians, trades unionists and employers. Although Sect. "1960–1970: Challenges to the Existing Regime" of the HSWA imposed a duty for the HSE to protect the public from workplace hazards, for the first decade or so of its operation, this was a relatively low priority. Partly this was an understandable lack of appreciation of the gravity of this new requirement, and partly a practical response to the need to continue day-to-day service under the new organisational arrangements. It was believed that the views of the public, where relevant, would be captured via the HSC's public-facing (if not directly public-representing) constituent members: "*The public interest is still well safeguarded by the local authority representatives on the Commission, especially the general public interest*" (Employment Committee 1982: para. 114). Gradually, though, the HSE came to take on more roles in which the health and safety of the public were a consideration, including, for example, domestic gas safety, the use of pesticides and the transport of dangerous goods by road. The political scrutiny to which health and safety regulation was subjected was of course reported in the media, further heightening the visibility of health and safety in general. So, for example, in 1988, *The Guardian* observed that:

[t]he Health and Safety Executive has been severely cut back [...] Public concern about industrial hazards - from the Chernobyl and Bhopal disasters to the alleged dangers of visual display units - has never been higher. But our capacity to study and control them is being progressively eroded. (Laurance 1988: 23)

One approach that the HSE developed, responding to public pressures around issues such as nuclear safety, was the Tolerability of Risk Framework (1988). This model explicitly included as part of its matrix justice-based societal values about the acceptability of hazards, to be balanced against scientific calculations of probabilities and costs of prevention (Bandle 2007; McQuaid 2007). It equipped the HSE with a tool with which to assess and respond to not just risks but also public attitudes, crucial around controversial issues like nuclear safety. Older, heavy industries which had long acted as the established 'front line' in terms of health and safety became a less prominent part of the British economy, meaning that the focus of health and safety concerns altered. This greater diversity of employment arguably meant that day-to-day health and safety issues garnered less immediate public attention. While new frontiers and issues did emerge (workplace stress, display screen equipment), these arguably commanded less unequivocal support as health and safety issues. Health also received increasing attention, including the introduction of COSHH (Control of Substances Hazardous to Health) Regulations in 1988. Beyond such changes in day-to-day health and safety concerns, in the late 1980s, a series of major workplace and public disasters focused public attention on health and safety and its regulation. These included the *Herald of Free Enterprise* ferry capsize and King's Cross fire (1987), Piper Alpha offshore platform explosion, Clapham Junction rail crash (1988), and sinking of the *Marchioness* on the Thames (1989). All had important ramifications, but the sense of crisis prevailing at the time seems to have given the HSE some degree of protection against the deregulatory agenda, stimulating demands for greater protection.

1990–2000: Managerialism and Governance

In many respects, the 1990s opened with a continued sense of profound uncertainty following the high-profile disasters of the late 1980s. The HSE accrued new regulatory responsibilities for the offshore (1991) and rail industries (1990); both of which had previously been regulated by officials based within the government departments that bore over-all responsibility for their governance (Energy and Transport, respec-tively). Following the disasters at Piper Alpha and Clapham Junction, this arrangement came to be seen by some as giving rise to conflicts of interest (Brotherstone and Manson 2011), and so was changed. These moves to shore up the constitutional legitimacy of regulation depoliti-cised health and safety, diminished public and political criticism (albeit only temporarily for the rail industry, with new concerns about regu-lation emerging at the end of the decade). Piper Alpha was viewed by some as a game-changer—for offshore oil and gas production, but also for health and safety more widely, as Peter Jacques, then Head of the TUC's Social Insurance and Welfare Department, recalled: "*Piper Alpha actually kicked health and safety back into the centre of [...] industrial life, and maybe [...] political life*" (Peter Jacques Interview, para. 139).

At the same time, the 1990s also witnessed a shift towards what polit-ical scientists have called the 'New Public Management' in relation to government more widely (Hood 1991; Baldwin and Cave 1999: Ch. 6). This approach to policy-making sought greater constitutional legitimacy, via accountability and efficiency, in the public sector, achieved particu-larly through a movement towards enhanced private sector involvement through deregulation and privatisation (including the creation of a Deregulation Unit within the Cabinet Office). Decision-making was to be based on principles of rational cost–benefit analysis, performance tar-gets and monitoring, and a culture of 'managerialism'. In practice, this meant that regulators found some of their capacity to act was reduced as businesses were expected to take further steps towards self-regulation. The continued decline of trade union membership (down from 9.8 mil-lion in 1990 to 7.9 million in 2000) and winnowing of heavy industries traditionally concerned with health and safety also meant that the politi-cal prominence of health and safety also decreased. This was not radically altered even with the election of the Labour government in 1997 (Beck and Woolfson 2000). Health and safety issues did receive some new sup-port, notably in the 'Revitalising Health and Safety' initiative announced in 1999, but the general suspicion of regulation continued.

The year 1992 turned out to be a crucial year, with the enactment into UK law of the 'six-pack' of regulations, wide-ranging regulatory tools that originated in European Community initiatives, and which in some ways reintroduced elements of regulatory prescriptiveness which were in tension with the more risk-based approach contained within the 1974 Act (Löfstedt 2011; Rothstein et al. 2015; Vogel 2012). With the increasing Euroscepticism that emerged from the mid-1980s, this set of regulations posed further challenges to the constitutional legitimacy of health and safety by establishing it as an area in which European 'interference' was threatening 'traditional' British freedoms. Conservative suspicion also manifested itself in 1992 when the HSC started a review of health and safety legislation, with a view to removing 'unnecessary' regulation (HSC 1994). Whilst the principles of health and safety regulation were found to be sound, seven pieces of legislation and 100 sets of regulations were recommended for removal, with further simplification of the health and safety regime also recommended. Finally, the tolerability of risk framework was revised in 1992, tailoring approaches developed in the nuclear industry to lower hazard sectors (Bandle 2007). This reflected the growing importance (in terms of democratic legitimacy) of public attitudes towards health and safety, as well as an increasing recognition of the duties towards the public that existed under HSWA Sect. "1960–1970: Challenges to the Existing Regime".

Some of the most significant challenges to health and safety during the 1990s came from the shift from a nationalised to a privatised railway system. Between 1993 and 1996 the entire industry was restructured, with over 100 companies becoming responsible for various aspects of operation, including, of course, health and safety of employees and passengers. As in the offshore industry, the 'safety case' regime was introduced to determine permission to operate: plans were submitted by potential service providers showing how risks would be managed, via goal-setting rather than via more prescriptive regulations. However, two major incidents in the closing years of the 1990s called the health and safety system, including the role of the HSE, into question: the crashes at Southall in 1997 (7 fatalities and 139 injuries) and Ladbroke Grove in 1999 (31 fatalities and 523 injuries). One further case, beyond the railway industry, was also significant: the Lyme Bay canoe tragedy of 1993, in which four schoolchildren drowned whilst taking part in a trip run by an outdoor activity centre (Wells 2001). The case attracted significant public concern about perceived risk aversion as well as the limited regulatory oversight of adventure activities. Such high-profile incidents temporarily raised the public visibility of health and safety and politicised health and safety further.

2000–2010: REGULATING IN PUBLIC

The period after 2000 saw a profound hardening of attitudes—public and political—towards health and safety. Despite some apparently positive moves, the New Labour governments retained a lingering suspicion of health and safety, possibly resulting from a combination of the continuing politicisation of the issues, a commitment to a business-friendly agenda, and growing media hostility to health and safety (Almond 2009; Dodds 2006; Tombs and Whyte 2010). The publication of the 'Revitalising Health and Safety' strategy in 2000 marked perhaps the most proactive attempt of the period to shape debate and practice, being championed by the Deputy Prime Minister and with the self-professed intention *of "injecting new impetus to better health and safety in all workplaces"* (DETR 2000: 1). Subsequent HSE enforcement strategies, such as the 'Fit3'[11] initiative, sought to proactively address areas and issues of particular quantitative concern (such as 'slips, trips and falls'). Nevertheless, health and safety continued to fall within the remit of wider reviews of regulatory practice, including the Hampton (2005) and Macrory (2006) reviews, which were couched in by-now familiar terms of reducing burdens upon businesses. The political suspicion of regulation that was established during the 1980s remained in place and even grew, particularly following the start of the economic crisis in 2007.

Intimately connected with this growing political hostility towards health and safety were the democratic and justice-based legitimacy challenges posed by an increasing public scrutiny, mainly expressed via an increasingly hostile media. Whilst media awareness of health and safety issues was not new, what was different was the seemingly relentless nature of the negative coverage, particularly in the tabloid press (Almond 2009; Ball and Ball-King 2013). So, in an early but extreme articulation, a feature in *The Times* in November 2000 railed against the *"Hitlerite absolutism"* of *"health and safety weirdos [who] sit in health and safety headquarters congratulating themselves on their patronising dogmas and total lack of accountability"* (Jay 2000: 9[s]). Such stories were readily seized upon by public and politicians alike where they confirmed existing beliefs about the appropriate role for the state. In opposition, elements within the Conservative party became increasingly vocal in their condemnation of

---

[11]'Fit for Work, Fit for Life, Fit for tomorrow'. http://www.hse.gov.uk/statistics/publications/fit3.htm.

'health and safety gone mad': leader of the party David Cameron stated in 2008 that *"this whole health and safety, human rights act culture, has infected every part of our life"*.[12] There was an increasing emphasis on the ways in which health and safety supposedly reduced discretion and increased risk aversion, ignoring the principles of sensible risk management which had been put forward by bodies like the HSE for many years (HSE 2006). The rhetoric became so widespread that the HSE felt it necessary to counter the negative press by releasing, from April 2007, a 'myth of the month'.

Major incidents continued to affect the public, posing significant questions about health and safety. These included the explosions at the Buncefield oil depot (2005), which, despite not inflicting anything like the same level of injury and harm as the Flixborough explosions 30 years before, had a dramatic impact on public perceptions and awareness of risk, thus demonstrating once again the capacity that major industrial hazard sites have to shape discourses about safety. The railway industry also continued to incur a number of high-profile incidents, including passenger crashes at Hatfield (2000), Potter's Bar (2002) and Ufton Nervet (2004), as well as the Tebay incident (2004), in which four workers were killed by a runaway wagon. These incidents were widely viewed as indicting both the privatised system and the regulatory regime, and in 2006 responsibility for the railway industry was removed from the HSE and handed to the Office for Rail Regulation (now known as the ORR). These incidents were one of the drivers behind the introduction in 2007 of the Corporate Manslaughter and Corporate Homicide Act, which criminalised companies which caused the death of workers or members of the public (Almond 2013). The other major change was the formal merging of the HSC into the HSE in 2008, signalling the end of the experiment with a separate corporatist policy-making body. Given the continuing strength of understandings that self-regulation was important, that businesses should play a role in shaping regulation, and declining trade union membership, the ways in which the health and safety agenda was shaped changed significantly in the 2000s. In addition, it was necessary for all actors involved—regulators, employers, trades unions, or health and safety professionals—to

[12] Speech to the Conservative Party Conference, 1 October 2008; Reported by *The Guardian*, at http://www.guardian.co.uk/politics/2008/oct/01/davidcameron.toryconference1.

take more formal account of public opinion (a trend seen in other areas of political life at this time: Bohmann 2000; Chambers 2003; Fishkin 2009).

2010 ONWARDS: RED TAPE AND REVIEW

The publicly expressed hostility towards health and safety that mounted from 2000 continued unabated in the period after 2010 (Almond 2015). Similarly, the existing pressures around regulation and the politicisation of health and safety continued to develop, particularly with the Conservatives leading the coalition government after the 2010 election and the ongoing economic crisis. As Prime Minister, David Cameron continued to express the view that health and safety regulation was a problem, declaring his 2012 "*New Year's resolution: to kill off the health and safety culture for good. I want 2012 to [be] the year we get a lot of this pointless time-wasting out of the British economy and British life once and for all*" (Kirkup 2012). Such extreme statements reinforced the prevailing negative tone surrounding health and safety. One of the early actions of the new coalition government was to commission a review (Young 2010) of the alleged compensation culture and the ways in which health and safety law was applied, a move which was seen to "*delight the Tory leader's spurned right wing, with the issue of over-restrictive rules filling many MPs' postbags*" (Stratton 2010). One of the significant preoccupations at the heart of the Young Review was over the *perception* of health and safety legislation, an acknowledgement of the extent to which its legitimacy was being challenged. At the same time, this and subsequent reviews appeared to open up the area of health and safety to still further questioning. Part of the Young Report also focused on the rise of the health and safety professional, seeing the need for appropriate professional validation, reflecting wider concerns about the proliferation of unnecessary advice (Young 2010: 15). As in the 1960s, legislation was perceived as having become over-complex, and a desire for simplification was expressed (Young 2010: 16).

This was followed in 2011 by the Löfstedt Review, commissioned *to "look into the scope for reducing the burden of health and safety regulation on business*" (Löfstedt 2011: 1). Recommendations which were subsequently enacted included exempting some self-employed workers from health and safety law, a review of all Advisory Codes of Practice, and the now-familiar call for further simplification of the regulatory framework.

Significantly, so far as the legitimacy of health and safety is concerned, Löfstedt expressed concern about the effects of a *"constant stream of stories in the press blaming health and safety [...] for preventing individuals from engaging in socially beneficial activity, overriding common sense and eroding personal responsibility"* (2011: 16). As in the 1980s and 1990s, both the Young and Löfstedt reviews gave a particular emphasis to business interests. Despite the health and safety system being found broadly fit for purpose by these two major reports, the process of review continued, raising questions about the degree to which health and safety was being targeted on the basis of ideological opposition to regulation. In 2011, for example, the 'Red Tape Challenge' was launched, including a specific strand on health and safety, and in 2014 the HSE was subject to a triennial review (Temple 2014). Reflecting the continued saliency of the issue, to this day, the (now Conservative) Government retains a policy webpage dedicated to 'Health and Safety Reform', with updates about previous and new initiatives.[13]

In terms of the day-to-day work of the HSE, there have been some changes as a result of having to operate in this hostile environment. In 2012 the HSE responded to perceptions of over-regulation by introducing the 'Myth Busters Challenge Panel',[14] intended to allow anyone concerned to raise cases in which they felt health and safety had been incorrectly applied. Economic pressures to reduce costs to taxpayers and to maximise revenue have further developed commercialisation of HSE services, including the controversial introduction of 'Fee for Intervention' (FFI), the cost recovery scheme under which those guilty of breaking health and safety laws might be charged for costs incurred by the HSE. The Temple Review, a triennial government review of HSE's operations, concluded that FFI ran a serious risk of undermining trust in the regulator because it brought the motives of the regulator into question (Temple 2014). The HSE has also lost responsibility for the nuclear industry, with the Office for Nuclear Regulation being established in 2011 to bring together all agencies concerned with the civilian nuclear industry and form a single body; it became formally independent of the HSE in 2014. As with the ORR, though, this new regulator has been criticised for handling both economic and safety aspects. Finally,

---

[13] https://www.gov.uk/government/policies/health-and-safety-reform.

[14] http://www.hse.gov.uk/contact/myth-busting.htm.

reflecting the push to greater self-regulation, particularly amongst low-risk enterprises, the number of inspections by the HSE has continued to fall, from around 33,000 in 2010–2011 to under 22,000 in 2011–2012 (DWP 2013). This has led to concerns about the enforcement capability of the HSE, particularly from the political left. In sum, health and safety and its regulation are now subject to intense challenges, some of which have remained the same since at least 1960, if not earlier; in particular, debates about the constitutional legitimacy of regulation, and the 'appropriate' role for the state, are long-standing, as is the predominance of the notion of self-regulation. At the same time, there are new aspects to these challenges, such as the level of media hostility directed at health and safety in recent times, and the increasingly dismissive tone of political debates about regulation.

CONCLUSION

Underlying this chapter is a key question: who has political influence? This has determined the scope and focus of health and safety regulation from the outset—just as it does to this day. The susceptibility of any particular government to public opinion, the changing nature of that 'public opinion', the rise and fall of organised labour, ideas about compensation and insurance: over the years the state has to respond to all of these factors when regulating health and safety. But such an awareness, whilst useful, may obscure those aspects that have often been missed: things such as gender (Harrison 1990; Johnston and McIvor 2004a), race and ethnicity, and individual agency (Bradley 2011; Brotherstone and Manson 2011; Johnston and McIvor 2000a, 2004a, b; McIvor 2001: 171; Melling 2003; Strangleman 2011; Walker 2011).

There are other gaps, too, in our understanding of the ways health and safety regulation has been shaped and continues to be shaped. As has hopefully become clear, whilst the visibility and immediacy of workplace safety and accidents initially occupied the gaze of contemporary regulators and later historians, health has in the last 25 years or so taken on increasing prominence. Similarly, for both regulators and historians, the early focus on traditional heavy industries, such as mining, shipping and manufacturing, has begun to shift to include 'newer' roles and workplaces, including office and service work. For historians, the post-1945 era is much less well researched than the pre-1945 period, though again this is beginning to change, and there is a wealth of research produced contemporaneously by social scientists from which to draw. All of this flags up the necessity of awareness both of who has been overlooked when considering occupational health and safety and its regulation, and of the impact of the past upon the present.

It is possible to identify some broad changes over time that might be helpful in understanding the regulation of health and safety since 1960. Occurring across the post-1960 period, though having started before this, there is a move from the specific to the open-ended: so, for example, a move from being concerned with precise places of work to a more general conception of the employment environment. As McIvor has suggested (2001: 173), this also involved a shift in location of contest over workplace issues including health and safety, from the traditional fora of industrial relations, and to the more overtly politicised arenas of Whitehall. In addition, in some regards, the state has been involved in a move from a relatively reactive approach (inspectors investigating after an event) to a more proactive approach (risk assessments to identify issues before they arise). At the same time, there are important continuities stretching from the nineteenth century to the present, demonstrating that the state and all actors involved in health and safety are constrained both by their immediate circumstances and by their pasts. Primarily, it is important to recognise that the bounds of acceptable standards of health and safety, and the means through which to achieve them, including by regulation, are constantly being renegotiated. For the most part, their frameworks for action have remained parliamentary or via negotiation, rather than anything more radical or disruptive. One key area of legitimacy has also continued to recur—the moral legitimacy claims for action (even if the morals of society around health and safety have changed over the years), including the 'right to be safe.'

The account provided here is not intended to suggest a straightforward trajectory of improvement over time: the past is too messy for that. There were and remain great geographic differences in experience, as well as movements 'backwards' after periods of seeming improvement. The boundaries presented between periods were permeable. Changes were contingent, the product of multiple factors, and were often contested as opposing schools of thought gained the upper hand, a messy process of ebb and flow which continues today. It is therefore vital to look for the hand of the past in the present.

REFERENCES

PRIMARY SOURCES

Allen, B. D. (1966, March 4). Almost Full Inspection Within 17 Months. *Municipal and Public Services Journal, 78,* 691.

Building Businesses…Not Barriers. (1986). Cmnd.9794. London: Crown.

BIS. (2014). *Trade Union Membership 2013 Statistical Bulletin.* London: Crown.

Chief Inspector of Factories. (n.d., c. 1969). *Untitled Memorandum, to Managing Directors and Managers of Firms Employing More Than 50.* Modern Records Centre, University of Warwick (MRC), MSS.292D/146/18/1.

Department of Employment and Productivity. (1970). *Annual Report of H.M. Chief Inspector of Factories, 1969.* Cmnd. 4461. London: Crown.

Department of Employment. (1973). *Annual Report, H.M. Chief Inspector of Factories, 1972.* Cmnd. 5398. London: Crown.

Department of Environment, Transport and the Regions (DETR). (2000). *Revitalising Health and Safety: Strategy Statement.* London: Crown.

Dock and Harbour Authorities' Association. (n.d., c. 1971). *Written Evidence to the Robens Committee.* The National Archives of the UK, London [TNA], LAB 96/57.

DWP. (2013, February). Common Sense Restored to Health and Safety. DWP Press Release. https://www.gov.uk/government/news/common-sense-restored-to-health-and-safety. Accessed 30 April 2015.

Employment Committee. (1982, June 16). *The Working of the HSC and E, Minutes of Evidence* (HC 400-ii). London: Crown.

Hammer, J. (2014). *How the HSW Act Came About* (Unpublished).

Hampton, P. (2005). *Reducing Administrative Burdens: Effective Inspection and Enforcement (The Hampton Review).* London: HM Treasury.

HSC. (1994). *Review of Health and Safety Regulation: Main Report.* London: HMSO.

HSE. (1980). *Health and Safety, Manufacturing and Service Industries, Report for 1978.* London: HMSO.

HSE. (1988). *The Tolerability of Risks from Nuclear Power Stations.* London: HMSO.

HSE. (2005). *Fit for Work, Fit for Life, Fit for Tomorrow.* http://www.hse.gov.uk/statistics/publications/fit3.htm.

HSE. (2006, April 13). *Sensible Risk Management* (HSE Board Paper HSE/06/46).

HSE. (2012). *Myth Busters Challenge Panel.* http://www.hse.gov.uk/myth/myth-busting/index.htm.

Industrial Health and Safety Bulletin. (1971). *Total Loss Control* (Spring), 7–8.

Jay, P. (2000, November 10). Join My Opera Chorus: Open Roads for All. *Times* (Suppl.), 9.

Kinnersly, P. (1976). The Right to Work and Live in a Safe, Healthy Environment. In J. Dale & T. Emerson (Eds.), *Jobs and the Environment* (pp. 3–7). West Kingsdown: SERA.

Kirkup, J. (2012, January 5). Coalition Plans to Kill Off 'Health and Safety Monster' with Limits on Lawyers' Fees'. *The Telegraph*. http://www.tele-graph.co.uk/news/politics/8995276/Coalition-plans-to-kill-off-health-and-safety-monster-with-limits-on-lawyers-fees.html. Accessed 30 April 2015.

Laurance, J. (1988, March 9). Paying the Wages of Peril. *The Guardian*, 23.

Lifting the Burden. (1985). London: HMSO.

Löfstedt, R. (2011). *Reclaiming Health and Safety for All: An Independent Review of Health and Safety Legislation* (The Löfstedt Review). London: Crown.

Lord Robens. (1972). *Safety and Health at Work: Report of the Committee 1970–72* (The Robens Report). London: HMSO.

Lord Young. (2010). *Common Sense, Common Safety* (The Young Review). London: Crown.

Macrory, R. (2006). *Regulatory Justice: Sanctioning in a Post-Hampton World* (The Macrory Review). London: Better Regulation Executive.

Ministry of Labour. (1962). *Annual Report of the Chief Inspector of Factories, 1961*. Cmnd. 1816. London: Crown.

Ministry of Labour. (1964). *Annual Report of the Chief Inspector of Factories, 1963*. Cmnd. 2450. London: Crown.

Paterson, A. (1975, February 7). Faults in Safety Laws. *Morning Star*, p. 4.

Royal Society for the Prevention of Accidents (RoSPA). (1964, June 12). *Sixth Annual Report of Head of Agricultural Safety Division*. National Archives Scotland, Edinburgh, AF59.152.

RoSPA. (1965, September 24). *Minutes of the National Agricultural Safety Committee*. RoSPA Archives, Birmingham, D/266/2/25.

RoSPA. (n.d., c. 1971). *Written Evidence to the Robens Committee*. TNA, LAB 96/75.

Spiegelberg, R. (1971, May). Chaos in British Industry. *Times*, p. 5.

Stratton, A. (2010, June 13). Tories Start Drive to Cut Health and Safety Rules. *The Observer*. http://www.theguardian.com/politics/2010/jun/13/tories-start-health-safety-rules-cuts. Accessed 30 April 2015.

Temple, M. (2014). *Triennial Review Report: Health and Safety Executive [The Temple Review]*. London: HMSO. https://www.gov.uk/government/publications/triennial-review-report-health-and-safety-executive-2014.

Times. (1973, January 31). Accidents Caused by Unguarded Minds, 7.

Times. (1980, July 26). Concern Over Staff Cuts at Health and Safety Body, 3.

Tubby, J. (1979, April). Safety Act Not Good Enough. *The Land Worker* (p. 6). Museum of English Rural Life, University of Reading.

TUC. (1964, November 17). National Joint Advisory Council, Industrial Safety Sub-Committee, Notes for Meeting. MRC, MSS.292B/146.17.1.

TUC. (1968, July 10). Representatives' Comments, Industrial Safety Advisory Council, Joint Safety Organisation. MRC, MSS.292B/146/17/2.

TUC. (1972, October 12). Verbatim Report of TUC Conference on the Robens Committee Report. TUC Library, HD 7273.

Watson, B. (2005). Interviewed by David Walker, Chemical Workers Project, Scottish Oral History Centre. University of Strathclyde Archives and Special Collections, GB249 SOHC7 (2004–2005).

Williams, R. (c. 1979). 'A Key Issue', *Work Hazards* 21 (c. 1979). Samuel Barr Collection, Glasgow Caledonian University (GCU) DC 140/2/1/2.

Secondary Sources

Ackers, P. (2014). Game Changer: Hugh Clegg's Role in Drafting the 1968 Donovan Report and Redefining the British Industrial Relations Policy-Problem. *Historical Studies in Industrial Relations, 35*(3), 63–88.

Aldrich, M. (1997). *Safety First. Technology, Labor and Business in the Building of American Work Safety, 1870–1939*. Baltimore: Johns Hopkins University Press.

Almond, P. (2009). The Dangers of Hanging Baskets: Regulatory Myths' and Media Representations of Health and Safety Regulation. *Journal of Law and Society, 36*(3), 352–375.

Almond, P. (2013). *Corporate Manslaughter and Regulatory Reform*. Basingstoke: Palgrave Macmillan.

Almond, P. (2015). Revolution Blues: The Reconstruction of Health and Safety Law as 'Common-Sense Regulation'. *Journal of Law and Society, 42*(2), 202–229.

Almond, P., & Esbester, M. (2016). Il/Legit.imate Risks? Public Opinion and Health and Safety Regulation in Post-1960 Britain. In T. Crook & M. Esbester (Eds.), *Governing Risks: Danger, Safety and Accidents in Modern Britain, c. 1800–2000* (pp. 295–314). Palgrave Macmillan.

Baldwin, R. (1987). Health and Safety at Work: Consensus and Self-Regulation. In R. Baldwin & C. McCrudden (Eds.), *Regulation and Public Law*. London: Weidenfeld and Nicholson.

Baldwin, R. (1996). Regulatory Legitimacy in the European Context: The British Health and Safety Executive. In G. Majone (Ed.), *Regulating Europe* (pp. 83–105). London: Routledge.

Baldwin, R., & Cave, M. (1999). *Understanding Regulation: Theory, Strategy, and Practice*. Oxford: Oxford University Press.

Ball, D., & Ball-King, L. (2013). Safety Management and Public Spaces: Restoring Balance. *Risk Analysis, 33*(5), 763–771.

Balleisen, E. J., & Brake, E. K. (2014). Historical Perspective and Better Regulatory Governance: An Agenda for Institutional Reform. *Regulation & Governance, 8*(2), 222–245.

Bandle, T. (2007). Tolerability of Risk: The Regulator's Story'. In F. Bouder, D. Slavin & R. Löfstedt (Eds.), *The Tolerability of Risk: A New Framework for Risk Management* (pp. 93–103). London: Earthscan.

Bartrip, P. (1982). British Government Inspection, 1832–75: Some Observations. *Historical Journal, 25*(3), 605–626.

Bartrip, P. (1983). State Intervention in Mid-Nineteenth Century Britain: Fact or Fiction? *Journal of British Studies, 23,* 63–83.

Bartrip, P. (1996). "Petticoat Pestering": The Women's Trade Union League and Lead Poisoning in the Staffordshire Potteries, 1890–1914. *Historical Studies in Industrial Relations, 2*(2), 3–26.

Bartrip, P. (1998). Too Little, Too Late? The Home Office and the Asbestos Industry Regulations, 1931. *Medical History, 42*(4), 421–438.

Bartrip, P. (2001). *The Way from Dusty Death: Turner & Newall and the Regulation of Occupational Health in the British Asbestos Industry 1890s–1970s.* London: Athlone.

Bartrip, P. (2002). *The Home Office and the Dangerous Trades: Regulating Occupational Disease in Victorian and Edwardian Britain.* Amsterdam: Rodopi.

Bartrip, P. (2014). "Enveloped in Fog": The Asbestos Problem in Britain's Royal Naval Dockyards, 1949–1999. *International Journal of Maritime History, 26*(4), 685–701.

Bartrip, P., & Fenn, P. T. (1983). The Evolution of Regulatory Style in the Nineteenth Century British Factory Inspectorate. *Journal of Law and Society, 10*(2), 201–222.

Beck, M., & C. Woolfson, C. (2000). The Regulation of Health and Safety in Britain: From Old Labour to New Labour. *Industrial Relations Journal, 31*(1), 35–49.

Black, J. (2008). Constructing and Contesting Legitimacy and Accountability in Polycentric Regulatory Regimes. *Regulation & Governance, 2*(2), 137–164.

Bohman, J. (2000). *Public Deliberation: Pluralism, Complexity, and Democracy.* Boston, MA: MIT Press.

Bowden, S., & Tweedale, G. (2003). Mondays Without Dread: The Trade Union Response to Byssinosis in the Lancashire Cotton Industry in the Twentieth Century. *Social History of Medicine, 16*(1), 79–95.

Bradley, D. (2011). Oral History, Occupational Health and Safety and Scottish Steel, c. 1930–1988. *Scottish Labour History, 46,* 86–101.

Braithwaite, J. (1985). *To Punish or Persuade: Enforcement of Coal Mine Safety.* Albany, NY: University of New York Press.

Brotherstone, T., & Manson, H. (2007). North Sea Oil, Its Narratives and Its History: An Archive of Oral Documentation and the Making of Contemporary Britain. *Northern Scotland, 27,* 15–41.

Brotherstone, T., & Manson, H. (2011). Voices of Piper Alpha: Enduring Injury in Private Memory, Oral Representation and Labour History. *Scottish Labour History, 46,* 71–85.

Burgess, A. (2011). The Changing Character of Public Inquiries in the (Risk) Regulatory State. *British Politics, 6*(1), 3–29.

Carson, W. G. (1979). The Conventionalisation of Early Factory Crime. *International Journal of the Sociology of Law, 7*(1), 37–60.

Carson, W. G. (1982). *The Other Price of Britain's Oil.* Oxford: L.M. Robertson.

Chambers, S. (2003). Deliberative Democratic Theory. *Annual Review of Political Science, 6*(1), 307–326.

Clegg, H. A. (1985). *A History of British Trade Unions Since 1889* (Vol. 2). Oxford: Clarendon Press.

Clegg, H. A., Fox, A., & Thompson, A. (1964). *A History of British Trade Unions since 1889* (Vol. 1). Oxford: Clarendon Press.

Cooter, R., & Luckin, B. (1997). 'Accidents in History: An Introduction'. In R. Cooter & B. Luckin (Eds.), *Accidents in History: Injuries, Fatalities and Social Relations* (pp. 1–16). Amsterdam: Rodopi.

Cox, P. (2007). Compulsion, Voluntarism, and Venereal Disease: Governing Sexual Health in England After the Contagious Diseases Acts. *Journal of British Studies, 46,* 91–115.

Crook, T., & Esbester, M. (2016). Risk and the History of Governing Modern Britain, c. 1800–2000. In T. Crook and M. Esbester (Eds.), *Governing Risks in Modern Britain: Danger, Safety and Accidents, c. 1800–2000* (pp. 1–26). London: Palgrave.

Dawson, S., Willman, P., Clinton, A., & Bamford, M. (1988). *Safety at Work: The Limits of Self-Regulation.* Cambridge: Cambridge University Press.

Dodds, A. (2006). The Core Executive's Approach to Regulation: From 'Better Regulation' to 'Risk-Tolerant Deregulation'. *Social Policy & Administration, 40*(5), 526–542.

Eichener, V. (1997). Effective European Problem-Solving: Lessons from the Regulation of Occupational Safety and Environmental Protection. *Journal of European Public Policy, 4*(4), 591–608.

Esbester, M. (forthcoming). *The Birth of Modern Safety: Preventing Worker Accidents on Britain's Railways, 1871–1948.* Abingdon: Taylor & Francis.

Evans, E. J. (1978). *Social Policy 1830–1914: Individualism, Collectivism and the Origins of the Welfare State.* London: Routledge.

Fishkin, J. (2009). *When the People Speak: Deliberative Democracy and Public Consultation.* Oxford: Oxford University Press.

Frazer, W. M. (1950). *History of English Public Health, 1834–1939.* London: Balilliere, Tindall, and Cox.

Gamble, A. (1994). *The Free Economy and the Strong State: The Politics of Thatcherism* (2nd ed.). London: Macmillan.

Gilad, S. (2011). It Runs in the Family: Meta-Regulation and Its Siblings. *Regulation & Governance, 4*(3), 485–506.

Greenberg, M., & Wikeley, N. (1999). Too Little, Too Late? The Home Office and the Asbestos Industry Regulations, 1931: A Reply. *Medical History, 43*(4), 508–513.

Greenleaf, W. H. (1983). *The British Political Tradition. Vol. 1, The Rise of Collectivism.* London: Methuen.

Harrison, B. (1989). "Some of Them Gets Lead Poisoned": Occupational Lead Exposure in Women, 1880–1914. *Social History of Medicine, 2*(2), 171–195.

Harrison, B. (1990). Suffer the Working Day: Women in the "Dangerous Trades", 1880–1914. *Women's Studies International Forum, 13*(1–2), 79–90.

Harrison, B., & Nolan, M. (2004). Reflections in Colonial Glass? Women Factory Inspectors in Britain and New Zealand 1893-1921. *Women's History Review, 13*(2), 263–288.

Harvey, D. (2005). *A Brief History of Neoliberalism.* Oxford: Oxford University Press.

Hawkins, K. (2002). *Law as Last Resort: Prosecution Decision-Making in a Regulatory Agency.* Oxford: Oxford University Press.

Higgins, D., & Tweedale, G. (2010). Oil on the Water: Government Regulation of a Carcinogen in the Twentieth-Century Lancashire Cotton Spinning Industry. *Business History, 52*(5), 695–712.

Higgison, A. (2005). Asbestos and British Trade Unions, 1960s and 1970s. *Scottish Labour History, 40,* 70–86.

Hood, C. (1991). A Public Management for All Seasons? *Public Administration, 69*(1), 3–19.

Humphries, J. (1988). Protective Legislation, the Capitalist State and Working-Class Men: The Case of the 1842 Mines Regulation Act. In R. E. Pahl (Ed.), *On Work: Historical, Comparative and Theoretical Approaches* (pp. 95–124). Oxford: Blackwell.

Hutchins, B. L., & Harrison, A. (1911). *A History of Factory Legislation* (2nd ed.). London: P. S. King & Son.

Hutter, B. (1997). *Compliance: Regulation and Environment.* Oxford: Clarendon Press.

Hutter, B., & Manning, P. (1990). The Contexts of Regulation: The Impacts Upon Health and Safety Inspectorates in Britain. *Law and Policy, 12*(2), 103–136.

Jackson, M. (Ed.). (2015). *Stress in Post-War Britain, 1945–85.* London: Pickering and Chatto.

Johnston, R., & McIvor, A. (2000a). *Lethal Work: A History of the Asbestos Tragedy in Scotland.* East Linton: Tuckwell Press.

Johnston, R., & McIvor, A. (2000b). Whatever Happened to the *Occupational Health Service? The NHS, the OHS and the Asbestos Tragedy on Clydeside.* In C. Nottingham (Ed.), *The NHS in Scotland: The Legacy of the Past and the Prospect of the Future* (pp. 79–105): Aldershot: Ashgate.

Johnston, R., & McIvor, A. (2004a). Dangerous Work, Hard Men and Broken Bodies: Masculinity in the Clydeside Heavy Industries, c. 1930s–1970s. *Labour History Review, 69*(2), 135–152.

Johnston, R., & McIvor, A. (2004b). Oral History, Subjectivity and Environmental Reality: Occupational Health Histories in Twentieth-Century Scotland. *Osiris, 19,* 234–249.

Johnston, R., & McIvor, A. (2007). *Miners' Lung. A History of Dust Disease in British Coal Mining.* Aldershot: Ashgate.

Jones, H. (1983a). *The Home Office and Working Conditions 1914 to 1940* (Unpublished PhD thesis). London University.

Jones, H. (1983b). 'Employers' Welfare Schemes and Industrial Relations in Inter-War Britain, *Business History, 25*(1), 61–75.

Jones, H. (1985). An Inspector Calls: Health and Safety at Work in Inter-War Britain. In P. Weindling (Ed.), *The Social History of Occupational Health* (pp. 223–239). London: Croom Helm.

Jones, H. (1994). *Health and Society in Twentieth-Century Britain.* London: Longman.

Kirby, P. (2013). Victorian Social Investigation and the Children's Employment Commission, 1840–42. In N. Goose & K. Honeyman (Eds), *Children and Childhood in Industrial England: Diversity and Agency, 1650–1900.* Aldershot: Ashgate.

Kostal, R. W. (1994). *Law and English Railway Capitalism 1825–1875.* Oxford: Oxford University Press.

Leka, S., Jain, A., Hollis, D., Andreou, N., & Zwetsloot, G. (2012). *The Changing Landscape of OSH Regulation in the UK: A Review.* Leicester: IOSH.

Livesey, R. (2004). The Politics of Work: Feminism, Professionalisation and Women Inspectors of Factories and Workshops. *Women's History Review, 13*(2), 233–262.

Löfstedt, R. (2005). *Risk Management in Post-Trust Societies.* London: Earthscan.

MacDonagh, O. (1958). The Nineteenth Century Revolution in Government: A Reappraisal. *Historical Journal, 1*(1), 52–67.

MacLeod, R. M. (1967). Law, Medicine and Public Opinion: The Resistance to Compulsory Health Legislation, 1870–1907. *Public Law, 107–128,* 189–211.

Majone, G. (1997). From the Positive to the Regulatory State: Causes and Consequences of Changes in the Mode of Governance. *Journal of Public Policy, 17*(2), 139–167.

McIvor, A. (1997). State Intervention and Work Intensification. The Politics of Occupational Health and Safety in the British Cotton Industry,

c. 1880–1914. In A. Knotter, B. Altena & D. Damsma (Eds.), *Labour, Social Policy and the Welfare State* (pp. 125–139). Amsterdam: Aksant Academic.

McIvor, A. (2001). *A History of Work in Britain, 1880–1950.* Basingstoke: Palgrave.

McIvor, A. (2013). *Working Lives: Work in Britain Since 1945.* Basingstoke: Palgrave Macmillan.

McLean, I., & Johnes, M. (2000). *Aberfan: Government and Disasters.* Cardiff: Welsh Academic Press.

McQuaid, J. (2007). A Historical Perspective on Tolerability of Risk. In F. Bouder, D. Slavin, & R. Löfstedt (Eds.),*The Tolerability of Risk: A New Framework for Risk Management* (pp. 87–92). London: Earthscan.

Melling, J. (2003). *The Risks of Working and the Risks of Not Working: Trade Unions, Employers and Responses to the Risk of Occupational Illness in British industry, c. 1890–1940s* (LSE Discussion Paper 12).

Melling, J. (2014). Making Sense of Workplace Fear: The Role of Physicians, Psychiatrists, and Labor in Reframing Occupational Strain in Industrial Britain, c. 1850–1970. In D. Cantor & E. Ramsden (Eds.), *Stress, Shock and Adaptation in the Twentieth Century* (pp. 189–221). Rochester, NY: University of Rochester Press.

Mess, H. A. (1926). *Factory Legislation and Its Administration 1891–1924.* London: P. S. King & Son.

Mills, C. (2010). *Regulating Health and Safety in the British Mining Industries, 1800-1914.* Aldershot: Ashgate.

Mitchell, B. R. (1988). *British Historical Statistics.* Cambridge: CUP.

Nardinelli, C. (1980). Child Labor and the Factory Acts. *The Journal of Economic History, 40*(4), 739–755.

Nichols, T. (1997). *The Sociology of Industrial Injury.* London: Mansell.

Pantti, M. K., & Wahl-Jorgensen, K. (2011). "Not an act of God": Anger and Citizenship in Press Coverage of British Man-made Disasters. *Media, Culture and Society, 33,* 105–122.

Paterson, J. (2007). The Evolution of Occupational Health and Safety Law on the UK Continental Shelf, 1964–2006. *Northern Scotland, 27,* 43–67.

Pearce, F., & Tombs, S. (1990). Ideology, Hegemony, and Empiricism. *British Journal of Criminology, 30*(4), 423–443.

Porter, D., & Porter, R. (1988). The Politics of Prevention: Anti-Vaccinationism and Public Health in Nineteenth-Century England. *Medical History, 32,* 231–252.

Price, K. (1993). Women of Steel: Health and Safety Issues During World War Two. *North East Labour History Bulletin, 27,* 1–13.

Rimington, J. (2008, October 9). *Health and Safety—Past, Present and Future.* Alan St. John Holt Memorial Lecture.

Roberts, D. (1960). *Victorian Origins of the British Welfare State.* New Haven, CT: Yale University Press.

Roberts, I. (2007). Women's War Work in the North East Shipbuilding Industry. *North East History, 38,* 67–102.

Rothstein, H., Beaussier, A., Borraz, O., Bouder, F., Demeritt, D., de Haan, M., et al. (2015). When 'Must' Means 'Maybe': Varieties of Risk Regulation and the Problem of Trade-offs in Europe. *HowSAFE WP, 15*(1), 1–21.

Rhodes, G. (1981). *Inspectorates in British Government: Law Enforcement and Standards of Efficiency.* London: Allen and Unwin.

Sirrs, C. (2016a). Risk, Responsibility and Robens: The Transformation of the British System of Occupational Health and Safety Regulation, 1961–1974. In T. Crook & M. Esbester (Eds.), *Governing Risks in Modern Britain: Danger, Safety and Accidents, c. 1800–2000* (pp. 249–276). Palgrave: London.

Sirrs, C. (2016b). *Health and Safety in the British Regulatory State, 1961–2001: The HSC, HSE and the Management of Occupational Risk* (Unpublished PhD thesis).

Strangleman, T. (2011). Writing Workers: Re-Reading Workplace Autobiography. *Scottish Labour History, 46,* 26–37.

Swuste, P., Gulijk, C., & Zwaard, W. (2010). Safety Metaphors and Theories: A Review of the Occupational Safety Literature of the US, UK, and the Netherlands, Till the First Part of the 20th Century. *Safety Science, 48*(8), 1000–1018.

Tombs, S., & Whyte, D. (2010). A Deadly Consensus: Worker Safety and Regulatory Degradation under New Labour. *British Journal of Criminology, 50*(1), 46–65.

Tucker, E. (1995). And Defeat Goes On: An Assessment of Third-Wave Health and Safety Regulation. In F. Pearce & L. Snider (Eds.), *Corporate Crime: Contemporary Debates* (pp. 245–267). Toronto: University of Toronto Press.

Tweedale, G. (2000). *Magic Mineral to Killer Dust: Turner & Newall and the Asbestos Hazard.* Oxford: Oxford University Press.

Tweedale, G., & Hansen, P. (1998) 'Protecting the Workers: The Medical Board and the Asbestos Industry, 1930s–1960s'. *Medical History, 42*(4), 439–457.

Vogel, D. (2012). *The Politics of Precaution: Regulating Health, Safety and Environmental Risks in Europe and the United States.* Princeton: Princeton University Press.

Walker, D. (2011). "Danger Was Something You Were Brought Up Wi'": Workers' Narratives on Occupational Health and Safety in the Workplace. *Scottish Labour History, 46,* 54–70.

Walkowitz, J. (1980). *Prostitution and Victorian Society.* Cambridge: Cambridge University Press.

Walters, D. (1996). Health and Safety Strategies in Europe. *Journal of Loss Prevention in the Process Industries, 9*(5), 297–308.

Ward, J. T. (1962). *The Factory Movement 1830–1855.* London: Macmillan.

Wells, C. (2001). *Corporations and Criminal Responsibility* (2nd ed.). Oxford: Oxford University Press.

Williams, J. L. (1960). *Accidents and Ill-Health at Work.* London.

Wilson, G. K. (1983). *The Politics of Safety and Health: Occupational Safety and Health in the United States and Britain*. Oxford: Clarendon Press.

Wrigley, C. (Ed.). (1996). *A History of British Industrial Relations. Vol. III 1939–1979*. Brighton: Edward Elgar.

Yarmie, A. H. (1984). British Employers' Resistance to "Grandmotherly" Government, 1850–80. *Social History, 9*(2), 141–169.

CHAPTER 4

A New Order? Constituting Health and Safety

INTRODUCTION

The regulation of risks to the health and safety of workers and the public involves many different actors engaging at many levels of society, both formal and informal. One recurrent feature of much scholarship—historical and socio-legal—on regulation is a tendency towards statism, a 'macro-fetish' which focuses attention on the relationship between citizens and central state agencies (Almond and Gray 2017; Baldwin et al. 2010; Bartrip 1982, 1983, 2002; Fraser 2009; Gray 1996), and which is implicit in the concept of the 'regulatory state' (Braithwaite 2000; Levi-Faur 2013; Majone 1997; Moran 2003; Sunstein 1990). While this book aims to account for change over time at as many levels of activity as possible, not just this 'macro' one, the state remains a good starting point for discussion, not least because the practices that originate there exert such an organising influence, changing practices, ideas, and expectations across the *'regulatory universe'* (Aalders and Wilthagen 1998: 427). In systems of 'decentred' regulatory practice (Black 2001) such as health and safety, where actors beyond the core agencies of the state are primarily responsible for providing solutions to the issues to be addressed, the state also acts as a 'backstop' for private or networked provision (Crawford 2006: 459). Certainly, within the sphere of health and safety, some of the most significant developments since 1960 have originated or occurred at the level of the state.

© The Author(s) 2019   79
P. Almond and M. Esbester, *Health and Safety in Contemporary Britain*, https://doi.org/10.1007/978-3-030-03970-7_4

This chapter will explore the ways in which the formal, constitutional elements of the health and safety 'system' (that is, its legal, bureaucratic, and governmental bases) have impacted on the perceived legitimacy of this broader concept. The ways in which authority has been founded and exercised, the legality of rules, and the procedural fairness of the use of powers are central to the constitutional legitimacy claims made on behalf of 'health and safety': does the regulatory regime have a clear source of authority to justify its actions (Baldwin 1995; Prosser 2010)? This question of constitutional legitimacy is also largely one of 'input legitimacy', relating to the basis on which the regulator is empowered to act by law and political mandate (Scharpf 1999); this also draws attention naturally to the state. This is where regulations are made (in the form of primary and secondary legislation),[1] where policy-making and enforcement take place (via regulatory agencies), and where processes of accountability are focused. As this chapter will show, over the last 60 years, specific changes within the British regulatory framework, broader developments within government and the political system, and influences from outside, via the European Community/Union, have all impacted significantly upon the ways in which health and safety is perceived by those who engage with it.

As Chapter 3 has shown, the health and safety 'system' underwent a significant reinvention in the early 1970s, and two conjoined changes made by the Health and Safety at Work Act (HSWA) 1974 are particularly relevant here. First, it introduced a new framework which changed the legal rules imposed, making 'health and safety' more flexible but also more variable and arguably, inconsistent, with significant consequences for perceptions of legality. Second, it created a single, unified state inspectorate (the Health and Safety Executive [HSE]) to implement this new law. This new body had to establish its authority, and this process both contributed to a modernisation of health and safety, and allowed the new regime to respond to external political pressures and demands. Relatedly, the third theme explored in this chapter is the changing relationship between health and safety regulators and central government.

---

[1] 'Primary legislation' refers to statutes and Acts of Parliament; 'secondary legislation' refers to legal instruments (regulations and other statutory instruments) made under powers conferred by Acts of Parliament; for example, the HSWA 1974 (primary legislation) confers a power on the relevant Secretary of State to create regulations to govern specific safety risks.

The institutional independence that the HSWA gave to the HSE reduced the capacity of government to influence policy and outcomes, just as their desire to do so was growing, raising issues of procedural account-ability. Finally, an external influence which has produced significant change is that of Britain's relationship with Europe, which has regularly been cited as a source of overregulation and illegitimate authority.

Together, these strands illustrate the centrality of ideas of 'constitutional legitimacy' to perceptions of health and safety. They also demonstrate the ways in which this area contributes to historians' increasing focus on notions of diffuse governance, beyond the central state as classically understood (Crook and Esbester 2016: 4–6) and has come to reflect what for socio-legal scholars are widely accepted contemporary regulatory orthodoxies, such as those of 'networked governance' (Crawford 2006; Schmidt 2004), pluralised and decentred regulation (Bartle and Vass 2007; Black 2001, 2008; Grabosky 2013; Parker 2008), and meta-regulation (Gilad 2010; Morgan 2003). Broadly speaking, these concepts connote a movement away from "*the assumptions that states are the main loci of control over social and economic life or that they ought to have such a position*" and towards "*regulatory control…as diffused through society with less emphasis on the sovereign state*" (Scott 2004: 166). But while these tendencies can be identified in the contemporary health and safety landscape, how longstanding are they? In addition, there are also countervailing trends towards political intervention and the external imposition of regulatory oversight in one form or another. How have conflicts over the constitutional basis of health and safety manifested? And what influence have they had on broader perceptions?

LAW, CERTAINTY, AND LEGITIMACY

As discussed in Chapter 3, the 'Factories Acts' model of legislative intervention had gradually extended the scope and intensity of legal control over workplace health and safety practices during the nineteenth and early twentieth centuries, albeit in a piecemeal and often quite limited way (Almond 2013; Bartrip 1983; Baldwin 1987). These laws tended to be prescriptive and narrowly focused, so that by 1970, there were nine Acts of Parliament and over 500 Statutory Instruments governing workplace health and safety (Baldwin 1987: 132; Hutter 1997: 22), giving rise to a perception that there was simply "*too much law*" (Robens 1972: para. 28). This framework of laws was fundamentally changed by the HSWA

1974, which introduced a principles-based, goal-setting regime to replace the existing system of rules (Baldwin 1995; Dawson et al. 1988; Gunningham and Johnstone 1999; Hutter 1997). The existing law was also perceived as being inflexible, unresponsive to the changing industrial context, and prone to lapsing into a state of *"regulatory fatigue"* (Gunningham and Johnstone 1999: 30). As one example, the Factories Act 1961 s.1(3) made it a legal requirement that factory walls had to be whitewashed or cleaned with soapy water every 14 months.[2] They largely ignored systemic issues of management in favour of ones of physical provision (Dawson et al. 1988; Robens 1972: para. 30), and left little room for worker engagement, reframing compliance as a disciplinary issue (as performance could be 'right' or 'wrong': Dawson et al. 1988: 9). Many interviewees (such as this nuclear industry worker) recalled this culture of rule-reliance:

> These little books told you everything you should know at the time to be safe [...] there was an element of faith, they would actually protect you, if you followed those rules you wouldn't go far wrong. (Chris Marchese Interview, para. 4)

A New Legal Model

Following Robens' recommendations, the HSWA introduced a new approach to imposing legal obligations via wide-ranging, open-ended 'portmanteau' duties to ensure the health, safety, and welfare of employees and others affected by work (HSWA 1974: ss. 2–3), which employers would be required to prove they had fulfilled 'so far as is reasonably practicable' (the 'SFAIRP' test). This test was designed to allow for the balancing of a quantum of risk against cost and difficulty in determining *"whether the time, trouble and expense of the precautions suggested are disproportionate to the risks involved"*,[3] but also to allow for flexibility in the methods used by duty-holders to fulfil their obligations. More specific requirements were to be imposed via 'lower-order' rules, in the form of regulations and delegated legislation, which offered greater flexibility to

---

[2] This provision remained in force until its repeal by the Workplace (Health, Safety and Welfare) Regulations 1992.

[3] Per Lord Oaksey in *Marshall v Gotham* (1954), AC 360 at p. 370; see also *Austin Rover v H.M. Inspector of Factories* (1989), 3 WLR 520.

rule-makers (Baldwin 1995: 131; Hutter 1997: 68; Hutter and Manning 1990: 112). In general terms, this new Act functioned as:

> a piece of enabling legislation through which [duty-holders] would be encouraged, within nationally enforced standards, to develop the forms of organisation and types of activities which they considered to be most appropriate to secure the safety and health of those at work. (Dawson et al. 1988: 14)

This placed the primary responsibility for ensuring compliance with these duties onto the regulated entity, not the state regulator (Almond 2013; Baldwin 1987; Dawson et al. 1988; Hutter 1997; Ogus 1995; Sirrs 2016).

To its supporters, the 1974 Act was a revolutionary, modernising step, introducing changes which have remained intact for 40 years.[4] By internalising the self-regulatory capacity of duty-holders, the Act extended and flexibilised the health and safety infrastructure, established the need to ensure safety as an ongoing obligation, underpinned a shift towards thinking of health and safety in terms of management and culture, and made the law more enforceable (Hutter and Manning 1990: 112; Sirrs 2016). This role was acknowledged by, for example, British Rail in 1977, who noted that the HSWA had *"already excited the imaginations of both managers and staff into a new awareness of the general safety problem [because]...employees themselves will be able to take the initiative"*.[5] For its critics, the Act was overly optimistic about the realities of business compliance, relying too heavily on voluntary action and downplaying the 'bounded rationality' of the regulated population (Gunningham and Johnstone 1999: 68–74; Wells 2001: 74). This self-regulatory orientation has also arguably facilitated deregulation, via the SFAIRP principle, by making compliance a matter of cost–benefit analysis, from where it can be rationalised away as costs change in times of reduced productivity (Almond 2013: 113–114; Bain 1997; Beck and Woolfson 2000; Dawson et al. 1988; Tombs and Whyte 2010). It also arguably 'normalised'

---

[4]An account of this sort is provided by HSE, who laud the Act as having *helped make Britain one of the safest and healthiest places in the world to work, [and] saving thousands of lives.* http://www.hse.gov.uk/aboutus/40/beyond.htm.

[5]C. A. Rose, *Memorandum to British Rail Board*, 16 August 1977, p. 10. The National Archive of the UK (TNA), AN 156/936.

workplace harms by framing them as risks (to be weighed and then taken) rather than dangers (to be avoided), 'conventionalising' the status of the law as subordinate to production imperatives (Carson 1979; Nichols and Armstrong 1973; Tombs and Whyte 2007).

Many of these issues have impacted on the perceived legitimacy of health and safety. For instance, the SFAIRP principle, which establishes that an employer can only be liable in relation to a failure of health and safety where they have not exercised appropriate due diligence in relation to that risk, was intended to mitigate the *"burning sense of grievance"* (Norrie 2001: 84) associated with the strict liability standard. Debate around the legitimacy of the SFAIRP test resurfaced in 2007 when the European Commission challenged its compatibility with the obligations enshrined in European law to introduce measures to encourage improvements in health and safety.[6] The concern had been that 'reasonable practicability' was a qualified standard that essentially tolerated a degree of non-compliance on the basis of cost; this was perceived as undermining the absolute nature of the commitment imposed at European level. Although the test was held not to breach these obligations, concerns over the legitimacy of this test (on due-process, legal certainty, and justice grounds) remain in some quarters, along with concerns that the law is under-inclusive and less easily enforceable as a result of the concession made to the cost–benefit calculation of regulatory 'need' (Baldwin 1990; Tombs and Whyte 2007).

'Deliberate Vagueness'

By far, the most significant legitimacy challenge posed by the new law, however, was that of the non-specificity introduced via the general duties, the effect of which was summarised by Jim Hammer, Deputy Director-General of the HSE 1984–1989:

> The great thing about the [HSWA] is that it doesn't specify the means. It specifies the objective. That means people have to think [...] The obligation is to do one's best to use one's imagination to assess the risks

---

[6]This obligation is contained within Articles 4(1) and 5(1) of Council Directive 89/391/EEC of 12 June 1989, and the issue was examined by the European Court of Justice in *Commission of the European Communities v United Kingdom*, 14 June 2007, C-127/05.

and then to plan for, and take, the relevant precautions. (Jim Hammer Interview, para. 49)

Rather than precisely setting out what must be done, the Act framed its requirements in terms of a duty to ensure health and safety in general terms. This constituted a shift from *specification* standards to *performance* standards; from precise proscriptions relating to defined hazards (such as the requirement to whitewash walls, mentioned above), to general duties and open-ended obligations to reach acceptable levels of provision (Gunningham and Johnstone 1999: 24–25). The reasons for such an approach were set out in the Robens Report, which envisaged this "*positive declaration of over-riding duties*" as establishing "*the preservation of safety and health at work...[as] a continuous legal and social responsibility*", and so encouraging employers to "*take a less narrow and more rounded view of their roles and responsibilities*" (1972: para. 130). The doctrinal reasons for taking this approach related to the need for the law to encompass classes of widely varying conduct, which themselves reflected multiple, and often incommensurate, dimensions (Asgeirsson 2015: 438). Adequate health and safety provision has never been reducible to a single unit of measurement but is instead dependent on the interplay of many elements of risk and performance; a general duty allowed for these to be reflected by a common norm.

This communicative aim linked health and safety to "*social citizenship*" rights to security and welfare, by ensuring that safety risks were not imposed with impunity by those who benefited from them, onto those who did not (Marshall 1992; also Almond 2013: Ch. 6; Prosser 2010: 103; Ramsay 2006). At the same time, such a principles-based approach provided for a breadth and flexibility of application that precise statutory rules could not; secondary legislation (regulations) and Approved Codes of Practice (ACoPS) would allow for more detailed rules to be developed responsively over time (Baldwin 1995: Ch. 5; Hutter 1997: 70; Prosser 2010: 95), and without necessarily being themselves directly actionable[7] (Robens 1972: para. 142). General duties have enforceability implications, in that they are thought to be harder to disprove, but more likely to be contested (Baldwin 1990: 323–324). However, some stakeholders, such as the General Secretary of the Trades Union Congress (TUC),

---

[7] In that, breaching a regulation might not be an offence in and of itself, but provide evidence that the general duties within the HSWA have been breached.

Vic Feather, who raised the issue in 1972, felt that abandoning precision would undermine the enforceability of the Act, as the "*threat of injury and death ought to be met by safety standards that are legally specific, not shadowy*".[8]

This aspirational approach reflected a shift of legal philosophy, from an Anglo-Saxon conception of law as a set of absolute commands, to a 'Roman Law' approach (Baldwin 1996: 98), "*an elaborately articulated system of principles abstracted from the detailed rules which constitute the raw material of law*", allowing for a greater flexibility and simplicity of formulation (Nicholas 1962: 1). This conception of 'law as principle not rule' informs contemporary continental legal systems (as well as, for example, human rights systems), but posed many legitimacy challenges when introduced in 1974 due to "*the essential tension between that Act and traditional and powerful Anglo-Saxon conceptions in favour of law clearly stated and limited in extent, so that, as it were, we know where we stand with it*" (Rimington 2008: 6). The law posed new challenges for duty-holders, who suddenly needed to exercise much more informed judgements about what the law required, based on professional expertise, rather than on a straightforward ability to fulfil mandated instructions. "*Giving meaning and substance to the principles of law*" required attending to "*broad considerations across a range of issues*" (Hutter 1997: 102), thus placing an emphasis onto decision-making that went beyond the 'disciplinary' approach:

> The Inspector's approach would [go from...] giving them a list of things to put right to saying [...] you need to introduce a system which [will...] achieve some reasonable standard of compliance. That's where the safety committees and safety representatives would come into help, and firms could employ consultants if they wished. (David Eves Interview, para. 27)

One outcome of this approach, then, was the proliferation of a 'safety profession', made up of 'sociological citizens' whose jobs involved interpreting and implementing legal requirements (Haines 2011; Huising and Silbey 2011). This is discussed further in Chapter 6, but for now, we can note that it introduced a complexity into regulatory dealings that did not exist before.

---

[8]V. Feather, quoted in 'Both Sides Welcome Alf's Report', *Safety and Rescue*, September 1972, p. 2. TNA, LAB 96/476.

This theme of accessibility and the ease of compliance under a principles-based regulatory system was also reflected in the change that the HSWA imposed onto the law-making process. Robens did not remove the need for any degree of precision in the law; rather, it stipulated that this should be located in secondary legislation (statutory instruments and regulations) and ACoPs, with the level of prescriptiveness increasing at each stage down this hierarchy of rules (Baldwin 1995: 131–132). Section 15 of the HSWA gives the relevant Secretary of State the power to make new regulations to impose specific, legally enforceable, requirements by laying them before Parliament, where they become law unless disapproved by either House within 40 days, but also to modify and repeal existing regulations and general statutory duties (so-called 'Henry VIII powers'). ACoPS were initially authored by the Health and Safety Commission (HSC), until its merger into the HSE in 2008, since when the latter has performed the role. They have not required approval by Parliament or the Secretary of State, and constitute a non-enforceable source of practical guidance. Following the provisions of an ACoP will usually mean that the general duties under the Act are, in that regard, met, however, a duty-holder who has not done so can defend against a prosecution under the Act by showing that they have complied in some other way (Löfstedt 2011: 55). Over time, each of these sources of rules has grown in number, so that in 2018, the HSE could list 101 pieces of secondary legislation that it enforced,[9] and 55 currently available ACoPS.[10]

The issues of accessibility, consistency, and predictability that this provoked constituted the basis of a significant legitimacy challenge to 'health and safety', culminating in a Government (and media) campaign against 'red tape' and over-regulation in this area during 2010–2015 (Almond 2015). The Löfstedt Review (2011) examined the legal framework in response to concerns about a proliferation of regulatory requirements; while it found relatively little evidence of regulatory overload (in the *scope* of the obligations imposed), it did highlight the tendency towards diffusion and proliferation in the sources of law (Löfstedt 2011: 70–77). This tendency is not new, and was highlighted in a 1982 review of health and safety in the railway industry: "*[t]he proliferation of instructions on*

[9]http://www.hse.gov.uk/legislation/statinstruments.htm.

[10]http://www.hse.gov.uk/pubns/books/index-legal-ref.htm.

safety is a matter of concern to some managers who quote instances where it is sometimes necessary to consult several separate documents in order to find out how a job should be done". [11]

These delegated forms of rule-making also pose constitutional challenges, not least via the perception that they lack proper constitutional accountability and an appropriate democratic mandate, as they have often been introduced without much participative engagement, and so carry a greater potential for abuse or partiality (Craig 2012: 433–435; McHarg 2006: 555). Scrutiny mechanisms are weak, and Parliamentary debates on delegated legislation rarely occur (Baldwin 1995: 64–68). The processes by which these sources are made and remade are subject to judicial review, but are often driven by expediency, and remain subject to procedural objections, not least that they give rule-makers a relatively unchecked form of discretionary power to make law, leading to a significant increase over time in the volume and complexity of material generated (Baldwin 1996: 87), as one former Health and Safety Commissioner (2005–2011) recalled:

> HSE was providing 60-odd different bits of guidance around manual handling, one was labelled golf clubs, one was labelled hairdressers…it was just ridiculous, that…overburdening. (Sayeed Kahn Interview, para. 123)

This also reflected a political tendency towards distancing the mechanisms of policy delivery from the arenas where that policy is created, as recognised by a former HSE Director of Health Policy (1970s and 1980s):

> [G]overnment starts with quite tight ideas of what should be done…It goes down and down and down from regulation, to Approved Code of Practice, to campaign with educational materials, to inspection campaign, to leaflet. (Tim Carter Interview, para. 97)

The consequences of this core/periphery distinction within the legal framework are both practical and psychological, relating to the possibilities for inconsistency and uncertainty that result. Some indication of the sense of illegitimacy that a non-prescriptive legal regime can produce is

---

[11] British Rail Board (BRB)/HSE, *An Outline Report on OSH by the Accident Prevention Advisory Unit*, November 1982, p. 17. TNA, AN 16/157.

provided by the 2008 case of *R v Chargot*,[12] where a dumper truck had tipped over on a construction site, crushing the driver. As there were no witnesses, the precise cause of the event could not be established; however, the HSE did find shortcomings in the management of health and safety at the site. The victim's employers were charged with a breach of s.2(1) of the Act. It was held that it was not necessary to prove the specific acts or omissions by which a defendant had breached s.2 in order to establish liability; the fact of death, the Court said, "*itself demonstrates that the employer failed to ensure [the victim's] health and safety at work*".[13] In upholding this results-based approach (the outcome establishes a prima facie breach of duty), the House of Lords paid particular attention to the social and economic purposes of the general duties, and to the role of the SFAIRP test in mitigating an otherwise absolute requirement.[14] But critics maintain that *Chargot* highlights the procedural unfairness associated with enforcing imprecise regulatory standards (Spencer 2009: 265). Regulators tend to prefer rules that are precise enough to be easily enforceable (Baldwin 1990: 323; Hawkins 2002: 356), but *Chargot* confirms that the general duties may run counter to this, introducing greater clarity for enforcers at the cost of increased uncertainty for the regulated, as a former HSC Chair (1993–1999) reflected:

> Almost always when we'd gone to Court, we'd use Section 2 or Section 3. You can't lose. 'He's hurt – you've lost'...And that got me thinking, why do we want all the others [offences]? Because the truth is that people think it's too complicated. (Sir Frank Davies Interview, para. 59)

Uncertainty, Anxiety, and Fear

The broader psychological effects of the open-endedness of the law in this area can perhaps be summarised as ones of uncertainty and anxiety. A recurrent feature of contemporary debate around health and safety law is that of the fear of red tape, civil liability, and prosecution (Almond 2009; Furedi 1997). For example, the Young Review mentioned 'fear' no fewer than 35 times (Young 2010), and the academic

---

[12] *R v Chargot Limited (t/a Contract Services) and Others* (2008), UKHL 72.

[13] Ibid., *per* Lord Hope of Craighead at para. 30.

[14] Ibid., paras. 21 and 29–31.

literature on compliance motivations has acknowledged the role that fear of punishment, mediated via considerations of reputation and normative duty, plays in prompting compliance (Genn 1993; Kagan et al. 2011; Mendeloff and Gray 2005). But Young did not seem to be discussing either the fear of punishment per se, nor the general sense of 'manufactured uncertainty' found in accounts of the 'risk society' of late modernity (Beck 1992; Giddens 1991). Rather, the Report articulated a broader anxiety among the regulated population about their status as decision-makers, and the possibility of failure attached to uncertainty about the scope of the law. As we have seen, the open-ended duties in the HSWA give rise to some lack of clarity in how duties should be interpreted (Hutter 1989, 1997: 77). As the former Chairman of Railtrack and the Olympic Delivery Authority put it:

> Many people didn't like [the new regime] because it required them to think. No longer could you say, 'oh, I'm doing it in accordance with the regulation' [...] you actually had to think about the specific task in hand. (Sir John Armitt Interview, para. 9)

Because the adequacy of health and safety provision has been assessed since 1974 on the basis of a sliding-scale of reasonable practicability, the management of risk has been explicitly framed as a matter of 'organized uncertainty' (Power 2007). Duty-holders have been required to exercise judgements about the gravity, likelihood, and tolerability of risks, and face some form of organisational risk if they have done so inadequately. This means the scope for individual choice that the HSWA brought was distinctly double-edged, being both empowering but also anxiety-provoking.

The practical and logical consequences of this new form of health and safety regulation are made clear in the observation of Bill Simpson, the first Chair of the Health and Safety Commission, who described the new approach as "*not so much 'Thou shalt not', as, 'Thou shalt'*".[15] On the one hand, there was a move away from explicit instruction; on the other, a move towards a positive obligation that placed a greater burden on the individual. This external demand was necessarily more onerous than what had come before, and acted as a direct source of negative

---

[15] B. Simpson, quoted in *New Powers Aid Inspectors, Safety and Rescue*, March 1975, p. 3.

freedom; you were free to pursue this goal however you like, but pursue it you must. Regulatory interactions, particularly those around issues of individual judgement, have been observed as anxiety-inducing emotional events; this emotional content is related to the issue of pursuing *control* as a source of certainty and order, but at the cost of having to serve particular external interests above one's own (Fineman and Sturdy 1999; Lange 2002: 209–211). As a result of this fundamental tension, the rational process of 'controlling' is exacerbated into a state of anxiety because of the relative inability of (or lack of space given to) duty-holders to exercise this freedom in a self-realising way. The organisational and economic imperatives of the business context, the status of regulations as enforceable legal commands, and the bureaucratic demands of compliance (Hutter 1997; Power 2007), all ensure that this is not appreciated as a free or self-individualising process.

The outcome of this anxiety within the regulated population has been a movement towards the self-imposed prescription and rule-proliferation that Young (2010) was responding to. But this process has also been relatively independent of the regulatory system. As the HSE's former Deputy Chief Executive (2008–2015) argued:

> the prescriptive legislative regime that we've tried to do away with [...] is now being replaced, within parts of our society by, in the name of self-regulation, people writing their own proscriptive lists of what you can and can't...do. (Kevin Myers Interview, para. 48)

A well-recognised response to external uncertainty in the regulatory sphere is self-regulation, in the sense of imposing controls that go 'beyond compliance' with the law (Gunningham et al. 2004; Prakash 2001). This is not usually viewed as a negative feature of the regulatory landscape, as 'beyond compliance' is often a state of best practice. So, for example, anxieties about the loss of reputational capital lead to a *"social expansion of the legal license"* via the observance of standards that go beyond the law (Gunningham et al. 2004: 331), including those that are self-imposed. This becomes problematic when the forms that are adopted are rigidly bureaucratic and target-driven; *"the significant driver of the managerialization of risk management is an institutional fear and anxiety, namely that which is associated with the demands of organizational and individual accountability"* (Power 2007: 180). This leads to 'social defence' via the pursuit of an 'auditable surface' of outcomes that are seen to embody and resolve the demands placed upon organisations in objective, systematic, measurable ways (Hoggett 2006, 2010: 206). In the context of systems of company audit, Power refers to these processes as *"rituals of verification"* (1997), performative actions that reduce uncertainty by substituting the 'knowability' of targets and performance management for the 'unknowability' and anxiety engendered by the law. This is also an example of "juridification," or the extension of regulation to an ever-wider range of areas (Habermas 1992; Teubner 1987).

Within the health and safety sphere, the primary legitimacy risk of the law's system of open-ended duties is that the freedom from prescription it provides encourages variations in practice by placing decision-making burdens onto individual duty-holders; some of the variation that results will sometimes tend to be excessive, ill-advised, or risk-averse. A wide-ranging public narrative about 'health and safety gone mad', or excessive regulatory over-reaching, has emerged in the United Kingdom since approximately 2000, and has been attributed in no small part to the proliferation of self-imposed regulatory burdens and the excessive interpretation of the law by duty-holders (Almond 2009, 2015; Ball and Ball-King 2013; Löfstedt 2011: 40–43). In particular, it has been suggested that the open-endedness of the law encourages duty-holders to replicate the prescriptiveness of the pre-1974 regime within their own organisations, leading to risk-aversion and poor decision-making. Oft-cited examples of this relate to the proliferation of formal but superficial, and unnecessary, written risk assessment processes:

> They're mainly about ergonomics and performance and comfort rather than about illness. Those are the things that lay the ground for everybody saying, 'there's over-officiousness'. Doing the risk assessment on the display screens becomes a bit of a joke. (Tim Carter Interview, para. 84)

One consequence of this is the reduction of these processes to superficial 'rituals of verification', centred on the production of surface level, demonstrable evidence that requirements and expectations have been met, rather than the deeper (anxiety-inducing) process of exercising informed judgement about technical risks and uncertainties (Hoggett 2010: 210).

The focus group data gathered as part of this book's research (see Chapter 2) suggest that this problem is now encountered widely;

inflexible written policies, rigid internal safety officers, and a sense of pervasive managerialism characterised many public accounts of their relationship with health and safety, and crucially, this was often located at local, not regulatory, levels. All of these mechanisms can be understood as forms of 'social defence', a risk-avoidant bureaucratic reaction to the uncertainties of the modern world, whereby 'virtual' organisational objectives that are easily defined are pursued as a means of containing the anxieties that are posed via a non-prescriptive legal regime (Hoggett 2010). As the HSWA had first envisaged, the duty-holder was to be the ultimate, and (in Robens' view) most preferable, decision-maker and source of rules. It may be argued that a degree of uncertainty is necessary in motivating ongoing safety 'mindfulness' (Weick et al. 1999), in rendering the regulatory process deliberative (Waldron 2011), and in making duty-holders aware of the weight of their obligations. It may also be argued that it is part of pursuing the broader goal of granting autonomy to duty-holders to fulfil a complex, multidimensional standard (Asgeirsson 2015). But arguably, the imposition of regulatory standards of this sort is no different from any other scheme of rules, in that it prompts a tendency towards prescription and rule-proliferation, and may do so to a greater degree as a result of the freedom from prescription that exists.

THE CREATION OF A UNIFIED STATE INSPECTORATE

As well as considering the changes that have taken place in the form and function of the law since the middle of the twentieth century, an account of the constitutional legitimacy of health and safety must also focus on the institutional means through which this law is regulated and enforced. The establishment of a functionally appropriate regulatory infrastructure that could respond to the challenges of a changing workforce has been a major challenge, particularly since, by the start of our period of study, the existing institutional arrangements were 140 years old (see Chapter 3 for a discussion of this history). Perhaps as a result of the contested origins of the Factory Inspectorate (Thomas 1948: 78; Ward 1962: 115), the approach to inspection and enforcement taken by that body was conciliatory and compliance-oriented in nature (Bartrip and Fenn 1983; Bartrip and Hartwell 1997; Carson 1979). This institutional structure and approach endured largely unchanged until the early 1970s (Baldwin 1987; Rhodes 1981), by which time both were well established; however, as Chapter 3 has shown, this coverage was uneven, and related to the enforcement of piecemeal statutes and regulations. It was also undertaken via a fragmented institutional arrangement of government inspectorates, with responsibility divided between the Factory Inspectorate (part of the Department of Employment), the Mines and Quarries Inspectorate and Nuclear Installations Inspectorate (parts of the Department of Trade and Industry), the Agriculture Safety Inspectors (part of the Ministry of Agriculture and Fisheries), the Explosives Inspectorate (part of the Home Office), and the Radiochemical, Alkali, and Clean Air Inspectorates (parts of the Department of the Environment).

The Robens Committee regarded this institutional fragmentation as a source of "*bewildering complexity*" (1972: para. 32), uncertainty, inefficiency, and bureaucratic inertia at a policy level (1972: paras. 33–39). Accordingly, it was recommended that regulatory activity be organised under the auspices of a single unified inspectorate, pooling technical expertise, maximising resources, and uniting the Inspectorates under a single set of organisational objectives (1972: para. 205). These changes would be the source of both constitutional and functional legitimacy for the new agency, which had to be a:

> self-contained organization clearly responsible for the area; it had to have day-to-day autonomy; it had to be organized in a manner consistent with responsible and accountable management; and, finally, those involved in the area [...] had to be fully involved in managing the new institution. (Baldwin 1995: 127; drawing on Robens 1972: paras. 111–114)

The HSE was this new arm's-length agency, created under s.10 of the HSWA 1974, which also created the HSC as an executive supervisory body, which would make relevant policy decisions. The establishment of the HSE as a non-departmental public body (NDPB) signalled a desire to prioritise the independence and clarity of focus that separation from immediate Government oversight allowed, over the increased political profile and accountability that would attach to it were it to be part of a Government Department (Baldwin 1987: 134; Rimington 2010; Robens 1972: para. 119). More will be said later about the subsequent realities of this relationship with wider government; for now, it is worth reflecting on the challenges posed by the process of creating this unified regulator.

'Where There Is Discord...'

On the face of it, the unification of health and safety regulation, and the promise of consistency and clarity that it brought, was welcomed by politicians and commentators, including Neil Kinnock, then a young backbench MP: *"A unified inspectorate is sensible...a national authority for safety and health at work will provide a much-needed expertise, long-needed co-ordination and a welcome expansion of research".*[16] The HSE was created as the operational arm of the HSC and was given the primary role of implementing and upholding the requirements laid out in that statute via its advisory, supervisory, and enforcement roles. The problem was that the breadth of the HSE's coverage, and the depth of resource and expertise needed to provide this, required the corralling of existing Inspectorate staff and functions away from other Departments in government (a process which generated much conflict, as discussed later in Chapter 7). To begin with, the HSE retained separate functional (research, planning) and specialist inspection (Factories, Mines and Quarries, Nuclear) divisions, coordinated by a central Management Board (the division heads, the Director-General, and his Deputy), in order to allow for a gradual standardisation of working arrangements (Rhodes 1981: 83). This did leave a number of problems of integration unaddressed, as a former Chief Inspector of Factories (1985–1988, 1992–2002), recalled:

> we'd now got this cluster of Inspectorates all working within their own silos...[John] Locke [first Director-General of the HSE] wanted all that changed and [...] he failed, because the Chief Inspector of Factories, for example, couldn't persuade his own senior staff that this was a good idea [...] the Mines Inspectorate didn't want [...] to be in any position where they might be seen to be ceding sovereignty to the Factory Inspectorate. It was like robins fighting over a stretch of hedgerow. (David Eves Interview, para. 45)

As a result, the HSE remained a federation of sorts for some time. Each of the specialist inspectorates retained its own culture and way of working, a product of the extremely high levels of technical expertise of its members. One commentator observed in 1979 that *"there is a new*

---

[16] 'Workers' Safety and Health', HC Deb 21 May 1973, Hansard Vol. 857 cc62–117, 65.

attitude and a new way of working in the Factory Inspectorate. Not all inspectors have yet been converted and the mines and railway inspectorates are untouched by new ideas",[17] while Eves estimated that it was only "*by about 1990, we were beginning to think we had at last achieved a unified HSE*" (para. 45). Efforts at integration conflicted with the broader political and institutional settings within which the HSE operated, frustrating efforts at reform and creating divisions that were only resolved via time-honoured diplomacy:

> Specialist inspectors [were] recruited from the industries they inspected… if they were confronted with a reform they did not like, or even if they happened to dislike the Director-General, they had many contacts in 'their' industry to which they could go so as to create pressure…Locke was having to eat humble pie…two very expensive bottles of whisky had been placed on the table as Locke's peace offering to his colleagues. (John Rimington Interview, paras. 37–38)

These different traditions and approaches to issues of compliance and decision-making among the specialist inspectorates had important procedural legitimacy implications because they led to inconsistency, variable treatment, and an appearance of subjectivity that undermined the HSE's claims to objective expertise (Dawson et al. 1988; Hutter 1989). Some observers, such as the industrialist (and later HSC Chair) Sir Frank Davies, suggested that these issues were overstated and related to local rather than institutional variation (also Hutter 1997: 38–39):

> I didn't notice any great difference, one industry to another. We did notice difference from one District to another […] it was difference in the inspector. (Frank Davies Interview, para. 12)

But it was clear that consistency was to remain a challenge, particularly in light of the discretion in the interpretation and application of particular measures which the new legislative framework allowed inspectors and inspectorates. Different priorities and modes of regulating would continue to exist at the same time within the HSE (Hawkins 2002: 323; Hutter 1997: 77). Over time, however, the "*robber barons*" within the

---

[17] B. England, 'Discussion', *Work Hazards* 21 (c. 1979), p. 18. *Samuel Barr Collection*, Glasgow, Caledonian University, GCU DC 140/2/1/2.

specialist inspectorates were *"gradually brought to heel"* (Kevin Myers Interview, para. 9). Subsequent changes, such as the creation of the generalist Field Operations and Hazardous Installations Directorates, brought a greater degree of joined-up practice, but arguably, this came at the price of a loss of autonomy and integration with the industries that the specialist inspectorates regulated. On the one hand, this was a concern to trades unions, who worried about the loss of access to knowledgeable inspectors[18]; on the other, those links created ongoing problems for the management of the unified HSE.

The Extension of Control

From the late 1980s onwards, the history of the unified inspectorate has been one of expansion and contraction, often in response to external events. In particular, a concurrence of major disasters in the late 1980s, most notably the Clapham Junction rail crash and the Piper Alpha offshore disaster (both in 1988), led to the HSE being given responsibility for HM Railways Inspectorate and for regulating offshore installations (in 1990 and 1991, respectively). In both cases, there was evidence of conflicts of interest around the management of safety by government departments which were also responsible for maintaining levels of productivity and service in those industries (Hidden 1989). For example, Carson's (1982) work on the offshore oil sector demonstrated how the Department of Energy's pursuit of a *"political economy of speed"*, in order to sustain a revenue flow on which the British national economy was increasingly dependent, led to an *"institutionalized tolerance"* among regulators of levels and types of risk that would ordinarily be unthinkable (1982: 291). Similarly, Hutter (2001) found that railway safety was, during this time-period, undergoing a major transition from traditional mode of regulating risk, focused on technical standards, to a newer, safety-case-based approach (in line with the new legal approach discussed previously). This exacerbated, to some degree, the tension between safety and funding (Hutter 2001: 67). In both cases, the incorporation of the regulatory body into the HSE was intended to address this conflict and modernise the regulatory approach, and on both counts, it

---

[18]Report of TUC Conference on the Robens Report, 12 October 1972, pp. 44–45. TUC Library, HD 7273.

proved unpopular, as Sir Bill Callaghan, who later became HSC Chair (1999–2007), observed:

> the rail industry did not like the Rail Inspectorate coming into HSE from the Department for Transport. And a decision was taken, in my view quite rightly, just as it was wrong for inspection of offshore oil platforms to be done by the Department of Energy, it was wrong for inspection of railways to be done by the Department for Transport. But the view that only railwaymen could look at the problems of railways was very strong. (Sir Bill Callaghan Interview, para. 23)

Evidence from other examples of offshore safety failures suggests that disasters often arise when a single accountability and decision-making frame comes to predominate within an industry setting, with assumptions and motivational drivers being shared between key players (including captured regulators: Bradshaw 2015) and used to resolve trade-offs (Mills and Koliba 2015), leading to a narrow focus on profit-maximisation as a goal. The HSE was seen as an expert, independent regulator who could provide a degree of impartiality, challenging this 'cognitive closeness' via the imposition of a new regulatory approach; it has also rarely acted as an 'economic' regulator (Prosser 2010: 98–101; 109), so offered a means of distancing controversial issues from the economic pressures brought to bear by government policy. Major disasters often constitute a source of distinct tension and crisis for both regulators and political actors (Almond 2007; Haines 2011; Hutter and Lloyd-Bostock 2017), and much of this pressure relates to the need to demonstrate the legitimacy, particularly in constitutional and procedural terms (impartiality, due process, legality: Baldwin 1995), of the regulatory response to the causes of such disasters. Regulatory oversight thus operates as a form of 'risk management' on behalf of governments, a 'steering mechanism' to dispel conflict and control issues that pose political risks to decision-makers (Almond 2013: 132–134; Carson 1979; Haines 2011: 48–50). But this can expose the regulator to a greater degree of political pressure, as was the case in these examples. As the *Guardian* noted in 1988, additional responsibilities demanded by the Government "*combined with a series of disasters […] has put worker and public safety high on the agenda and the role of the HSE is being closely examined*" (Brown 1988: 1). The pursuit of procedural objectivity also posed challenges, particularly in ensuring the appropriate balance of competence and independence in specialist sectors, as the then-Deputy Director General of the HSE recalled:

> It was after Piper Alpha and the Cullen Inquiry that the offshore inspectorate came into the HSE [...] who do you use to inspect a very specialist industry? All mines inspectors had to have been mine managers [...], poacher turned game keeper. It used to be similar for the railways. But the offshore inspectorate certainly depended on people's specialised knowledge [...] we did our best to train them in a 'control mode', to be independent. But of course they inspected their former colleagues and friends. It's not easy. (Jim Hammer Interview, para. 73)

But risk is also about offsetting danger with rewards. The management of these sources of social and political risk presented the HSE with an opportunity to enhance its own position. Post-disaster reforms tended to strengthen the regulators' hands, as they rendered it difficult for even a deregulatory government to take action that might be construed as exposing the general populace to risk; as one Labour MP recalled:

> for virtually all of Margaret Thatcher's government, I think that they would have liked to have done things with the Health and Safety at Work Act but there were so many disasters, it was very difficult. (Frank Doran Interview, para. 46)

In the immediate aftermath of Clapham Junction, Piper Alpha, and other disasters, HSE was able to make a play for greater responsibility (and hence resources) by taking responsibility for offshore and railway safety (Employment Committee 1988: paras. 20–24). The invitation of John Rimington, the then-Director-General, to Government (delivered via a Select Committee hearing) was clear: "*if one has the hands-on experience and is doing the job oneself one knows what is going on [but...] that was not the situation created by Parliament [...] we do of course have immense experience of inspection*" (ibid.: para. 33). This invitation was heeded, and by 1992, once railway and offshore safety had been brought within the HSE, the same Committee welcomed the HSE's management "*back here with responsibilities which we thought you should have back under your belt*" (Employment Committee 1992: para. 2). The political capital of such processes won the HSE valuable friends at a time of budgetary restraint; as trade unions, public debate, and the Labour party put pressure on the government over the HSE's resourcing, the Director-General of the HSE was reported as smiling *"discretely and welcom[ing] the political weight they are able to put behind his requests for more money and resources"* (Brown 1988: 1). While the newly amalgamated HSE continued for some years to struggle to integrate specialist bodies like the Railway Inspectorate into its ways of working (Hutter 1997, 2001), and would see responsibility for nuclear installations and railway safety taken away once more in the 2000s and given to independent agencies located within the relevant government Departments (the Office for Nuclear Regulation [ONR], and the Office of Rail and Road [ORR]), it has, on the whole, secured broad legitimacy benefits from its status as a unified inspectorate.

HSC and HSE: Happy Bedfellows?

One of the fundamental innovations introduced under the 1974 Act was the creation of a two-tier regulatory organisation. The Robens Report had argued for the creation of both a new national Authority (inspectorate), and a Managing Board to control its operations (1972: para. 117–118). This Board would be participatory and representative of the regulator's stakeholders, and so embody the 'corporatist' policy ideals of the time (Hutter 1997: 23). Section 10 of the HSWA implemented this by creating two new bodies; the HSE to implement and enforce health and safety law, and the HSC to make policy and direct the actions of the Executive. The Commission would consist of three members representing employer organisations (primarily the Confederation of British Industries [CBI]), three representing employee organisations (primarily via the TUC), and up to three other members from professional or related bodies (such as the medical profession), plus a Chairperson appointed by the Secretary of State. This led to the HSC being characterised as *"the most corporatist body in Britain"* (Wilson 1985: 113; also Hutter 1997: 24), the most fully formed expression of a policy ideal centred on *"interest representation [via...] a limited number of singular, compulsory, non-competitive, hierarchically ordered and functionally differentiated [actors...], recognised [...] by the state"* (Schmitter 1974: 93), which only ever partly displaced other forms of pluralism within the UK policy context (Jordan 1981; Marsh and Rhodes 1992). These interest groups were placed at the centre of the regulatory process; the functional separation of Commission and Executive was in order to prevent these interests appearing too closely involved in day-to-day enforcement decision-making (Baldwin 1987: 136). At the same time, the

HSC was intended to act as a buffer between the HSE and the political sphere by providing a forum within which different interests could be heard without becoming embroiled in the politics of industrial relations more broadly. Regulatory arrangements of this sort perform a *'steering role'*, redirecting conflicts away from the economic and political system, and into more manageable forms (Almond 2007, 2013; Carson 1979; Habermas 1973).

In practice, the HSC proved a reasonably constructive oversight body, though it possessed a pronounced tendency towards consensus-seeking, with disagreements usually internally resolved and a unified front presented on policy issues, as senior HSE figures recall:

> In my experience, the Commission never, ever voted. What they did if there was a disagreement, they said we'll refer this back to officials and the officials used to then go and have a word with the employer representative and they'd have a word with the trade unionists and it was either a one-lunch or a two-lunch job to resolve the impasse. (Jim Hammer Interview, para. 39)

The HSC was also criticised for being less representative than Robens intended; the members *reflected* their respective constituencies, rather than *representing* them in any direct sense, and so were heavily reliant on the backing of organisational constituents (CBI and TUC briefing papers) rather than wider consultation. The HSC also tended to be deferential towards the policy agenda that the HSE provided, as one former CBI-nominated Commissioner recalled: "*We as Commissioners had [...] very little initiative, we followed what the Executive, Rimington's people, suggested should be done, because they were following it all the time*" (Rex Symons Interview, para. 62).

This agenda-setting involved coordination between the leaders of the two bodies ("*we had meetings, the Executive and the Chairman, a small group, a few days ahead of each Commission, so that he was on side*": John Rimington Interview, para. 45), something which made the personal relationships between HSE Director-Generals and HSC Chairs centrally important; some, like that between John Locke (HSE) and Sir Bill Simpson (HSC) in the 1970s were very difficult ("*an absolute disaster. They weren't on speaking terms*": John Rimington Interview, para. 25), while others, like that between Sir John Cullen (HSC) and John Rimington (HSE) in the 1980s, and Sir Frank Davies (HSC) and Jenny Bacon (HSE) in the 1990s, were more constructive. At times, personal distrust between post-holders created conflict, who sought to pressurise, contain, or brief against one another (Neal Stone Interview, para. 56; Sir Bill Callaghan Interview, para. 54). In general, however, the HSC functioned alongside the HSE in a consensual manner, noting in 1982 that:

> [w]e have to reach our decisions by consensus because we are aware that once we have agreed on a solution, we have committed those whom we represent to that solution and have to stick by it. Differences of view must, of course, arise and we resolve them through continuous and persistent discussion. (Employment Committee 1982: 74)

Such accounts position HSC within the institutional framework of Moran's 'club government' model of regulation; interpersonal, 'mutualised', and colonised by powerful, established interests (2003: 137). The HSC was envisaged as the key corporatist forum through which interest groups would coordinate, but over time, the nexus of these processes arguably shifted away from the HSC and into the broader political and institutional context around it, as the political climate turned away from corporatism and towards neoliberalism (*"Trades Unions' participation in government had been fashionable in 1972, but no longer was in 1992"*: John Rimington Interview, para. 63). This new climate encouraged a more diverse 'market-place' of influence and involvement, with a much more diverse, complex, and less tightly defined, range of interrelated interests and actors playing a role in health and safety policy, such as the EU (discussed below), commercial providers such as the emergent safety professions (discussed in Chapter 6), public interest groups, and so on. The regulation of safety was better understood as a site of *networked governance*, that is, of policy as an outcome of interactions within clusters of interdependent institutions and actors (Hancher and Moran 1989; Marsh and Rhodes 1992; Rhodes 2001; Jordan 1981). On this view, *policy networks* provide the settings for decision-making, rather than the centralised apparatus of the corporatist state, not least because they are more permeable, flexible, and multidirectional, and less restrictive of the agency of their constituent members. The movement towards pluralism identified here is consistent with notions of a shift from a semi-corporatist *policy community* (typified by stability, limited membership, and vertical interdependence) to a more marketised *producer network* arrangement, with a more open membership, more economic relationships, and more reliance of the core on the periphery

(Marsh and Rhodes 1992: 13–15), driven by a political climate of economic liberalism, privatisation, arm's-length regulation, and the 'New Public Management' (NPM) of private sector engagement in the delivery of public goals (Hood 1991; also Black 2005; Gamble 1994; Moran 2003).

With health and safety regulation thus occurring increasingly 'beyond the state', the HSC became less central to the *delivery* of safety, and subsequently struggled to function as a political buffer, attracting rather than deflecting criticism in the aftermath of major railway disasters between 1999 and 2002, as the then-Chair recalled:

> I was castigated. I was asked to see the Prime Minister [...] this series of high profile crashes was politically very toxic. Lots of public interest. I think that's the time when people were questioning openly whether we had the right approach. (Sir Bill Callaghan Interview, paras. 26–27)

While the involvement of individuals within the HSC who had a strong grounding in particular constituencies was still valuable (Podger 2015: para. 12), it was clear that by the late 1990s/early 2000s the Commission was less adept at depoliticising health and safety than in the past:

> [In] getting both sides in industry to agree, [HSC] was eliminating one source of potential political difficulty to government. Government became much less interested in having a tripartite HSC once they believed that they were in charge [...] the various rail crashes, that was the great disaster of HSE [...] the point at which Ministers lost confidence. (Senior HSE Source Interview, para. 22)

At the same time, the residual role of a public body like the HSC was, under the 'NPM', one of ensuring accountability and managerial oversight of regulatory delivery (Hood 1991; Moran 2003). The HSC came to be seen as unable to fulfil these contemporary governance norms, even by those running it:

> [HSC] was not a proper non-executive board [...] The job of the Commission is to hold the Executive to account, and that needs proper and robust procedures for corporate governance, and we didn't have that [...] we weren't really getting to grips on resource allocation and the Executive. (Sir Bill Callaghan Interview, paras. 54–56)

This led to a loss of trust between the HSC and HSE, with the latter, by the late-2000s, having come to see the former as an obstacle to work around:

> the Commission had completely lost confidence in the Executive, the Executive thought the Commission rather tiresome, irrelevant, it ought to go away. It is rather unfortunate that the organisation [...] was completely divided [prompting...] legitimate concerns as to what wasn't getting done as a result. (Senior HSE Source Interview, para. 82)

In 2007, the Labour Government proposed the merger of the HSC and HSE, as the existing structure was "*outdated and...not as effective as it need[ed] to be in managing the financial and performance monitoring aspects*" of the HSE (DWP 2007: 3). This was supported in principle by most respondents to a consultation, and the subsequent Legislative Reform (HSE) Order 2008[19] abolished the HSC, passing some of its functions to a new HSE Board and Chairperson; a degree of continuity was ensured as the last HSC Chair, Dame Judith Hackitt, became the first HSE Chair. Initial signs were that the merger was fulfilling the hope of Lord McKenzie, the Government Minister behind it, that it would "*give more focus to the organisation*" (Lord McKenzie Interview, para. 22), but the subsequent triennial Temple Review of the HSE concluded that the new HSE Board was not "*dynamic or an example of best practice*", and lacked the commercially and strategically "*necessary competencies/attributes common to all Boards*" (Temple 2014: paras. 4.8 and 4.12). The HSE Board is still not necessarily regarded as containing either the necessary political influence, nor the competence at interrogating management, to be fully effective (Podger 2015: para. 12). Perhaps as part of an effort to address these concerns, Martin Temple, the author of the triennial review, became the new HSE Chair in 2015, with Dame Judith Hackitt succeeding him in his prior role as Chair of the EEF, the UK's largest sectoral employers' organisation. This pursuit of 'business-like' leadership via a 'revolving-door' link (Makkai and Braithwaite 1992) to one particular constituency body suggests that the legitimatory role of the HSC, as a representative body and 'policy buffer', has been displaced by a focus on procedural governance. The unified HSE may eventually be more equipped to fulfil new governance norms, but may be less

---

[19] https://www.legislation.gov.uk/uksi/2008/960/article/2/made.

well equipped to represent external political constituencies as a result of its reduced capacity to present regulation as the outcome of cooperative processes of consultation. How significant this may be, in relation to legitimacy, remains to be seen, especially given the changes to have taken place in Britain's economic structure and trades union membership, discussed in Chapter 5.

A 'NATURAL IDENTITY OF INTEREST'? REGULATORS AND CENTRAL GOVERNMENT

The relationship that a health and safety regulator has with the rest of government functions as a fundamental source of its legitimacy. Just as Robens was criticised for his perception of a "*natural identity of interest*" between the different sides of industry (Robens 1972: para. 66; criticised by Tombs and Whyte 2010; Woolf 1973), so we must also be critical of accounts of regulation that too readily assume an identity of interest between 'regulators' and central government. The Government of the day is one of the key audiences that a regulator must present its legitimacy credentials to, and so the relationships between the two cannot be assumed to be either easy or straightforward (Almond 2007: 299). In particular, procedural values of "*efficiency and effectiveness*" (Baldwin 1995: 46), implicit in a governmental commitment to a minimised state apparatus and economic approach to policy as was the case for much of the period after approximately 1980, must be accorded with, via evidence that the regulator offers value for money and so deserves support. A regulator must also demonstrate that it is accountable to central government, and that it possesses the requisite expertise to perform its role, if it is to establish its organisational legitimacy. The HSE's interactions with other parts of government therefore had a major impact on the legitimacy of health and safety more broadly.

A Sort of Independence

The HSE sits in what was, at the time of its creation, an unusual place, constitutionally, within government; as a NDPB, it had a high degree of functional autonomy and a significant, expertise-led technical capacity to implement and develop practice within its area of regulation. At the time it was created, the HSE was part of the civil service and a separate entity from the HSC, which was a genuinely arms-length body tasked with the pursuit of statutory objectives, and empowered to develop policy on behalf of government, and given a very significant degree of functional autonomy to exercise in doing so (Prosser 2010: 90). In 1974, this 'hyper-modern' (Moran 2003) and innovative model of regulatory practice was novel, and was viewed with considerable hostility (Rimington 2010), not least because the creation of the HSE had taken areas of resource and responsibility away from existing government Departments:

> John Locke [...] who really created the HSE, he'd gone around arguing the toss with Permanent Secretaries in various Departments [...] so there was a lot of hostility in the early days from other Departments towards the creation of this new-fangled QUANGO. (David Eves Interview, para. 45)

The HSE now is one of 392 NDPBs in the United Kingdom,[20] indicating how far the model of 'distributed public governance' has spread since then (Flinders 2004). The Robens Report, in proposing the creation of the HSE, had argued that the benefits of having the new inspectorate form part of a central Government department (joined-up administration, direct accountability, and political profile) were outweighed by the loss of independence and diffusion of responsibility that this would entail (Robens 1972: para. 119). The intention was that a 'managing' body like the HSC could drive forward the health and safety agenda in a more dynamic way than traditional Whitehall policy mechanisms, and that the HSE would remove the day-to-day operations of health and safety from the politics of industrial relations policy (Moran 2003).

At the same time, however, there was a realisation that this arrangement would pose considerable accountability challenges. For as regulatory agencies become devolved from central government, their connections to the democratic processes that legitimise them necessarily become more tangential (Baldwin 1995; Black 2000, 2008; Scharpf 1999), something that can lead to accusations of regulatory capture (as narrow interests dominate decision-making in place of popular input: Makkai and Braithwaite 1992), technocracy (as inward-facing rationalities dominate decision-making in place of public agendas: Radaelli 1999), and illegitimacy (as regulators come to be seen as immune to public input: Smismans 2004). As the HSWA 1974 was moving through its second reading in the Commons, the then-Secretary of State for

---

[20]They are listed at https://www.gov.uk/government/organisations.

Employment, Michael Foot, moved to address such concerns by assuring Parliament that *"it was not the Government's intention to set up a body which was remote from questioning in the House of Commons"*.[21] In order to bolster this accountability, the HSE was positioned within the purview of a sponsoring department of state, initially the Department of Employment, which would exert budgetary control over the agency and also ensure appropriate Ministerial responsibility (Prosser 2010: 92).

Being functionally independent, the HSE was sometimes required to intervene around issues where government was, in one form or another, the relevant duty-holder, for example, publicly owned industries or those where approval has to be given for work (offshore oil or mining). In many of these sectors, there was a tendency for conflicts of interest to arise between the competing responsibilities of Secretaries of State, who often prioritised production over safety (Carson 1982), were *"lobbied by the industry to make [regulations] as easy-going as possible"*, and tended *"to under-resource the inspectorates"* (Helen Leiser Interview, para. 31). This role exposed regulators to particular scrutiny; when the HSE investigated an incident (the Bilsthorpe Colliery collapse of 1993) that fell within the Department of Employment's oversight, there were calls for an independent inquiry, which *"scared the Health and Safety Commission, who were terrified that their independence of government would be compromised"* (Benn 2002: 225–226). When the Secretary of State, Michael Forsyth, gave evidence to Parliament on the matter, he referenced the HSC's independence as a means of defending the Government from accusations of improper interference: *"We have listened to a disgraceful speech, a vicious and unwarranted attack on a tripartite body, the Health and Safety Commission [...] It is a disgrace that Opposition Members should seek to challenge their integrity"*.[22] Independence from government was clearly regarded as something that protected both parties.

Rimington has subsequently argued (2010: 608) that the devolution of the HSE's functions in this way had a double effect; it was necessary to be a 'third party' if existing Departments were to surrender their existing staff and functions (in which they would retain some interest, for instance, the Department of Energy in nuclear safety), but

---

[21] Hansard (1974), Vol. 871, Col. 1290, 3 April.

[22] 'Mines (Health and Safety)', HC Deb 26 October 1993, Hansard Vol. 230 cc698–747, 703.

this independence also exposed the HSE to powerful political forces. It was answerable to Ministers, but could not rely on a particular Minister or Department to deflect criticism or pressure on its behalf. The HSE was thus compelled to work independently to establish its own credentials within government; the Chief Inspector of Factories, Jim Hammer, suggested in 1975 that it was pursuing a strategy of "*defusing criticism before it came by 'selling' the organisation to MPs in times when it was not under attack due to the occurrence of a particular incident*".[23] Rather than functioning as a 'steering body', able to depoliticise issues of health and safety, the HSE has instead found itself playing a political role of sorts at different times. On the one hand, as a former HSE Director of Health Policy observed, the high-profile nature of the issues that it handled gave it authority within government:

> HSE actually kept its credibility better than a lot of other regulatory bits of government, largely because it was focused mainly on what everybody saw as reasonably high risk bits, and generally had quite a lot of support from people close to it in industry. (Tim Carter Interview, para. 134)

At the same time, however, this visibility has brought health and safety regulators within the political realm, where they are subjected to shifting governmental preferences, and hence differing demands. This was noted, for example, in the early 1980s by trade unionists when the HSE's funding was reduced: "*it is in line with present Government policy that the health, safety and working conditions of all workers, but especially those in farming, count for less than debating points scored in the political area*".[24]

On the one hand, Government poses problems when it has strong ideological views, which it seeks to impose onto the regulator, particularly when they are deregulatory. The HSE was constrained by the outlook of Government in making and implementing law and policy (Baldwin 1987: 140–141; Hawkins 2002: 118–119), and an anti-regulation stance on the part of Government had a significant impact on decision-making, as did demands for particular responses to events.

---

[23] HSE Management Board Minutes, 19 June 1975, Min. 2,a,i. The National Archives of the UK [TNA], London, EF 10/1.

[24] J. Hose, President of the National Union of Agricultural and Allied Workers, quoted in 'Safety Is Expendable', *The Land Worker*, March 1980, p. 7. MERL.

Each of these tendencies had the potential to be short-termist, incoherent, and highly disruptive, as a former Director-General lamented:

> Whitehall is run by Ministers [...] and Ministers have ideas all the time [...] they want answers to this, that, and the other, and a brief has rapidly to be prepared, the Minister persuaded that it won't work or the Minister has to be obeyed. (John Rimington Interview, para. 55)

This often meant that pressures came from multiple sources within Government. Indeed, one Senior HSE Source (Interview, para. 149) identified three separate policy agendas coming from different parts of the 2010–2015 coalition Government (the Department of Work and Pensions [DWP], the Department for Business, Innovation, and Skills, and the Cabinet Office), leading to incoherent institutional and policy relationships. On the other, Government is a source of pressure when it wants to intervene and control the health and safety agenda, as the Director-General of the HSE at the start of the New Labour era (1995–2000) recalled:

> [John Prescott] wanted a far stricter regime and more Ministerial involvement, he didn't want [HSC] doing things, he wanted to be in charge of everything. So that was a different kind of challenge, it wasn't one that said, 'let's cut the resources'...it was the opposite. But in many ways it was just as dangerous. (Jenny Bacon Interview, para. 45)

The result of this has been an uneasy, and ongoing, process of 'give and take', as the HSE's functional regulatory independence has been balanced against an increasing desire on the part of government to control the agency's work via 'regulation inside government' (Hood et al. 1998). Major disasters where the HSE was seen to have failed in its regulatory duties, such as the 1999 Ladbroke Grove rail crash, fundamentally challenged the procedural legitimacy of the HSE, at least in the eyes of government, leaving the agency "*shaken*" and lacking wider support (Jenny Bacon Interview, para. 72; Bill Callaghan Interview, para. 27; David Eves Interview, para. 58; Senior HSE Source Interview para. 20). Incidents of this sort have fuelled a broader scepticism about the functional arrangement that the HSE represents, as central government has recognised that it is not necessarily shielded from public criticism following such incidents (Pantti and Wahl-Jorgensen 2011). This "*growing anti-QUANGO animus in government [...] a view that the centre of government should do this, not arm's length bodies*" (Sir Bill Callaghan Interview, para. 50) has meant that there is now a greater tendency for control to be exercised via budget-setting (Almond and Esbester 2018: 8; Baldwin and Black 2016: 573). This sort of secondary 'framework' oversight (Hood et al. 2001; Moran 2003) has long provided a means for political priority-setting (Jim Hammer Interview, para. 55), and is the primary means of holding the regulator accountable (Hood et al. 1998; Prosser 2010; Rimington 2010), with performance targets and financial audits used to analyse the HSE's performance, as a former HSE Financial Manager recalled:

> you had to be able to forecast accurately what your expenditure was going to be month to month, you had to forecast accurately what your income was going to be […] If you didn't get that forecast right it could result in questions being asked which are very much along the lines of, 'do you know what you're doing?' HSE is constantly under the spotlight when it comes to the sponsor department. (Neal Stone Interview, para. 31)

The variability in the HSE's political accountability is reflected in the changing identity of the sponsoring Department. Although functionally independent, different government Departments have had oversight of the HSE (controlling its budgets and scrutinising its plans of work) at different times, and the power to make regulations under the HSWA 1974 lies with the Secretary of State of that Department, who also directs the HSE's work and holds it to account (Hawkins 2002: 159–161). Three things can be observed about these relationships. First, as with any process of institutional change, the discontinuity associated with moving from one Department to another (for instance, when the parent Department is dissolved) is disruptive, and functions as a source of institutional anxiety for the regulator (Hoggett 2006):

> [In 1995] we lost our parent department, the Department of Employment disappeared, and we had to form a whole lot of new relationships. And it took ages and was very resource intensive to build a relationship with the Department of the Environment…nobody really knew what Health and Safety Executive and Commission were and did, nor had much interest in it. (Jenny Bacon Interview, para. 88)

Second, these changes serve as indicators of different pressures that were brought to bear on the HSE at different times. The Department of the Environment (after 1997, the Department of the Environment, Transport, and the Regions [DETR]) was always a policy-oriented department with diverse responsibilities (Newman 2001), and (as observed above), the Secretary of State (John Prescott) was able to exert outcome-focused pressures on the HSE. Within the fiscally-oriented DWP (the successor to DETR as sponsoring Department), on the other hand, the focus of oversight was perhaps different, as the Government Minister with responsibility for the HSE at the time recalled:

> The DWP and the HSE…it was a bit of a battle at times on budgets, but that was not unique to the HSE. No, from a Ministerial point of view I think that that worked well. (Lord McKenzie Interview, para. 68)

Thirdly, these changes were in part reflective of the fluctuating fortunes and declining political legitimacy of the HSE. After the *'great institutional disaster'* of the rail crashes of 1999–2002 (Senior HSE Source Interview, para. 22), the HSE was passed around Whitehall, seemingly because it was viewed within Government as a liability rather than an asset:

> [There was] a big discussion when DETR was broken up about where we should go. We ended up in DWP. [**Did you have a period where you were Department-less?**] More or less, yeah. Well, the Department for Transport was then looking after us, and they obviously didn't want us. (Sir Bill Callaghan Interview, paras. 60–62)

The question of procedural accountability is central to questions about the legitimacy of a regulator like the HSE, which changes within government (Almond 2007) just as it does in the eyes of wider audiences. These fluctuations arguably have greater implications for the agency than almost any other form of legitimacy judgement.

Sharing Authority: The Role of Local Government

A final regulatory relationship within government, and one that has a key bearing on health and safety as a policy area, is that between the national regulatory authority, the HSE, and the 410 Local Authorities (LAs) in the UK who, as part of their local government function, perform key elements of the regulatory role in relation to approximately 1.1 million workplaces in mainly lower risk sectors (Work and Pensions Select

Committee 2008: para. 187; Löfstedt 2011: 80).[25] This role for local regulatory oversight goes back to the Factory Acts of the nineteenth century. Prior to 1833, oversight of factories was the responsibility of locally appointed visitors (usually Justices of the Peace and clergymen), who reported to the local courts.[26] After the 1833 Factory Act instigated the new, national Factory Inspectorate, local institutions retained a role (via certifying surgeons), but became more directly involved following the Factory and Workshop Act 1901, which gave LA inspectors the same powers of inspection and enforcement as the national Factory Inspectors. LAs were required to inspect and register workshops and pass on details of their inspection regime to the Secretary of State, a system of 'dual control' which expanded the machinery of regulation by co-opting public authorities which were not predisposed towards stringent enforcement (Hutchins and Harrison 1911: 242–248; Mess 1926: 29).

This sharing of the regulatory role gained impetus as a result of the Offices, Shops, and Railway Premises Act 1963, which brought many 'non-industrial' worksites within the scope of the legislative regime. Section 25 of that Act placed LAs under a duty to enforce requirements against the one million non-industrial workplaces that it brought within the scope of the regulatory system, allowing those authorities to appoint inspectors with the same sort of powers as the national Inspectorate. This constituted a *"rather haphazard compromise"* over the division of responsibility; the *"experience…prestige and authority"* of the central inspectorate would be applied to high-hazard industries, while the locally engaged and *"friendly"* LAs would handle low-hazard workplaces like offices and shops (Samuels 1963: 541). This division of responsibility was subsequently endorsed by Robens for the post-1974 Act era (1972: para. 246), and continues to this day, with LAs extending health and safety inspection into workplaces that are not picked up as part of the HSE's risk-targeted inspection programme (Löfstedt 2011: 82).

---

[25] In 2013–2014, LAs brought 92 prosecutions http://www.hse.gov.uk/statistics/prosecutions.htm compared to HSE's 588, and conducted 6300 proactive inspections and 37,000 'other visits'—where health and safety arises via other engagements with businesses (such as food hygiene inspections, licensing processes, and advisory visits) http://www.hse.gov.uk/aboutus/meetings/committees/hela/310714/data-collection%E2%80%93analysis.pdf. HSE conducted 23,740 proactive inspections (of high-risk or poorly performing workplaces) in the same year http://www.hse.gov.uk/aboutus/reports/1314/ar1314.pdf.

[26] A requirement imposed under the Health and Morals of Apprentices Act 1802, IX.

More controversially, the language of 'high' and 'low' risk workplaces has continued to shape the rationing of inspection and enforcement (Almond and Esbester 2018; see also Chapter 6), with the DWP publishing a strategy document in 2011[27] which set out three classes of industrial sector ('high risk areas', 'areas of concern', and 'low-risk areas'; 2011: 9), and restricted inspection to the first of these. The HSE also fulfils a coordinating role in relation to the health and safety-related activities of LAs, via a HSE/Local Authority Liaison Committee (HELA), and the operation of its own Local Authority Unit (established in 1983). As such, while LAs are functionally autonomous, the national regulator has a significant centralising role (receiving statistics, issuing guidance/training, promoting strategic goals), and exerts significant influence over their practice. A desire for closer working saw the Löfstedt Review recommend (2011: 83) greater central coordination, resulting in a new National LA Enforcement Code being produced in 2012.

The relationship between the HSE and LAs has given rise to a number of legitimacy-related challenges. In particular, questions of consistency and proportionality, and of accountability and control, have dogged the regulatory work of LAs and, by proxy, affected the perceived legitimacy of the HSE and of health and safety as a whole. There has long been a perception among some observers that LA inspectors are over-zealous (Dawson et al. 1988: 250; Robens 1972: paras. 233–234; Young 2010: 26–27) and variable in the standards they apply (Podger 2015: 6), tending to deviate from the understandings of reasonable practicability and compliance that characterise the HSE's work (Hawkins 2002; Hutter 1997). LAs are routinely portrayed as 'risk bullies' who take an overly restrictive approach to the imposition of standards (Löfstedt 2011: 82–83; also Halliday et al. 2011), although the evidence for this assertion is both limited and contestable. One former HSE Director-General attributed these differences to the risk culture within LAs, and the different expectations attaching to other regulatory functions which they perform:

> Environmental Health Officers [...] operate to standards of perfection. Most of what they do is inspect food shops and restaurants, where it really is important that things should be exactly right [...for] 80% of the

---

[27] *Good Health and Safety, Good for Everyone*. http://www.dwp.gov.uk/docs/good-health-and-safety.pdf.

economy, the people [...] administering the Health and Safety at Work Act are not imbued with the idea of reasonable practicability. (John Rimington Interview, para. 75)

It may be that the risk culture surrounding food safety inspection really is more precautionary than that around health and safety, not least because they are oriented towards the control of ongoing outbreaks (Hutter 2011).

Such concerns are longstanding; oral evidence presented to the Robens Committee in 1970 claimed that "*enforcement is shared between too many authorities, some of who are completely ineffectual. This is particularly the case with the Local Authorities, few of whom are sufficiently competent to understand what is required of them*".[28] Similarly, the National Federation of Professional Workers argued that greater centralisation was needed.[29] These persistent concerns have prompted a move towards greater central control of LA regulatory functions. For example, the Rogers Review of Local Authority enforcement (2007) tried to ensure consistency of approach by establishing a hierarchy of enforcement topics according to the level of risk, plus consideration of public and political priority (2007: 36–43), and the Löfstedt Review recommended that the HSE direct all LA enforcement activity to ensure both efficiency and consistency (2011: 83). A new National LA Enforcement Code was subsequently introduced to embed this recommendation, and to restrict the range of possibilities open to regulators.[30] The degree of practical control exerted over LAs has thus increased, but has done so, paradoxically, in response to criticism that the HSE was already receiving in relation to these functions (even when it was not formally responsible for them). The HSE has tended to bear the brunt of public and political blame for decisions that all regulatory actors, LAs included, make (Almond 2015); these developments have brought the actual constitutional legitimacy (accountability and oversight) of health and safety regulation into line with its *perceived* legitimacy.

[28]Written Evidence of P. O'Gorman to Robens Committee, 8 October 1970, p. 10. TNA, LAB 96/36.

[29]Written Evidence of National Federation of Professional Workers to Robens Committee, 15 March 1971, p. 2. TNA, LAB 96/697.

[30]National Local Authority Enforcement Code, *Health and Safety at Work: England, Scotland & Wales*. http://www.hse.gov.uk/lau/la-enforcement-code.htm.

Another reason for any inconsistency is the multiplicity of functional demands that LAs fulfil, and the limited resources they have to do so, something that has worsened as LA budgets have been cut (Steve Sumner Interview, para. 46):

> in the Local Authority sector we [HSE] didn't have effective ways of ensuring consistency…it's much more difficult to achieve consistency when you've got an inspectorate which is relatively weak because it's got all sorts of other pressures on it. Individual Local Authorities don't have the collective resources HSE has. (Sir Bill Callaghan Interview, para. 39)

When budgets are cut, but responsibilities remain multiple, public bodies naturally tend to consolidate their work around their core functions, which health and safety has perhaps never been for LAs, as one local government policy-maker recalled:

> health and safety was always the Cinderella part of the service, if you like, in terms of the job of [LA Environmental Health Officers]…we were always perceived as being not as important as food hygiene. (Steve Sumner Interview, para. 10)

The consequence of the shift towards centralised control has thus been a decline of 95% in rates of LA inspection since 2009–2010 (HSE 2014: 2). Central control has largely involved the restriction of LA activity, and financial budget setting has proved a more effective lever than direct policy oversight. Here, as elsewhere, central government has found that its desire to control the work of regulators has conflicted with its reduced capacity to do so within an arms-length regulatory arrangement (Baldwin 1995; Black 2008; Moran 2003). For central government decision-makers, the functional independence of local government has, like the independence from government afforded to the HSE as a NDPB, gone from acting as a welcome 'policy buffer' (protecting government from blame and regulators from interference), to a liability, obscuring formal accountability and exposing both parties to political risks, particularly in an era of risk-intolerance and scepticism towards government (Haines 2011). Central government has arguably become less supportive of health and safety regulation, and has been more directly influenced in its decision-making by its perception of public attitudes around this issue (Almond 2015). At the level of both NDPBs and

LAs, the responsibility attributed to government for health and safety policy and practice, has outstripped its capacity to exercise that control. The attempt to retrieve that sense of control has involved a shift towards demonstrating more direct functional legitimacy (accountability and control) on behalf of health and safety, but perhaps at the cost of the indirect elements of constitutional legitimacy (independence and impartiality).

HEALTH AND SAFETY AND THE EUROPEAN DIMENSION

One of the most fundamental constitutional legitimacy challenges to have arisen over the last fifty years is that of the influence of the EU. The broader political and social legitimacy of the EU has been widely debated, not least because of the legal and political forms and functions that it pursues. It is often argued that the EU is *legitimate* when judged upon its own standards, institutions, and legitimation needs as a transnational legal institution, but may appear *illegitimate* when assessed by the expectations that apply to traditional nation states (Beetham and Lord 2014; Gibson and Caldeira 1995; Greenwood 2007; Moravcsik 2002; Scharpf 1999; Smismans 2004). Many of the perceived problems arise from a sense that opaque institutions, heavily reliant on technocratic modes of decision-making, are less well suited to responding to the wishes of the populations of member states than more directly engaged local political institutions. Many of these debates are directly relevant to consideration of the role that the EU plays in regulating health and safety at the level of member states (Baldwin 1996; Smismans 2004); certainly within the United Kingdom, the common perception, reproduced by some politicians, business leaders, and in much of the mainstream media, has been that the EU is implicated in the growth of an illegitimate, overly interfering regulatory framework (Anderson and Weymouth 2014; Coulter and Hancké 2016).

An Emerging European Interest in Health and Safety

The EU's interest in health and safety goes back to the early days of the EC project, such as Article 118 of the 1957 Treaty of Rome, which prompted the Commission to promote *"close collaboration"* between member states in areas including *"labour law and working conditions"* and the *"prevention of occupational accidents and diseases"* (Smismans 2004: 88–89). This 'social dimension' to the EC's initially trade-based role was initially one of guidance and coordination rather than rule-creation (Nielson and Szyszczak 1992), but it became more activist in nature in the mid-1970s, prompted by pressures to ensure equal pay for women, to respond to new technologies, and by a recognition of the need to avoid a 'race to the bottom' in terms of safety standards within the single market area (Smismans 2004: 96–97). At this time (1974), a Community Advisory Committee on Safety, Hygiene, and Health Protection at Work was created to promote Social Action Programmes around emergent hazards and chemical risks (Baldwin 1996: 94). This spirit of activism was regarded by observers such as John Rimington, who worked for the Diplomatic Service in Brussels and sat on the EC Social Questions Working Group, as politically motivated:

> the Commission were thinking, 'what a good idea if we became much more involved in health and safety', because [...] the more the Commission could get hold of standards, the more they were smack in the middle of the domestic industries of the member states [...] the Commission were already, in 1974, interesting themselves in this possibility. (John Rimington Interview, paras. 13–16)

Health and safety was arguably a means of allowing elements of the EC Social Charter, which was created in 1989 to engage with social well-being in general terms (and was otherwise rejected by the UK), to be implemented via a functional 'spill-over' effect, whereby the delegation of powers in one policy field creates pressure to expand that authority into adjacent areas (Majone 1997). In this way, the close relations between 'workplace' health and safety, and 'social' health and safety, which the HSWA foreshadowed in 1974, allowed for the exertion of EC control over an expanding range of issues.

Health and safety thus had a strategic value as an issue that straddled both the common market (economic) and social chapter (rights) policy areas (Smismans 2004: 88–89); unlike other areas of industrial policy, which were resistant to harmonisation, the EC felt that progress could be made relatively easily via intervention in this area (Wright 2008). For the most part, the EC's influence on UK policy and practice at this time was slow to take effect, and limited to areas of emergent hazard like carcinogens, and uncontroversial issues like safety signage (Dalton 1998:

40).[31] The Iron and Steel Trades Confederation in 1998 attributed this historical fact to the Conservative Government's use of codes of practice as places *"where the inconvenient aspects of European legislation could be hidden"*.[32] But this changed after the Single European Act 1986s.118A introduced qualified majority voting (QMV) among member states, rather than unanimous agreement, as the basis for introducing measures to harmonise conditions around health and safety. This area was the first social policy sector where QMV was introduced, largely because the welfare protections involved were directly related to preserving a level playing field for competition within the single market, and because the cost of doing so was largely passed off by member states to employers and other actors (Smismans 2004: 98). The effect of this was described by the Director-General of the HSE at the time:

> All directives up till that time had been by unanimity. In other words, if we didn't like something […] we could veto it. That had given us enormous influence. […After QMV] our influence over the shape of European Directives disappeared […] I told [the Minister] what would be the effect of conceding health and safety to qualified majority voting. This was at the Madrid summit of 1987, and he understood that there would be a flood of health and safety regulations which we would have little control over […] the Government used health and safety as a 'loss leader' – an area where they were prepared to make concessions and to accept European proposals that they might otherwise have resisted. We were thrown to the European wolves. (John Rimington Interview, paras. 77–79)

The introduction of QMV meant that the UK was suddenly less able to block measures that it opposed (Baldwin 1996: 98), meaning that a more proactive approach could be taken by the Commission in relation to health and safety, as they no longer had to overcome British scepticism before a measure could be introduced.

---

[31] I vividly remember when the Commission proposed a directive on fire signage [in 1975…] the only controversial question, curiously enough, was whether the sign […] should be a running man, as it turned out to be, or a walking man. [The British thought] it utterly wrong that anybody should run if there was a fire, you walked! (John Rimington Interview, para. 13)

[32] M. Leahy, Assistant General Secretary Iron and Steel Trades Confederation, 'A Step in the Right Direction', *Focus*, OSH Newsletter for Reps of the Iron and Steel Trades Confederation, February 1998, p. 2. MRC, MSS.36.2000.72.

The late 1980s saw a shift towards a more activist policy mode, with the start of a third Action Programme, a number of directives on issues such as construction safety, and the creation of a Framework Directive for the Introduction of Measures to Encourage Improvements in Safety and Health of Workers.[33] This was a radical step away from a piecemeal approach to promoting measures on specific hazards, and towards the adoption of a comprehensive framework of general overall directives and specific daughter directives that would impose wide-ranging requirements upon all areas of public and private activity (Baldwin 1996: 95; Smismans 2004: 98–101). Commentators on European policy-making tend to regard this as the period when the EC started to become a 'regulatory state' (Caporaso 1996; Majone 1996, 1997), driven by a growing interest in social policy questions at the level of the EC, but also by an awareness of the relatively limited institutional capacity of traditional policy-making (Majone 1996: 56), and the limits of the EC's own perceived legitimacy as a rule-maker (Löfstedt and Vogel 2001; Vogel 2012). Intervention via the 'framing' mechanisms of regulation was thus intended to 'steer' a route towards a greater social policy influence, via a method that was less impactful than 'rowing' there directly (Braithwaite 2000). It was also a victory for consensual EC policy-making, part of a 'race to the top' driven by the tendency for collaborative working at a Community level to lead to the incorporation of 'best practice' from across the member states (Smismans 2004: 104).

The Framework Directive and its five subsidiaries were implemented in the UK on 1 January 1993 via six new regulations (the 'six-pack').[34] This was the high-water mark of the interventionist approach to health and safety at the EU level, and these directives imposed wide-ranging precautionary, welfare-oriented 'minimum requirements' which member states had to implement via their own systems of law-making (Baldwin 1996: 95; Smismans 2004: 102–105). The directives did, however, impose high standards and absolute expectations (part of this 'race to the top'), and so the six-pack came to be regarded as emblematic of issues of

---

[33] Directive 89/391/EEC.

[34] The Management of Health and Safety at Work Regulations; Workplace (Health, Safety and Welfare) Regulations; Provision and Use of Work Equipment Regulations; Personal Protective Equipment at Work Regulations; Manual Handling Operations Regulations; and Health and Safety (Display Screen Equipment) Regulations.

'red tape' and European over-regulation, and posed an immediate legit-imacy threat for regulatory agencies such as the HSC/E (HSC 1994: paras. 50 and 137; also Beck and Woolfson 2000; Senior Government Source Interview, para. 16). As the *Times* noted in 1994, "*[g]iven the political sensitivity of Europe, it is perhaps not surprising that the HSE has been accused of being over-zealous*" (Walker 1994: 22). Why was this pro-cess of implementation entered into by the UK Government at all, given this political climate? One HSE insider of the time recalled the reasoning behind it:

> Ministers gave very clear briefs to HSE staff to agree those six pack regu-lations [...] they wanted to concentrate all their fight on opposing hours of work being a health and safety directive, but then let all the others through. Now the real trouble is that the politicians are very good at for-getting those instructions [...] they then got highly critical of these unnec-essary burdens, having instructed officials at an earlier stage to just say yes to it all. (Tim Carter Interview, para. 134)

The Working Time Directive, which was intended to form part of the body of EU health and safety provision, and which was perhaps the most 'social' element of European policy in this area, was resisted by the UK until its implementation in 1998 by a New Labour government which made a policy commitment to social chapter implementation (Blair et al. 2001). The Directive aimed to mandate a maximum 48-hour work-ing week across Europe, on the basis of staff health and welfare. But the social citizenship and inclusion implications of this policy (Threlfall 2003: 123) meant that it impacted upon the wider legitimacy of health and safety, as it seemed to represent a significant expansion of that par-ticular agenda (from the economic to the social sphere). The HSE Director-General at the time of its introduction perceived it to be more influential in terms of perceptions of health and safety than the six-pack:

> despite QMV, we did win quite a lot of the arguments about approaches to health and safety regulation [...] things that were potentially quite oner-ous, we were able to moderate. The big exception, and the one which I think poisoned the atmosphere a great deal, was the Working Time Directive, which arguably should not have come under health and safety legislation. (Jenny Bacon Interview, para. 39)

This measure reflected some of the other themes of this book: the 'overspill' of regulation into the private sphere, the open-endedness of the law, and its relevance to new employment sectors. As it has moved from an approach based upon 'negative integration' (removing barriers to trade) towards one rooted in 'positive integration' (embedding markets in common social and institutional frameworks: Coulter and Hancké 2016: 150), the EU has been viewed by British policy commentators as a source of non-accountable legislative unreasonableness (Anderson and Weymouth 2014: 133). This has damaged the legitimacy of health and safety more generally (Baldwin 1996: 99; Löfstedt and Vogel 2001), although public attitudes towards Europe's role in this regard remain much less negative (see Chapter 2).

Europe as a Legitimacy Challenge

One key reason that 'Europe' has impacted on the legitimacy of health and safety is the type of regulatory intervention it has tended to be associated with. The EU appears to take a much more risk-averse and precautionary approach to risk than the 'Anglo-Saxon' jurisdictions of the UK and USA, and to favour absolute forms of prohibition over more 'proportional' and risk-based modes of regulation (Löfstedt 2011; Rothstein et al. 2015; Vogel 2012):

> Europe [...] has a lot of declaratory law, saying 'this is what you should do', rather than 'it's an offence if you don't do this' [...] Here, the law tends to have reasonable practicability, but within that framework, judges interpret it fairly strictly, whereas the European system [...] has a concept of proportionality that means that judges can look at the seriousness in a particular circumstance [...] an awful lot of the discussion in the EU has been about wording and how you bridge that gap. (Tim Carter Interview, para. 167)

The different approaches to enforcement taken within the UK and Europe meant that the same regulations could have different effects in each place, as one former HSC Chair recalled:

> I said to [a] French inspector, 'how are you getting on with CDM [the Construction, Design, and Management Regulations]?' He says, 'we won't do much for a couple of years and then we will gradually enforce it'. But that's not our way. We enforced it the day it came in. (Sir Frank Davies Interview, para. 117)

Over time, British policy-makers have sought to mitigate a perceived tendency on the part of the EU towards imposing regulatory measures that impose unrealistic burdens or are difficult to enforce (Baldwin 1996: 101). As far back as 1981, the HSC was designing a work programme to allow it "*to respond to, and where appropriate, seek to change European Community or other international initiatives, and to stimulate such initiatives where this will help to ensure that **realistic and effective** health and safety policies can be adopted in this country*".[35] For domestic policy-makers, there has been a long-standing concern about the translation of this precautionary, hazard-focused European approach into British law, particularly, for instance, in relation to health risks:

> The EU regulated [Display Screen Equipment] based on the fact that it was new and different rather than because there were known risks [...] There's much greater acceptance of the subjective as a basis for action [...] whereas the threshold for action here tends to be a proven accident or health risk. (Tim Carter Interview, paras. 167–168)

Much has changed in the last twenty years, however. The EU's activism and influence has receded since the high water mark of the early-1990s (Smismans 2004: 105–118; Tim Carter Interview, para. 178; Richard Jones Interview, para. 18). Inconsistency between member states within Europe (north and south, or now, west and east) has prevented the development of further regulation:

> the expansion of the European Union has put a brake on the system. There has been [...] a huge tranche of legislation interpreted in very different ways across Europe. It's coming to some sort of commonality now but even so the new accession states are struggling with it. (Janet Asherson Interview, para. 20; also Senior Government Source Interview, para. 37)

This difficulty in implementing directives in secession states has also meant that soft law approaches and consolidatory policy instruments have become more central to EU practices in the area (Leka et al. 2012: 55–56; Wright 2008); the EU Strategic Framework 2014–2020 refers to 'social dialogue' and 'awareness raising' as the key measures to be used. There has also been a shift towards a more pronounced deregulatory agenda at European level, with the High-Level (Stoiber) Group

---

[35] HSC Press Release, *Plan of Work for 1981–1982 and Onwards*, 18 June 1981. Glasgow, Mitchell Library, TD914/18/2 (emphasis added).

on Administrative Burdens (and the Molitor Report on legislative simplification before it: Smismans 2004: 108) recommending the adoption
of key principles of regulatory minimalism, non-universality, and economic rationality, to tackle red tape (all of which have shaped UK health
and safety policy-making since 2010: Almond 2015). One of the seven
strategic objectives of the EU's Health and Safety at Work Strategic
Framework 2014–2020 is now to simplify *"existing legislation where
appropriate to eliminate unnecessary administrative burdens"* (European
Commission 2014). Those arguing for a 'push back' to bring EU practice in line with British norms (Löfstedt 2011: 63) would seem to be
pushing at an open door.

Despite this, concerns over EU 'red tape' and excessive burdens
(Löfstedt 2011: 59–60) remained a key point of conflict right up until
the 2016 Brexit referendum (Buttonwood 2016). There is a need to
keep these issues in context, however; as far back as 1992 (when the
EC was a source of much regulatory activity, and much political conflict in the UK), the then-Chair of the HSC, Sir John Cullen, told a
Parliamentary Committee that the European model was *"very similar to
our Health and Safety at Work Act. It tends to be a little more prescriptive
and detailed than we would choose to be…[it should not be] a big task to
conform with the regulations"* (Employment Committee 1992: para. 18).
Perhaps unexpectedly, particularly given the negative nature of much
media and political discussion of health and safety over the preceding ten
years (Almond 2009), and the similarities in terms of tone and style to
the abrasive media debate over Brexit, the issue of health and safety burdens did not feature prominently within the referendum process. When
the UK voted to leave the EU, it seemed to turn its back on many of the
features of the EU highlighted within debates about the issue of health
and safety, and for the reasons given there, but 'health and safety' itself
was not singled out as a reason for doing so. It seems inevitable that the
EU's influence on the UK's health and safety policy will decline following Brexit, but it remains unclear how far the Government's strategy for
withdrawal from existing EU regulatory arrangements (DEEU 2017)
will actually alter the legal framework in the UK, especially given that
many EU regulations have been introduced into UK law under the delegated legislation powers contained within the HSWA 1974, rather than
the European Communities Act 1972.

Overall evaluation of the EU's impact on legitimacy is challenging.
On the one hand, it has had positive impacts in terms of standards; on

the other, elements of this have been less welcome. Leka et al. (2012: 65–67) highlight that 63% of UK statutory instruments originated at an EU level, at a cost of £2.5 billion to the UK, but that 'soft' EU laws have brought practical benefits, despite being perceived as burdensome. There are arguments that the EU has brought legitimacy deficits in terms of democratic accountability, and led to greater inefficiency (Baldwin 1996: 99). At the same time, analysis of the relevant legislation, and of the experience of policy-making in Europe, suggests that the UK has had much more agency in terms of EU law-making than might be supposed, and has not necessarily been dragged in directions it did not want to go (Rothstein et al. 2015; also Löfstedt 2011: 60). This might be considered a symptom of what Smismans (2004) refers to as the EU's 'functional participation' source of legitimacy; engagement by member states in a range of institutional, administrative, advisory, and coordinating arrangements is a means of grounding the constitutional legitimacy of those arrangements in what they *do*, not in what they formally *say*. This appraisal was echoed by Jenny Bacon, who saw the six-pack era first-hand while at HSE:

> we contributed quite a lot to the debate in Europe. If you look across the safety regimes in Europe, they owe quite a lot to our approach. But I also think, in particular in the occupational health area and environmental protection, Europe stimulated us and that was good. (Jenny Bacon Interview, para. 28)

The evidence suggests that, throughout the UK's engagement with European policy-making, the EU has been *given* competence to impose health and safety laws via political trade-offs and via the UK's functional participation in EU health and safety policy processes (Smismans 2004), rather than unilaterally *imposing* this upon the UK. In its drawing of the lines between areas of 'market-related', negative integration (where the EU was regarded as having legitimacy to intervene to remove barriers to trade) and 'social', positive integration (where it was not seen to have legitimacy to change the social framework), the UK Government placed health and safety on the market side of the equation. As such, it has generally been accepted, and concessions made; even the issue of QMV and the introduction of the six-pack were calculated political trade-offs rather than policy 'defeats'. It is probably the case that the toxic political climate around Europe within UK politics has 'poisoned the well' for domestic regulators, undermining their wider legitimacy, rather than actions taken within the field of health and safety itself.

CONCLUSION

The four areas of development set out in this chapter all share a grounding in the notions of constitutional legitimacy, that is, the sources of legality and procedural validity that provide a source of authority to justify the actions of those who pursue the goals of 'health and safety' (Baldwin 1995: 43). On the one hand, the presence of a formal legislative mandate (of the sort introduced by the HSWA 1974), the establishment of an effective authority (in the form of the new, unified inspectorate) able to follow due process, and the establishment of proper modes of accountability and oversight (in the form of a strong relationship with central government), have all contributed to the legitimation of health and safety since 1960 (Baldwin 1995; Hutter 1997; Prosser 2010). On the other hand, the complex, and sometimes fraught, relationship between the British health and safety system and the EU has challenged that legitimacy, as its perceived externality from domestic political controls has chafed against the constitutional certainties provided under the other three areas discussed in this chapter. Together, they illustrate the complexity and reach of the modern health and safety 'system', and its proximity to the mechanisms and actors who populate the political realm. The relative stability of these features (the regulatory framework has endured since 1974) is also testament to the capacity of those institutions to cope with change; the flexibility of the law, the insulating role of the HSC, the arm's-length relationship to Government, and the cooperative relationship with the EU, have all allowed the health and safety system to resist external pressures that might otherwise have fundamentally damaged its standing.

As the discussion in this chapter makes clear, however, the reality of each of these sources of legitimacy is more complex than that. Taken together, they all demonstrate many of the features of the contemporary regulatory state; the emergence of 'new' governance-based models of regulation (Braithwaite 2000; Levi-Faur 2013; Majone 1997; Sunstein 1990) that shift responsibility for regulatory control away from the state and towards a more devolved, embedded, arms-length and network-based approach to risk-management, based on the empowerment of local decision-makers, the establishment of policy networks which include public and private actors, and conformity to the core principles of the NPM (Diefenbach 2009; Hood 1991; Hood et al. 1998). Each of these developments has been double-edged; there has been a greater public and political demand for regulatory accountability, a greater emphasis on proceduralised compliance and the juridification (or 'legalisation' of social issues: Black 2001; Teubner 1987) of regulatory processes, a side-lining of deliberative interest-group input in favour of bureaucratic public management, and a growing social anxiety about regulation, engendered in part by the greater freedom and indeterminacy of these new arrangements.

As Diefenbach (2009) argues, many of these problems are inherent to the rationalities of the NPM, which has fundamentally shaped regulatory practice since the 1980s. But it is also a unique outcome of the specific trade-offs attaching to health and safety. The flexibility of the legal framework has allowed it to evolve, but has also created a 'fear of freedom', leading to a growth in the requirements placed onto decision-makers. The unification of the inspectorate has professionalised and streamlined regulatory delivery, but has also resulted in a more managerialised regulatory structure that is less embedded in the constituencies it serves. The arm's-length relationship with central government has given HSE a degree of functional independence but has problematised issues of accountability, leaving the agency exposed to criticism from politicians and the public. And the EU's influence has become less pronounced, and been less of an imposition, in reality than in the perception of many audiences, who have viewed a largely consensual relationship as more coercive than it actually is. Across all of the areas of constitutional legitimacy, a balance has had to be drawn between the values of accountability and freedom; the new regulatory state decentres decision-making capacity, thus empowering a wider range of actors and institutions (inside and outside of the public sphere), but in doing so, it exposes a greater number of those actors and institutions to different forms of accountability for the decisions and actions they take (Bartle and Vass 2007). The development of a more open and pluralistic system of health and safety regulation has thus introduced both legitimacy benefits, and legitimacy risks, for those involved in the process.

REFERENCES

PRIMARY SOURCES

Benn, A. (2002). *Free at Last!: Diaries 1991–2001*. London: Arrow.

British Rail Board and HSE. (1982, November). *An Outline Report on OSH by the Accident Prevention Advisory Unit* (AN 16/157). London: The National Archives of the UK [TNA].

Brown, P. (1988, November 16). Balance Sheets and Shrouds. *Guardian*, p. 1.

Buttonwood. (2016, March 3). Drowning in Red Tape? *The Economist*. http://www.economist.com/blogs/buttonwood/2016/03/brexit-debate-0.   Last accessed 23 June 2017.

Department for Exiting the European Union (DEEU). (2017). *Legislating for the United Kingdom's Withdrawal from the European Union* (Cm. 9446). London: HMSO.

Department of Work and Pensions (DWP). (2007). *Report Following the Ministerial Consultation on the Proposed Merger of the Health and Safety Commission (HSC) and the Health and Safety Executive (HSE)*. http://webarchive.nationalarchives.gov.uk/20080205134503/;   http://www.dwp.gov.uk/publications/dwp/2007/hsmerger/. Accessed 9 November 2017.

Department of Work and Pensions (DWP). (2011). *Good Health and Safety, Good for Everyone*. http://www.dwp.gov.uk/docs/good-health-and-safety.pdf.

Employment Committee. (1982). *The Working of the Health and Safety Commission and Executive: Achievements Since the Robens Report* (Minutes of Evidence, HC 400-iii). London: Crown.

Employment Committee. (1988). *Health and Safety Matters [The Health and Safety Commission and the Health and Safety Executive]* (Minutes of Evidence, HC 704-i). London: Crown.

Employment Committee. (1992). *The Work of the Health and Safety Commission and Executive* (Minutes of Evidence, HC 263). London: Crown.

European Commission. (2014). *Communication From the Commission [...] on an EU Strategic Framework on Health and Safety at Work 2014–2020* (COM [2014] 332). https://eur-lex.europa.eu/legal-content/EN/TXT/PDF/?uri=CELEX:52014DC0332.

Framework Directive for the Introduction of Measures to Encourage Improvements in Safety and Health of Workers (1989), Directive 89/391/EEC.

Hansard. (1974, April 3). Vol. 871, Col. 1290.

Hidden, A. (1989). *Investigation into the Clapham Junction Railway Accident* (Department of Transport). London: HMSO.

Hose, J. (1980, March). Quoted in 'Safety Is Expendable', *The Land Worker*: 7. Museum of English Rural Life, University of Reading.

HSC. (1981, June 18). *Plan of Work for 1981–82 and Onwards* (TD914/18/2). Glasgow: Mitchell Library.

HSC. (1994). *Review of Health and Safety Regulation: Main Report.* Sudbury: HSE Books.

HSE. (1974–1976). *Management Board Minutes (December 1974–December 1976)* (EF 10/1). TNA.

HSE. (2014). *HELA Data Collection–Analysis of LAE1 2013/14 Data from Local Authorities* (Paper H17/01). http://www.hse.gov.uk/aboutus/meetings/committees/hela/310714/data-collection%E2%80%93analysis.pdf.

Leahy, M. (1998, February). A Step in the Right Direction. *Focus,* OSH Newsletter for Reps of the Iron and Steel Trades Confederation. MRC, MSS.36.2000.72.

Löfstedt, R. (2011) *Reclaiming Health and Safety for All: An Independent Review of Health and Safety Legislation* (The Löfstedt Review). London: Crown.

'Mines (Health and Safety)', HC Deb 26 October 1993, Hansard Vol. 230 cc698–747, 703.

National Federation of Professional Workers. (1971, March 15). *Written Evidence to Robens Committee* (LAB 96/697). TNA.

National Local Authority Enforcement Code. (2013). *Health and Safety at Work: England, Scotland & Wales.* http://www.hse.gov.uk/lau/la-enforcement-code.htm.

O'Gorman, P. (1970, October 8). *Written Evidence to Robens Committee* (LAB 96/36). TNA.

Rimington, J. (2008, October 9). *Health and Safety—Past, Present and Future.* Alan St. John Holt Memorial Lecture.

Robens, L. (1972). *Safety and Health at Work: Report of the Committee 1970–72* (The Robens Report). London: HMSO.

Rogers, P. (2007). *National Enforcement Priorities for Local Authority Regulatory Services* (The Rogers Review). London: HMSO.

Rose, C. A. (1977, August 16). *Memorandum to British Rail Board* (AN 156/936). TNA.

Temple, M. (2014). *Triennial Review Report: Health and Safety Executive* (The Temple Review). https://www.gov.uk/government/publications/triennial-review-report-health-and-safety-executive-2014. London: HMSO.

Work and Pensions Select Committee. (2008). *The Role of the Health and Safety Commission and the Health and Safety Executive in Regulating Workplace Health and Safety: Third Report of Session 2007–08* (Vol. 1, HC 246). London: Crown.

'Workers' Safety and Health', HC Deb 21 May 1973, Hansard Vol. 857 cc62–117, 65.

Young, L. (2010). *Common Sense, Common Safety* (The Young Review). London: Crown.

SECONDARY SOURCES

Aalders, M., & Wilthagen, T. (1998). Moving Beyond Command-And-Control: Reflexivity in the Regulation of Occupational Health and Safety and the Environment. *Law & Policy, 19*(4), 415–443.

Almond, P. (2007). Regulation Crisis: Evaluating the Potential Legitimizing Effects of 'Corporate Manslaughter' Cases. *Law and Policy, 29*(3), 285–310.

Almond, P. (2009). The Dangers of Hanging Baskets: Regulatory Myths' and Media Representations of Health and Safety Regulation. *Journal of Law and Society, 36*(3), 352–375.

Almond, P. (2013). *Corporate Manslaughter and Regulatory Reform*. Basingstoke: Palgrave Macmillan.

Almond, P. (2015). Revolution Blues: The Reconstruction of Health and Safety Law as 'Common-Sense' Regulation'. *Journal of Law and Society, 42*(2), 202–229.

Almond, P., & Esbester, M. (2018). Regulatory Inspection and the Changing Legitimacy of Health and Safety. *Regulation & Governance, 12*(1), 46–63.

Almond, P., & Gray, G. C. (2017). Frontline Safety: Understanding the Workplace as a Site of Regulatory Engagement. *Law & Policy, 39*(1), 5–26.

Anderson, P., & Weymouth, T. (2014). *Insulting the Public? The British Press and the European Union*. London: Routledge.

Asgeirsson, H. (2015). On the Instrumental Value of Vagueness in the Law. *Ethics, 125*(2), 425–448.

Bain, P. (1997). Human Resource Malpractice: The Deregulation of Health and Safety at Work in the USA and Britain. *Industrial Relations Journal, 28*(3), 176–191.

Baldwin, R. (1987). Health and Safety at Work: Consensus and Self-Regulation. In R. Baldwin & C. McCrudden (Eds.), *Regulation and Public Law* (pp. 132–158). London: Weidenfeld & Nicholson.

Baldwin, R. (1990). Why Rules Don't Work. *Modern Law Review, 53*(3), 321–337.

Baldwin, R. (1995). *Rules and Government*. Oxford: Clarendon Press.

Baldwin, R. (1996). Regulatory Legitimacy in the European Context: The British Health and Safety Executive. In G. Majone (Ed.), *Regulating Europe* (pp. 83–105). London: Routledge.

Baldwin, R., & Black, J. (2016). Driving Priorities in Risk-Based Regulation: What's the Problem? *Journal of Law and Society, 43*(4), 565–595.

Baldwin, R., Cave, M., & Lodge, M. (2010). *The Oxford Handbook of Regulation*. Oxford: Oxford University Press.

Ball, D., & Ball-King, L. (2013). Safety Management and Public Spaces: Restoring Balance. *Risk Analysis, 33*(5), 763–771.

Bartle, I., & Vass, P. (2007). Self-Regulation Within the Regulatory State: Towards a New Regulatory Paradigm? *Public Administration, 85*(4), 885–905.

Bartrip, P. (1982). British Government Inspection, 1832–75: Some Observations. *Historical Journal, 25*(3), 605–26.

Bartrip, P. (1983). State Intervention in Mid-Nineteenth Century Britain: Fact or Fiction? *Journal of British Studies, 23*, 63–83.

Bartrip, P. (2002). *The Home Office and the Dangerous Trades. Regulating Occupational Disease in Victorian and Edwardian Britain.* Amsterdam: Rodopi.

Bartrip, P., & Fenn, P. T. (1983). The Evolution of Regulatory Style in the Nineteenth Century British Factory Inspectorate. *Journal of Law and Society, 10*(2), 201–222.

Bartrip, P., & Hartwell, R. (1997). Profit and Virtue: Economic Theory and the Regulation of Occupational Health in Nineteenth and Early Twentieth Century Britain. In K. Hawkins (Ed.), *The Human Face of Law: Essays in Honour of Donald Harris* (pp. 45–84). Oxford: Oxford University Press.

Beck, U. (1992). *Risk Society: Towards a New Modernity* (M. Ritter, Trans.). London: Sage.

Beck, M., & Woolfson, C. (2000). The Regulation of Health and Safety in Britain: From Old Labour to New Labour. *Industrial Relations Journal, 31*(1), 35–49.

Beetham, D., & Lord, C. (2014). *Legitimacy and the European Union.* London: Routledge.

Black, J. (2000). Proceduralizing Regulation: Part I. *Oxford Journal of Legal Studies, 20*(4), 597–614.

Black, J. (2001). Decentring Regulation: Understanding the Role of Regulation and Self-Regulation in a 'Post-Regulatory' World. *Current Legal Problems, 54*(1), 103–146.

Black, J. (2005). The Emergence of Risk-Based Regulation and the New Public Risk Management in the United Kingdom. *Public Law,* 512–549.

Black, J. (2008). Constructing and Contesting Legitimacy and Accountability in Polycentric Regulatory Regimes. *Regulation & Governance, 2*(2), 137–164.

Blair, A., Leopold, J., & Karsten, L. (2001). An Awkward Partner? Britain's Implementation of the Working Time Directive. *Time & Society, 10*(1), 63–76.

Bradshaw, E. A. (2015). "Obviously, We're All Oil Industry": The Criminogenic Structure of the Offshore Oil Industry. *Theoretical Criminology, 19*(3), 376–395.

Braithwaite, J. (2000). The New Regulatory State and the Transformation of Criminology. *British Journal of Criminology, 40*(2), 222–238.

Caporaso, J. A., (1996). The European Union and Forms of State: Westphalian, Regulatory or Post-Modern?. *JCMS: Journal of Common Market Studies, 34*(1), 29–52.

Carson, W. G. (1979). The Conventionalisation of Early Factory Crime. *International Journal of the Sociology of Law, 7*(1), 37–60.

Carson, W. G. (1982). *The Other Price of Britain's Oil*. Oxford: L.M. Robertson.

Coulter, S., & Hancké, B. (2016). A Bonfire of the Regulations, or Business as Usual? The UK Labour Market and the Political Economy of Brexit. *The Political Quarterly, 87*(2), 148–156.

Craig, P. (2012). *Administrative Law* (7th ed.). London: Sweet & Maxwell.

Crawford, A. (2006). Networked Governance and the Post-Regulatory State? Steering, Rowing and Anchoring the Provision of Policing and Security. *Theoretical Criminology, 10*(4), 449–479.

Crook, T., & Esbester, M. (2016). Risk and the History of Governing Modern Britain, c. 1800–2000. In T. Crook & M. Esbester (Eds.), *Governing Risks in Modern Britain: Danger, Safety and Accidents, c. 1800–2000* (pp. 1–26). London: Palgrave.

Dalton, A. J. P. (1998). *Safety, Health and Environmental Hazards at the Workplace*. London: Cassell.

Dawson, S., Willman, P., Clinton, A., & Bamford, M. (1988). *Safety at Work: The Limits of Self-Regulation*. Cambridge: Cambridge University Press.

Diefenbach, T. (2009). New Public Management in Public Sector Organizations: The Dark Sides of Managerialistic 'Enlightenment'. *Public Administration, 87*(4), 892–909.

Fineman, S., & Sturdy, A. (1999). The Emotions of Control: A Qualitative Exploration of Environmental Regulation. *Human Relations, 52*(5), 631–663.

Flinders, M. (2004). Distributed Public Governance in Britain. *Public Administration, 82*(4), 883–909.

Fraser, D. (2009). *The Evolution of the British Welfare State: A History of Social Policy Since the Industrial Revolution* (4th ed.). Basingstoke: Palgrave.

Furedi, F. (1997). *Culture of Fear: Risk-Taking and the Morality of Low Expectation*. London: Cassell.

Gamble, A. (1994). *The Free Economy and the Strong State: The Politics of Thatcherism* (2nd ed.). London: Macmillan.

Genn, H. (1993). Business Responses to the Regulation of Health and Safety in England. *Law and Policy, 15*(3), 219–233.

Gibson, J. L., & Caldeira, G. C. (1995). The Legitimacy of Transnational Legal Institutions: Compliance, Support, and the European Court of Justice. *American Journal of Political Science, 39*(2), 459–489.

Giddens, A. (1991). *Modernity and Self-Identity: Self and Society in the Late Modern Age*. Cambridge: Polity Press.

Gilad, S. (2010). It Runs in the Family: Meta-Regulation and Its Siblings. *Regulation & Governance, 4*(3), 485–506.

Grabosky, P. (2013). Beyond Responsive Regulation: The Expanding Role of Non-State Actors in the Regulatory Process. *Regulation & Governance, 7*(1), 114–123.

Gray, R. (1996). *The Factory Question and Industrial England, 1830–1860*. Cambridge: Cambridge University Press.

Greenwood, J. (2007). Organized Civil Society and Democratic Legitimacy in the European Union. *British Journal of Political Science, 37*(2), 333–357.

Gunningham, N., & Johnstone, R. (1999). *Regulating Workplace Safety: Systems and Sanctions.* Oxford: Oxford University Press.

Gunningham, N., Kagan, R., & Thornton, D. (2004). Social License and Environmental Protection: Why Businesses Go Beyond Compliance. *Law & Social Inquiry, 29*(2), 307–341.

Habermas, J. (1973). *Legitimationsprobleme im Spatkapitalismus* [Legitimation Crisis] (T. McCarthy, Trans.). London: Heinemann.

Habermas, J. (1992). *Faktizitat und Geltung* [Between Facts and Norms] (W. Rehg, Trans.). Cambridge: Polity Press.

Haines, F. (2011). *The Paradox of Regulation: What Regulation Can Achieve and What It Cannot.* Cheltenham: Edward Elgar.

Halliday, S., Ilan, J., & Scott, C. (2011). The Public Management of Liability Risks. *Oxford Journal of Legal Studies, 31*(3), 527–550.

Hancher, L., & Moran, M. (1989). *Capitalism, Culture, and Economic Regulation.* Oxford: Oxford University Press.

Hawkins, K. (2002). *Law as Last Resort: Prosecution Decision-Making in a Regulatory Agency.* Oxford: Oxford University Press.

Hoggett, P. (2006). Conflict, Ambivalence, and the Contested Purpose of Public Organizations. *Human Relations, 59*(2), 175–194.

Hoggett, P. (2010). Government and the Perverse Social Defence. *British Journal of Psychotherapy, 26*(2), 202–212.

Hood, C. (1991). A Public Management for All Seasons? *Public Administration, 69*(1), 3–19.

Hood, C., James, O., Jones, G., Scott, C., & Travers, T. (1998). Regulation Inside Government: Where New Public Management Meets the Audit Explosion. *Public Money and Management, 18*(2), 61–68.

Hood, C., Rothstein, H., & Baldwin, R. (2001). *The Government of Risk: Understanding Risk Regulation Regimes.* Oxford: Oxford University Press.

Huising, R., & Silbey, S. S. (2011). Governing the Gap: Forging Safe Science Through Relational Regulation. *Regulation & Governance, 5*(1), 14–42.

Hutchins, B. L., & Harrison, A. (1911). *A History of Factory Legislation* (2nd ed.). London: P.S. King & Son.

Hutter, B. (1989). Variations in Regulatory Enforcement Styles. *Law & Policy, 11*(2), 153–174.

Hutter, B. (1997). *Compliance: Regulation and Environment.* Oxford: Clarendon Press.

Hutter, B. (2001). *Regulation and Risk: Occupational Health and Safety on the Railways.* Oxford: Oxford University Press.

Hutter, B. (2011). *Managing Food Safety and Hygiene: Governance and Regulation as Risk Management.* Cheltenham: Edward Elgar.

Hutter, B., & Lloyd-Bostock, S. (2017). *Regulatory Crisis: Negotiating the Consequences of Risk, Disasters and Crises.* Cambridge: Cambridge University Press.

Hutter, B., & Manning, P. (1990). The Contexts of Regulation: The Impacts Upon Health and Safety Inspectorates in Britain. *Law and Policy, 12*(2), 103–136.

Jordan, A. (1981). Iron Triangles, Woolly Corporatism and Elastic Nets: Images of the Policy Process. *Journal of Public Policy, 1*(1), 95–123.

Kagan, R. A., Gunningham, N., & Thornton, D. (2011). Fear, Duty, and Regulatory Compliance: Lessons from Three Research Projects. In C. Parker & V. L. Nielsen (Eds.), *Explaining Compliance: Business Responses to Regulation* (pp. 37–58). Cheltenham: Edward Elgar.

Lange, B. (2002). The Emotional Dimension in Legal Regulation. *Journal of Law and Society, 29*(1), 197–225.

Leka, S., Jain, A., Hollis, D., Andreou, N., & Zwetsloot, G. (2012). *The Changing Landscape of OSH Regulation in the UK: A Review.* Leicester: IOSH.

Levi-Faur, D. (2013). The Odyssey of the Regulatory State: From a 'Thin' Monomorphic Concept to a 'Thick' and Polymorphic Concept. *Law & Policy, 35*(1), 29–50.

Löfstedt, R., & Vogel, D. (2001). The Changing Character of Regulation: A Comparison of Europe and the United States. *Risk Analysis, 21*(3), 399–416.

Majone, G. (1996). The Rise of Statutory Regulation in Europe. In G. Majone (Ed.), *Regulating Europe* (pp. 83–105). London: Routledge.

Majone, G. (1997). From the Positive to the Regulatory State: Causes and Consequences of Changes in the Mode of Governance. *Journal of Public Policy, 17*(2), 139–167.

Makkai, T., & Braithwaite, J. (1992). In and Out of the Revolving Door: Making Sense of Regulatory Capture. *Journal of Public Policy, 12*(1), 61–78.

Marsh, D., & Rhodes, R. (1992). *Policy Networks in British Government.* Oxford: Clarendon Press.

Marshall, T. H. (1992). *Citizenship and Social Class.* London: Pluto.

McHarg, E. (2006). What Is Delegated Legislation? *Public Law, 10*(2006), 539–561.

Mendeloff, J., & Gray, W. B. (2005). Inside the Black Box: How Do OSHA Inspections Lead to Reductions in Workplace Injuries? *Law & Policy, 27*(2), 219–237.

Mess, H. A. (1926). *Factory Legislation and Its Administration 1891–1924.* London: P.S. King & Son.

Mills, R. W., & Koliba, C. J. (2015). The Challenge of Accountability in Complex Regulatory Networks: The Case of the Deepwater Horizon Oil Spill. *Regulation & Governance, 9*(1), 77–91.

Moran, M. (2003). *The British Regulatory State: High Modernism and Hyper Innovation.* Oxford: Oxford University Press.

Moravcsik, A. (2002). Reassessing Legitimacy in the European Union. *Journal of Common Market Studies, 40*(4), 603–624.

Morgan, B. (2003). The Economization of Politics: Meta-Regulation as a Form of Nonjudicial Legality. *Social & Legal Studies, 12*(4), 489–523.

Newman, J. (2001). *Modernizing Governance: New Labour, Policy and Society.* London: Sage.

Nichols, B. (1962). *An Introduction to Roman Law.* Oxford: Clarendon Press.

Nichols, T., & Armstrong, P. (1973). *Safety or Profit? Industrial Accidents and the Conventional Wisdom.* London: Falling Wall.

Nielsen, R., & Szyszczak, E. (1992). *The Social Dimension of the European Community* (2nd ed.). Copenhagen: Handelshojskolens Forlag.

Norrie, A. (2001). *Crime, Reason, and History* (2nd ed.). Cambridge: University Press.

Ogus, A. (1995). Rethinking Self-Regulation. *Oxford Journal of Legal Studies, 15*(1), 97–108.

Pantti, M. K., & Wahl-Jorgensen, K. (2011). "Not an Act of God": Anger and Citizenship in Press Coverage of British Man-Made Disasters. *Media, Culture and Society, 33,* 105–122.

Parker, C. (2008). The Pluralization of Regulation. *Theoretical Inquiries in Law, 9*(2), 349–369.

Podger, G. (2015, October 6). *Health and Safety: Is Britain Going Backwards?* Alan St John Holt Memorial Lecture.

Power, M. (1997). *The Audit Society: Rituals of Verification.* Oxford: Oxford University Press.

Power, M. (2007). *Organized Uncertainty: Designing a World of Risk Management.* Oxford: Oxford University Press.

Prakash, A. (2001). Why Do Firms Adopt 'Beyond-Compliance' Environmental Policies? *Business Strategy and the Environment, 10*(5), 286–299.

Prosser, T. (2010). *The Regulatory Enterprise: Government, Regulation, and Legitimacy.* Oxford: Oxford University Press.

Radaelli, C. M. (1999). The Public Policy of the European Union: Whither Politics of Expertise? *Journal of European Public Policy, 6*(5), 757–774.

Ramsay, P. (2006). The Responsible Subject as Citizen: Criminal Law, Democracy and the Welfare State. *Modern Law Review, 69*(1), 29–58.

Rhodes, G. (1981). *Inspectorates in British Government: Law Enforcement and Standards of Efficiency.* London: Allen & Unwin.

Rhodes, R. (2001). *Understanding Governance: Policy Networks, Governance, Reflexivity and Accountability.* London: Open University Press.

Rimington, J. (2010). Devolved Functions in Government: Some Realities. *The Political Quarterly, 81*(4), 602–615.

Rothstein, H., Beaussier, A., Borraz, O., Bouder, F., Demeritt, D., de Haan, M., et al. (2015). When 'Must' Means 'Maybe': Varieties of Risk Regulation and the Problem of Trade-Offs in Europe. *HowSAFE WP, 1,* 15.

Samuels, A. (1963). Offices, Shops and Railway Premises Act, 1963. *Modern Law Review, 26*(5), 539–542.

Scharpf, F. W. (1999). *Governing in Europe: Effective and Democratic?* Oxford: Oxford University Press.

Schmidt, P. (2004). Law in the Age of Governance. In J. Jordana & D. Levi-Faur (Eds.), *The Politics of Regulation: Institutions and Regulatory Reforms for the Age of Governance* (pp. 273–295). Cheltenham: Edward Elgar.

Schmitter, P. (1974). Still the Century of Corporatism? *Review of Politics, 36*(1), 85–96.

Scott, C. (2004). Regulation in the Age of Governance: The Rise of the Post-Regulatory State. In J. Jordana & D. Levi-Faur (Eds.), *The Politics of Regulation: Institutions and Regulatory Reforms for the Age of Governance* (pp. 145–174). Cheltenham: Edward Elgar.

Sirrs, C. (2016). Risk, Responsibility and Robens: The Transformation of the British System of Occupational Health and Safety Regulation, 1961–1974. In T. Crook & M. Esbester (Eds.), *Governing Risks in Modern Britain: Danger, Safety and Accidents, c. 1800–2000* (pp. 249–276). London: Palgrave.

Smismans, S. (2004). *Law, Legitimacy and European Governance: Functional Participation in Social Regulation.* Oxford: Oxford University Press.

Spencer, J. R. (2009). Criminal Liability for Accidental Death: Back to the Middle Ages? *The Cambridge Law Journal, 68*(2), 263–265.

Sunstein, C. R. (1990). *After the Rights Revolution: Reconceiving the Regulatory State.* Cambridge, MA: Harvard University Press.

Teubner, G. (1987). Juridification: Concepts, Aspects, Limits, Solutions. In R. Baldwin, C. Scott, & C. Hood (Eds.), *A Reader on Regulation* (pp. 389–440). Oxford: Oxford University Press.

Thomas, M. W. (1948). *The Early Factory Legislation: A Study in Legislative and Administrative Evolution.* Leigh-on-Sea: Thames Bank.

Threlfall, M. (2003). European Social Integration: Harmonization, Convergence and Single Social Areas. *Journal of European Social Policy, 13*(2), 121–139.

Tombs, S., & Whyte, D. (2007). *Safety Crimes.* Cullompton: Willan.

Tombs, S., & Whyte, D. (2010). A Deadly Consensus: Worker Safety and Regulatory Degradation Under New Labour. *British Journal of Criminology, 50*(1), 46–65.

Vogel, D. (2012). *The Politics of Precaution: Regulating Health, Safety and Environmental Risks in Europe and the United States.* Princeton: Princeton University Press.

Waldron, J. (2011). Vagueness and the Guidance of Action. In A. Marmor & S. Soanes (Eds.), *Philosophical Foundations of Language in the Law* (pp. 58–82). Oxford: Oxford University Press.

Walker, D. (1994). 'Safe and Never Sorry'. *The Times,* (8 December 1994): 22.

Ward, J. T. (1962). *The Factory Movement 1830–1855.* London: Macmillan.

Weick, K. E., Sutcliffe, K. M., & Obstfeld, D. (1999). Organizing for High Reliability: Processes of Collective Mindfulness. *Research in Organizational Behavior, 21,* 81–123.

Wells, C. (2001). *Corporations and Criminal Responsibility* (2nd ed.). Oxford: Oxford University Press.

Wilson, G. K. (1985). *The Politics of Safety and Health: Occupational Safety and Health in the United States and Britain.* Oxford: Clarendon Press.

Woolf, A. (1973). Robens Report—The Wrong Approach? *Industrial Law Journal, 2*(1), 88–95.

Wright, F. (2008). Effective Regulation and Sanctioning in Health and Safety Law in the European Union. *Web Journal of Current Legal Issues, 1.*

CHAPTER 5

Health and Safety in a Changing World

INTRODUCTION

The previous chapter opened the substantive analysis of the moral legitimacy of health and safety (Suchman 1995), starting to explore how far health and safety has been perceived as having been carried out in accordance with core *procedural legitimacy* values, specifically those relating to issues of *constitutional legitimacy* (Beetham 1991); this includes ideas touching upon the legal mandate for official action, including questions of accountability and the rule of law. In doing so, Chapter 4 was particularly centred upon the role and function of the state (broadly conceived), exploring the political changes which had an impact upon health and safety after 1960. This included consideration of the changing relationship between central government and health and safety regulators, noting a trend towards an increasing decentralisation of decision-making. Whilst this was not an unproblematic development, by the end of the time period covered by this book, it had led to a much wider range of actors and organisations becoming involved in health and safety than had previously been the case. On the one hand, this was a democratising development, opening the doors for participation by a more diverse set of stakeholders and actors; but it also had consequent implications for issues of accountability and legal mandate, and for broader perceptions of legitimacy.

This chapter will expand upon this theme of participation, discussing changes and developments in relation to who has been involved in health

© The Author(s) 2019
P. Almond and M. Esbester, *Health and Safety in Contemporary Britain*, https://doi.org/10.1007/978-3-030-03970-7_5

and safety since 1960, and viewing this as a component of Beetham's notion of *normative legitimacy* (1991), the other component of moral legitimacy, as defined by Suchman (1995) and others. Rather than considering procedural questions about *how* health and safety is undertaken, normative legitimacy relates to *why* it is carried out: the 'rightness' of action taken in the name of health and safety. Black (2008) has further split normative aspects into *justice* claims (explored in Chapter 7) and *democratic* claims, explored here. These focus on ideas about who has been involved in and able to participate in health and safety, which interests are represented within processes of policy- and decision-making, and how far health and safety has been responsive to the concerns of those who might be affected by it. This therefore forms a type of 'input legitimacy', speaking to the *reasons for* regulating health and safety, but one that is especially concerned with questions of representation and participation.

As a counterpoint to the previous chapter's focus on the mechanics of the state role in health and safety, this chapter will explore aspects intimately connected with, but broader than, the high politics, legalism, and institutionalism that lies at the heart of those constitutional questions. It focuses on the major changes that have occurred in relation to the British economy and occupational structure since 1960, all of which have had knock-on effects on the practice and legitimacy of health and safety, especially with regard to who played a part and how. The period from 1960 has seen a major realignment of the British economy and workplace; traditional, manual, heavily unionised industries (manufacturing, heavy production, extraction/mining) have declined in terms of their scale and in the share of the workforce employed, and this has led to a number of effects, not least the reduction in the influence of traditional trade unions. At the same time, other areas (notably office, retail and service sectors) have increased their share of the economy. Who has worked, and where and how they have done so, have all changed significantly since 1960, and this has had important democratic legitimacy implications. By considering the economics of health and safety, the changed occupational structure, who was working and how they were represented (including the changing fortunes of organised labour), and lastly how health has largely remained at the periphery of health and safety initiatives, this chapter will demonstrate how health and safety has faced and responded to multiple legitimacy challenges concerning its scope and focus.

THE ECONOMICS OF HEALTH AND SAFETY

Before considering the implications of changes in Britain's economy and workforce for health and safety and its legitimacy, it is worth noting some more general relationships between economic performance and health and safety. These are certainly not specific to the period covered by this book, having been discussed from at least the nineteenth century onwards, and it is difficult to draw direct causal links between, for example, the state of the economy or a particular firm's performance and the number of workers killed or injured, as so many other factors are involved. However, it is clear that there are strong associations between financial pressures and health and safety (Genn 1993; Nichols 1999; Quinlan 1999; Quinlan et al. 2001; Viscusi 1983), something particularly important in the period covered here in terms of the declining heavy industries which might lack capital to invest in new equipment or processes. The need to sustain a profit margin has acted as a driver in relation to health and safety in many industries, including nationalised concerns. As one former HSE inspector recalled about the construction industry of the 1970s: *"there was an element of anarchy in the industry and the pursuit of the cheapest possible input to the construction process, so if someone could do the job cheaper by cutting corners, they would get the job"* (David Morris Interview, para. 42)—often with negative consequences for workers. Morris later noted:

> employers will always see the upfront cost of a risk control measure rather than the longer term benefit [...] And it's natural that they should because it's their money that's being spent [...] unless there is a countervailing force to say that, "yes, it's going to cost you X million pounds to do this, but over a ten year period it will save you Y million pounds, and isn't that a good thing?" then you only hear about the [...] cost. (David Morris Interview, para. 135)

Politics and ideology have, of course, lain behind all of the economic strategies adopted by governments, but in the 1980s in particular, the pursuit of profit was noticed as having an impact upon health and safety. In 1988, in the midst of strong economic growth, a 'senior government official' was quoted in the *Guardian* newspaper as saying *"A culture which is about enterprise, competition and profitability doesn't want to concern itself too closely with issues of health and safety"* (Laurance 1988: 23). Whilst ideology and the profit motive might have been important considerations at all times, they became more pronounced during periods of economic downturn. The impact upon business attitudes towards health and safety could be profound, as was noted by RoSPA in the early 1980s: "*It is a matter of concern that so many organisations, in a reaction to the depressed economy, have stopped or severely cut back on training and particularly safety training*".[1] These effects might be intensified in those industries that were already in decline; in its annual report for 1982–1983, RoSPA recorded that "*[t]he economic recession and financial climate has probably hit the occupational safety division harder than the other divisions because so much of its work is with those very industries under most pressure*".[2] Pressures of this sort were, of course, always likely to recur in future times of economic recession or crisis, or in pressured industries, as one former head of health and safety recalled of his time working for a privatised energy and utilities provider in the early 1990s:

> We were asked for cost reductions in all departments of 25%, which is quite a big number [...] we did some deep analysis on it with my team and we said "without changing the risk profile we think we can do 7%". And that got met with, "what is it about 25% you don't understand?" And you thought 'OK, we're not going to win this, we need to look at a completely different way of working' [...] it decimated the organisation, and I wasn't particularly proud of that, but if I said I wouldn't do it, I'd have been turfed out. (Graeme Collinson Interview, para. 39)

The economic climate also had an impact upon the ability of regulators to pursue their objectives, in terms of both their own resources and the responsiveness of those they regulated. Chief Inspector of Factories Jim Hammer started his 1979 report with a discussion about the economy, including the recognition that in light of a six per cent cut in budget in the coming years "*what the Factory Inspectorate can achieve is obviously constrained by the resources at its disposal*" (HSE 1981: vii). A significant policy pressure to ensure efficiency and avoid waste formed a key part of the broader 'New Public Management' culture that emerged within the public sector in the 1980s and early 1990s (Hood 1991). This necessitated a radical new approach, including cost–benefit analysis of where

---

[1] Royal Society for the Prevention of Accidents (RoSPA) Annual Report, 1981–1982, p. 12. RoSPA Archives, D.266/1/2/5/2.

[2] RoSPA Annual Report, 1982–1983, p. 10. RoSPA Archives, D.266/1/2/5/3.

inspection, in particular, should be applied (Ibid.: v). The link between the economy and the HSE's regulatory strategies was, to Hammer, clear, and econometric approaches such as cost–benefit analysis were understood by the regulated as well as (in some eyes) providing legitimisation for particular enforcement priorities. These ideas are engaged with in more detail as part of the discussion of 'enforcement, inspection, and the regulatory role' in Chapter 6 of this book. Public bodies, whether they be regulators, employers, or other organisations, remain subject to the same pressures from Government to both control and account for expenditure on health and safety. For HSE, this has meant existing *"very much under the spotlight from the Treasury"* (Neal Stone Interview, para. 31), while for other organisations, like the NHS, the need to prove the value for money of the initiatives and investments in safety that they make, as the head of health and safety for NHS Employers testified:

> We are a public organisation. The money is public money. We need to prove that what we're doing is making a difference to the patients and to the service. (Ruth Warden Interview, para. 66)

CHANGING EMPLOYMENT STRUCTURES

Returning to the overall structure of employment across this period, Tables 5.1 and 5.2 present indicative data which give an impression of the changes to the economic structure of Britain. The broad trend was one of contraction of heavy industries and expansion of the service sector. Construction is one notable exception to this, as was recognised by the HSE in its 1977 comment that *"it seems possible for there to be an increase in activity and in the numbers at risk. The prudent view is that*

Table 5.1 Numbers employed in selected sectors, Great Britain 1961–2009 (in 000s) (*Aggregated from* Mitchell [1988: 107] and ONS [2002: Table 7.5; 2010: Table 7.5])

Sector	1961	1971	1981	1996	2001	2009
Mining and quarrying	527	257	164	73	69	55
Agriculture	948	718	598	264	282	243
Textiles	466	309	176	174	134	57
Construction	571	554	621	898	1160	1241
Services, sport and recreation	2496	2952	3293	18,125	20,109	22,329

Table 5.2 Numbers employed in selected sectors as percentages of economically active population, Great Britain 1961–2009 (*Aggregated from* Mitchell [1988: 107] and ONS [2002: Tables 7.1 and 7.5; 2010: Tables 7.1 and 7.5])

Sector	1961	1971	1981	1996	2001	2009
Mining and quarrying	2.2	0.9	0.7	0.3	0.2	0.2
Agriculture	6.1	4.1	2.5	0.9	1.0	0.8
Textiles	1.9	1.1	0.7	0.6	0.5	0.2
Construction	2.4	2	2.6	3.1	3.9	4.0
Services, sport and recreation	10.4	10.8	13.9	63.1	68.2	71.2

unless there is a radical change in the effectiveness of accident prevention in the industry, the accident experience of the past 10 years may be repeated" (HSE 1979: 33).

These changes were clear to contemporary observers, as was seen in the 1981 HSE report: "*It is by now clear that there has been a long term, not to say permanent, change in the pattern of British industry*". The report went on to note that not only had the once major traditional industries been losing employees, but that "*[t]he safety departments of major companies have felt the same chill wind as other parts of the organisation and in many cases the health and safety function has been amalgamated with other duties*". At the same time, "*cutbacks in labour have led to a noticeable deterioration in maintenance and housekeeping*" (HSE 1982: v).

As well as impacting on the ways in which regulators could act, these changes also had an effect upon perceptions of what regulators should be doing. Evidence to the Robens Committee in 1970 noted that "*[t] he point I wish to establish is the naive assumption that an industrial code formulated in the nineteenth century is applicable or adequate to the needs of British industry and commerce*".[3] This was picked up by Peter Jacques of the TUC's Social Insurance and Industrial Welfare Department (1971–2000):

> much of the reasons for having a lot of health and safety [...] legislation, have disappeared, broadly meaning coalmining, ship building [...] in terms of the hundreds of thousands or tens of thousands of people involved in

---

[3]Written evidence of P. O'Gorman to Robens Committee, 8 October 1970, pp. 6–7. The National Archives of the UK, London [TNA], LAB 96/36.

those industries, steel [...] I think that quite a lot of that has disappeared. Quite a lot of the work, heavy health and safety, the safety, occupational health stuff has disappeared. (Peter Jacques Interview, para. 192)

One additional result of these changes has been a reduction in the perceived legitimacy of health and safety provision, as industries where there are clear and established risks, and long traditions of risk regulation, serve to reinforce and validate the idea of regulation as socially important and valuable. This perception was widely reflected in the focus groups discussed in Chapter 2. The decline of such industries, it is suggested, has lessened the public and political visibility of occupational ill-health, injury and death from high hazard processes by diffusing these outcomes (although they still continue, albeit at a lower rate than in the past). As Jim Hammer remarked:

> The other reason why people are not so aware of the real importance of health and safety nowadays is that the big industries that used to cause the dramatic, serious accidents, the ship building, chemical works, and big steelworks, in general terms, don't exist anymore. I know I'm exaggerating, but people don't have this perception that it's important to look after working people [any more]. (Jim Hammer Interview, para. 67)

If the visibility of risk has validated regulation in the past, then the apparent safety of newer sectors may have posed—and continue to pose—problems in terms of the legitimacy of health and safety.

Further, one of the strengths and enduring legacies of the traditional industries had been that they could act as 'leaders' for the health and safety agenda, developing and disseminating new modes of risk management and expertise. This included not just the role of major employers or sectoral organisations, but also the prominence given to health and safety issues by the major unions serving workers in those industries. As industries emerged, they discovered the pitfalls of their practices; these became sites around which debates about safety and risk were contested. So, for example, in the early nineteenth century, this included the textile industry; in the mid-nineteenth century, the railway industry (see, for instance, Hutter 2001); and in the early- to mid-twentieth century, the chemical industry, particularly in relation to synthetics (Otter 2016; Walker 2007). Leading industries of the day played a role in generating proposed solutions to the problems they were creating, sometimes ahead of the regulators (including at times in an effort to deflect further regulation). The evolving knowledge base of the railway industry provided a key example of this, as a former British Rail safety manager recalled:

> From 1990 to about '95 [...] we actually knew more about railway safety management systems than the Railway Inspectorate did [...] I was having to go to the Health and Safety Executive to explain it and convince them. (David Maidment Interview, para. 101)

Well into the period covered by this book, traditional heavy industries could represent themselves as carriers of a weighty and substantial body of expertise and influence, and thus engage with policy-makers and regulators as equal partners within a broadly cooperative health and safety regime. Roger Bibbings, formerly of the TUC and RoSPA, commented:

> Britain has ceased to be a big manufacturing nation. We don't have coal mining, we don't have ship building, we don't have these big centres of economic and political influence which could be deployed in this area to tell ministers, well don't worry too much, we know what we're talking about, we'll come up with a set of regulations, we'll advise the Factory Inspectorate and if there are any dissenting voices off stage we'll be able to sort them out. (Roger Bibbings Interview, para. 29)

In the absence of these 'leader industries', it may be harder for those working in health and safety to show and establish the value of what they do; the gap has not yet been filled to the same extent by existing and emerging issues in newer sectors.

New Sectors—New Workers and New Issues (?)

The changing nature of the workplace has, unsurprisingly, had an impact upon *who* has undertaken paid employment and how that employment has been structured. Table 5.3 outlines the changing gendered pattern of employment since 1960, demonstrating a trend which has continued into the present, with increasing numbers of women entering the formal paid economy and women forming an increasing proportion of the workforce. McIvor (2013: 11) notes that as a part of this, the general trend has been that women experienced an alteration in their modes of working, from hourly paid work to salaried roles, though this of course obscures big differences often based around class. Modes of working

Table 5.3 Numbers of men and women in employment (figures in 000s), Great Britain 1961–2009 (*Aggregated from* Mitchell [1988: 107] and ONS [2002: Table 7.1; 2010: Table 7.1])

Year	M	%	F	%	Total
1961	16,232.47	67.6	7781.85	32.4	24,014.32
1971	17,357.77	63.4	10,006.63	36.6	27,364.4
1981	15,526.71	61.1	9877.68	38.9	25,404.39
1991	14,887	56.4	11,512	43.6	26,399
2001	15,459	55.1	12,607	44.9	28,066
2009	15,497	53.5	13,482	46.5	28,979

were also gendered, with women more likely to be employed in part-time roles across the period after 1960 (McIvor 2013: 24). The work undertaken by women and men often remained bound by particular codes and conventions, even towards the end of the twentieth century (for example, Clarke and Wall 2009; McIvor 2013: 24), which had the effect of exposing women and men to different risks and different levels of risk (McIvor 2013: 21). These impacts were perhaps lessened by the decline of the heavier industries, with their concentration of male workers and adverse health and injury outcomes. That said, the health implications, in particular, of more insecure employment practices (for example, zero hours contracts) that have become increasingly prevalent in recent years (McIvor 2013: 40–42), and which still disproportionately impact upon women (Fudge and Owens 2006), are as yet unclear.

In addition to changes in the gendered nature of work, the ethnicity of Britain's workforce has also changed—though it is difficult to give a precise form to this, as information about ethnicity and work was inconsistently collected until the late twentieth century (McIvor 2013: 25). It would be fair to say, however, that following post-war immigration into Britain (and the associated tensions that this produced at times), the numbers of minority ethnic employees increased, typically concentrated in those locations around which immigration was focused. What impact this had in terms of health and safety and legitimacy is uncertain, though in terms of its democratic aspects, it created new audiences and needs for health and safety messages. Some working in health and safety showed some awareness of the potential need. A 1963 British Medical Association conference, for example, discussed the possibilities of offering "*a short course in occupational hygiene, safety and health*" to immigrants, as well as creating leaflets to help new arrivals and *"for the British people on how to receive immigrants"*, as *"it was up to us to set a good example"*. The question of language was raised, as it was suggested that *"many immigrants [...] were unable to speak English"*, though this view was challenged.[4] Needless to say, this topic was one in which existing prejudices manifested themselves: in 1960, one Wolverhampton-based group advised employers to treat immigrants *"as though they were children, kindly but firmly"* and not to employ them *"on jobs which involve the use of discretion"*. Whilst these views are, by today's standards, expressed in rather inflammatory terms, they reflect commonly held views of the time; yet other points that the document went on to note were perhaps more constructively articulated, and were applicable to all workers—*"they should be fully and properly instructed in the job"*, *"the direct answer to the problem is – Education"*.[5] Translation of material into other languages was also suggested, as was involving community leaders to convey health and safety information. Key concerns remain over the challenges and health and safety risks that associate with issues of ethnicity, particularly in industries like construction, where rates of migrant labour employment are high, and the training needs and opportunities are particularly demanding (Tutt et al. 2011). Generally, however, ethnicity has not been engaged with or given explicit consideration as a topic in relation to health and safety, with potentially negative implications for the democratic legitimacy of health and safety.

These changes in the make-up of the British economy and workforce have also seen the emergence of new industries and sectors, which have their own challenges and political contexts (Mayhew and Quinlan 2002; Quinlan 1999; Quinlan et al. 2001). The retail and service sectors have grown in terms of both their importance to the economy, and the size of the workforce they employ; office work is also much more significant today than in 1960. Whilst concerns about conditions in offices had existed since the early years of the twentieth century (see, for example, the files of the National Federation of Professional Workers

[4]British Medical Association (BMA) Conference 'The Immigrant in Industry' (1963), Minutes, pp. 4–5. Modern Records Centre, University of Warwick [MRC], MSS.292B/141/3/3.

[5]RoSPA Industrial Group Advisory Council Meeting, 29 January 1960, Minute 6. RoSPA archives, D.266/2/21/1.

from c.1913–1963),[6] the issue gradually became more significant as the number of office-based workers increased during the second half of the twentieth century.[7] Health and safety issues were often considered under the more general term of 'welfare', and were frequently tied to questions of productivity and workforce morale.[8] Following the Offices, Shops and Railway Premises Act 1963, greater attention was paid to this area by employers, unions and regulators (Local Authorities, in the main). Yet, it was also clear that regulation here was under-resourced. It was claimed in 1966 that "*Many local authorities are fully aware that their staff of inspectors falls short of the complement desirable to act as enforcement officers. Their argument is that they cannot afford to appoint more*".[9] Yet, the unions were also criticised for their failure to take the Act, and safety in the premises it covered, seriously, as a former British Rail safety manager, recalled:

> trade unions would fight for compensation but they never said anything to me about the state of the station [...] They used to moan about broken steps and windows and draughts [...] all the housekeeping things, but equally they knew nobody ever did anything about all the minor safety things. (David Maidment Interview, para. 87)

There is limited evidence as to how regulation was perceived in these sectors which were relatively unfamiliar with regulation. Some groups saw the focus as at odds with the priorities of higher risk workplaces, at least in the 1960s. The Institute of Directors spent some time in the 1960s considering office conditions, commenting in 1960 that "*[i]n factories conditions can quickly become appalling and of course extremely dangerous. They are kept under control by first class legislation and a first class inspectorate which is a model for the world [...] inspection is of course expensive both in money and trained manpower. It is simply not worthwhile*

---

[6] For instance, paperwork relating to office conditions in MRC, MSS.239/3/1/103.

[7] See, for example, the note in the *Observer*, specifically focusing on office conditions for female workers (29 December 1949). Brighton Design Archive, 1685/1.

[8] See, for example, Institute of Directors 'outline for aide memoire on working conditions in offices' (1960: p. 1). Brighton Design Archive, 1685/1.

[9] *Justice of the Peace and local Government Review*, 22 January 1966. See also *The Municipal and Public Services Journal*, 4 March 1966, p. 691. MRC, MSS.239/3/1/95.

expending the money on office conditions".[10] Early in 1966, white-collar unions pushed for greater enforcement action by local authorities.[11] At the same time, some employers did take action: a British Rail memo of March 1970 notes that safety exercises had been carried out at head-quarters in relation to the typing services.[12] The change in the economy was recognised by the regulators and legislators, as was noted in 1986 by the retailer WH Smith: "*In the face of new and proposed legislation and a tougher attitude by enforcement authorities, it is important that staff are aware of fire and safety regulations. This is now imperative*".[13] Significantly, there is also some suggestion that the extension of regulatory activity to office work from the 1980s, in particular, increased the numbers of people in contact with the HSE and contributed to later concerns about the legitimacy of health and safety:

> lower-risk 'white-collar'-type employers had new duties of care towards the safety of members of the public who might also be affected, while visiting shops, offices, hospitals and so on [...] They were what we called the 'new entrants' when HSE was being set up [...] the risks were different in kind, lower in general, but capable of much panic and disproportionate public and media reaction if there was a case where something went wrong. (Helen Leiser Interview, paras. 67–69)

Particular issues of consistency and proportionality have recently been debated in these contexts (Löfstedt 2011; Young 2010); there have also been suggestions that the levels of regulatory expertise in these sectors are lower than elsewhere, and lower than may be desirable (Peter Kinselley Interview, para. 32), in part because this was not always seen as a regulatory priority, as one former HSE inspector recalled:

> There was a tendency to put people who were less effective...into the 'less important' new-entrant activities [...] there was an unwillingness to

---

[10]Correspondence from B. O'Donovan to Mr Farr, 'An Office Workers' Charter', 6 January 1960, p. 1. Brighton Design Archive, 1685/1.

[11]*Justice of the Peace and Local Government Review*, 22 January 1966. MRC, MSS.239/3/1/95.

[12]'British Railways Board Accident Prevention Policy', Internal Memorandum, 9 March 1970. TNA, AN 171/550.

[13]WH Smith Retail Group, Management Executive Minutes, 2 January 1986, Minute 16. Museum of English Rural Life [MERL], WHS 1415.1–4.

recognise that there were things we [HSE] had to learn. (David Morris Interview, paras. 97–101)

The process of mapping traditional notions of 'health and safety' across into newer sectors such as office work remains an ongoing one, in terms of making the normative case for intervention, and showing that the delivery of effective health and safety is efficient, proportionate, and of value. The service sector is one area to which the HSE has begun to pay particular attention since the late 1980s. Despite the growth of the sector and the HSE's interest in it, the 1990–1991 HSC report observed that "*[t]he Commission continues to be concerned about the rising number of accidents in the service sector over the past few years*" (HSC 1991: ix). A range of new concerns followed from workplaces beyond heavy industry, something Peter Jacques, formerly of the TUC, reflected upon:

> health and safety is not only about old industry it's about new industries and moving into areas which previously I don't think people like me ever took account of. Things like stress…they're new areas and I think some employers perhaps, and not only employers, are a bit sceptical about whether there is such a thing as stress…a lot of the issues in health and safety now are much more difficult to deal with, I think, than the ones when I first started. (Peter Jacques Interview, para. 192)

Of the new high-risk areas that grew during this period, it is worth dwelling upon the offshore industry. Oil production, in particular, increased massively—up from 87,000 tons in 1974 to 77,671,000 tons in 1980 (Mitchell 1988: 270). Yet, whether it adopted a profit-centred motivation from older heavy industries or from the American industry from which it took its lead, production was viewed as privileged over safety (Brotherstone and Manson 2007: 35; Carson 1982; Paterson 2007): "*a lot of [casualties] were because of poor practice, probably brought about by the urgency to get the job done that little bit quicker and the culture at that time was […] get the installation producing, that's the important thing, everybody was focused on production of hydrocarbons and health and safety perhaps didn't take its rightful place*".[14] At least in the early days of health and safety regulation (whilst inspectors fell under the

---

[14] Martin Thompson, Interviewed as Part of the University of Aberdeen's 'Lives in the Oil Industry' Project. British Library, C963/13, tape 2, side b.

Department of Energy), there was an acknowledgement that the industry was leading. One inspector recalled *"our management were quite enlightened and said "you know, if you actually work to the rule book, then, you know, you'll never keep up with the oil industry", so we were given quite a loose rein [...] as long as we got the work done to the required quality'.*[15]

Yet, with the changes in the economy in more recent years, it is important to consider what this might mean for health and safety and for leadership. Jenny Bacon, former Director-General of the HSE, noted of the 1990s:

> the approach to health and safety in small businesses and self-employment did become much more important, partly because that was the way in which the labour market was changing, far more jobs coming out of that sector than were coming out of major manufacturing. (Jenny Bacon Interview, para. 79)

In terms of leadership, however, the capacity of the self-employed or smaller businesses to influence health and safety regulation, policy and practice is perhaps limited. In recent years, the rise of zero hours contracts and the 'gig' economy (Prassl and Risak 2015) also presents significant challenges to health and safety of workers, to ideas of legitimacy in relation to health and safety, and to potential sources of leadership. Instead it is suggested that some larger employers (major retail, supermarkets), as well as SMEs, are now leaders in influencing the agenda of government (Senior Government Source Interview, para. 56), and in setting the agenda for issues that need to be addressed. This was something observed by Bill Callaghan, HSC Chair 1999–2007, as having an impact:

> As we change from being a manufacturing country to becoming effectively a retailing economy, economic power is there amongst big supermarkets and DIY chains and all the rest of it. And they certainly had the ear of government. And a lot of the discussions about proportionality of health and safety enforcement came from complaints from the big retailers. And that of course was the local authority enforcement sector for the most part. But they certainly had the ear of the Prime Minister. (Bill Callaghan Interview, para. 30)

[15] Ibid., tape 2, side a.

The changing nature of employment has posed challenges for regulators. As early as 1981, the Chief Inspector of Factories noted that the jobs then being created were tending to be temporary roles or for sub-contractors, and in smaller firms, and that businesses were focusing only on aspects 'directly' related to the product or service they were sell-ing: all reasons why health and safety might not be considered (HSE 1982: v). These aspects have continued in the years since, including fur-ther complications of supply chains and increasingly mobile workforces (James et al. 2007). With such transient work, it was difficult for employ-ees to accrue experience that would serve them well in relation to health and safety. Jenny Bacon saw this in the 1990s:

> [you] need something slightly different when you've got a much more mobile and short term workforce on contracts for services rather than con-tracts of employment [...] also the instability of the job market, people now have [...] portfolios, they don't have careers, they're not in one job for a long time, they're moving around. It's much more difficult to get application of health and safety legislation in those circumstances. (Jenny Bacon Interview, para. 29)

Similarly, it might be the case that within small firms—and even within larger industries, for specific jobs—there might be insufficient work for employees to gain experience of the role. This was even the case in some aspects of the nuclear industry in the 1990s, where asbestos removal work was outsourced because "*we actually didn't have enough work to produce a competent workforce, it was better to employ contractors*" (Chris Marchese Interview, para. 45.).

At the same time, there has also been some awareness of health and safety as an issue with porous boundaries: where 'workplace' ended and 'home' or 'leisure' started have not been as clear as it might be imagined. Homeworking is certainly not new—it was a marked feature of pro-duction pre-dating industrialisation and lasted well into the years often thought of as characterised by factory-scale production (Simonton 2002: ch. 9). Indeed, homeworking exists to this day, in terms of production and service roles undertaken where individuals live and in terms of flex-ible working arrangements for those otherwise regularly based in office or other environments (Allen and Wolkowitz 1987). Understanding the potential for health and safety to apply expansively, and beyond the confines of traditional conceptions of occupational locations, is not necessarily a recent phenomenon. In the early 1960s, according to RoSPA, "*[m]ore and more, progressive and enlightened organisations are taking an active interest in their workers welfare at all times. They see safety on the roads, in the home and at work as part of the same problem*".[16] This cross-over was often related to road risks, but it also applied to asbestos, with the understanding that it spread beyond the factory gate (for example, on workers' clothing) and killed those at home (for example, Gillie and Gillman 1976). Some thought has been paid in recent years to pre-existing conditions that the population have *before* they enter the workforce:

> The mental health issues in the general population are finding their way into the workplace [...] I think there's a great shift in what are the issues in health and safety. (Peter Jacques Interview, para. 192)

Evidently, then, the last 60 years have seen huge changes in not only the economic structure of Britain, but also the ways in which occupational health and safety have been conceived, regulated and discussed.

THE CHANGING NATURE OF TRADE UNIONISM AND WORKER REPRESENTATION

The period since 1960 has seen great changes in the nature of trade unionism in general and worker representation on health and safety matters in particular, perhaps summarised as a rise and then a fall in rates of participation and involvement. This has brought with it challenges in terms of legitimacy—including responding to long-standing (since at least the nineteenth century) managerial concerns about prerogative in the workplace and trade union power through to more recent issues about the impact of changes upon claims that can be made about the democratic legitimacy of health and safety regulation. Although containing some gaps, Table 5.4 gives a statistical picture of the position of the unions across the period covered by the book. The key trends are a rise in membership and density of coverage (percentage of working

---

[16] RoSPA, 'Off the Job Safety Planning' Leaflet (c.1963), p. 2. MRC, MSS.292B/146/2/2.

Table 5.4 Trends in trade unionism, 1960–2013. Figure for total number of unions for 2010 is actually 2008–2009 (*Aggregated from* Bain and Price [1980: 38], BIS [2014: 21–25], Mitchell [1988: 137], and ONS [2002: 118; 2010: 116])

Year	Total no. of unions	Members (000s)					
		M	Density (%)	W	Density (%)	Total	Density (%)
1960	664	7884	54.2	1951	25.4	9835	44.2
1970	543	8444	58.2	2743	32.1	11,187	48.5
1980	438	9162		3790		12,952	
1990	327					9810	
2000	237	3752	30.4	3367	29.1	7119	29.8
2010	185	2969	23.8	3561	29.5	6530	26.6
2013		2928	22.9	3522	28.3	6450	25.6

population in a union) until 1979 and a decline thereafter, and the decline of male trade unionism and rise of female union membership. This decline of trade union membership has also occurred in other developed European economies during the same period (Ebbinghaus and Visser 1999; Visser 2012). This decline has been reflected in the absence of organised labour as a leading influence from policy-making processes since the 1980s. There have been wider cultural shifts in how labour protection is viewed, how policy is made, and how representation and worker engagement is undertaken.

In 1960, there was considerable variation in workforce participation in trade unionism as well as variation in trade union activity on health and safety matters. As might be expected, unionisation was higher in older established heavy industries such as mining, iron and steel, shipbuilding and railway work, with resulting geographic distributions (Bradley 2012; Johnston and McIvor 2000; McIvor 2013: 171); some sectors, such as agriculture, construction and shop work were weakly unionised (Jim Hammer Interview, para. 65; Peter Jacques Interview, paras. 4–18; McIvor 2013: 207). It is likely that this impacted on the prominence with which health and safety issues were addressed within respective industries and workplaces, as they were seen by many as matters of 'natural' interest to the unions. Whilst it is not possible to establish that this is a cause and effect relationship, given the other potential factors in play (such as fragmented and diffused workforces), it is possible to suggest that sectors with relatively restricted union membership were also those in which health and safety issues received relatively limited attention.

Unions acted on health and safety in a number of ways. On a day-to-day basis, shop stewards might raise concerns with foremen (and they were usually men) or management, though the ability to do so was located within the local context of industrial relations at a particular workplace; this at a time which one interviewee, a former inspector, saw that *"management/ trade union conflict [...] was very much part of the industrial culture in the late '60s, early '70s"* (David Morris Interview, para. 34). Union and worker representatives might be found sitting on voluntary safety committees, though this tended to be at the discretion of the management who constituted the committees. Indeed, the existence of those voluntary safety committees was not guaranteed; between 1956 and 1964, they declined in number in the largest industries, which might have been better placed to institute them (Sirrs 2016b: 77). Beyond individual workplaces, a series of regional (often RoSPA-affiliated) health and safety councils existed, though it is clear that, in the 1960s, unions occupied an ambivalent place: the Teesside Industrial Accident Prevention Committee, for example, decided to allow unions to join the Committee in 1968 only after some discussion, as they felt *"that many district union officials were out of touch with their members at works level"*.[17] Unions provided representation at hearings and during investigations into incidents, including discussing or raising cases with state inspectors, coroners' inquests and legal proceedings, and lobbied MPs or raised issues through union-sponsored/affiliated MPs, or approached relevant Ministers directly. This latter role was threatened in 1962, for example, when the TUC was unhappy at the slow rate of progress in providing washing facilities at ports for workers who might come into contact with anthrax: *"unless this was accelerated it was likely that direct representation would be made to the Minister of Labour"*.[18]

Calls for worker representation in health and safety issues within industries were long-standing, having been made since at least the start of the twentieth century, although with relatively little progress—particularly as employers were concerned about employee involvement

---

[17] Teesside Industrial Accident Prevention Committee, Minutes 3149c (20 December 1967) and 3165 (20 March 1968). RoSPA Archives, D.266/2/18/11.

[18] Note of Meeting Under the Committee of Inquiry on Anthrax, Between Ministry of Labour and Worker and Trade Union Representatives, 10 January 1962, p. 5. TNA, LAB 14/957.

in 'managerial' matters. From the 1950s onwards, a number of unions started to push more vocally for the appointment of safety representatives and, in 1962, the TUC voted for a resolution calling for statutory safety representatives, a departure from the previous support for voluntary arrangements (Grayson and Goddard 1975: 18). Perhaps more immediately, union representatives (including the TUC) had contact with regulators. Peter Jacques recalled his early days at the TUC's Social Insurance and Industrial Welfare Department in the late 1960s/early 1970s:

> we spent a lot of time with Ministers [...] we would get whatever we wanted in a way, but it was like talking to a sponge bag. 'What do you want? Yeah, right-ho Peter.' And then pass them over [...] if you wanted anything real you had to talk to the people who really knew. And initially it was the Factory Inspectorate but later on with the Executive it was exactly the same. (Peter Jacques Interview, para. 40)

As well as securing a place at the table through the tripartite structure of the HSC, the role of trades unions was formalised through provisions for worker consultation (Dawson et al. 1988). The unions had wanted the appointment of worker representatives made compulsory and they continued to campaign on the issue after the 1974 Act noted this as 'desirable'. This generated considerable opposition from employers, on the grounds that it would have an adverse effect upon industrial relations.[19] This long-standing employer opposition to union or worker involvement was noted by the TUC in 1975: "*Most commentators saw 'worker participation in safety effort as solely a matter of discipline'. This fundamentally paternalist view is certainly not dead*" (Grayson and Goddard 1975: 17). The depth of the contest over worker representatives demonstrates the significance to employers and unions of safety representatives, which were eventually introduced under the Safety Representatives and Safety Committees Regulations 1977.

The issue of worker representation and consultation was also a matter of significant concern to the newly formed HSC, possibly as it offered

---

[19] See, for example, the comments of the Cement Makers Federation: 'Safety Representatives and Safety Committees', 16 February 1976, pp. 1–3, in the files of the Confederation of British Industries relating to the HSC's Consultation on Safety Representatives. MRC, MSS.200/C/3/EMP/4/40.

a route to enhancing that body's democratic legitimacy. As David Eves, later Deputy Director-General of the HSE, recalled:

> you want [...] employers, and employees' representatives, usually through the trade unions, to agree that there is a way forward that both sides can agree to. So this is why the Safety Committee and Safety Representatives Regulations were one of the Commission's first priorities in the 1970s, and they struggled with it at first, it took them three years [...] These for the first time gave safety representatives who were elected by their fellow workers, powers to inspect workplaces themselves, investigate, bring matters to the attention of their bosses. (David Eves Interview, para. 21)

The unions gave the regulations a cautious welcome, while recognising the challenges that remained:

> Looking around at all the legislation supposedly affording protection to working people, it might be comfortable to conclude that old-style union organisation and activity are no longer needed. Nothing could be further from the truth. A healthy and safe workplace is not going to be given to you. YOU must negotiate it! (Cunningham 1978: 28)

In addition, it is worth noting that these new powers applied only to representatives of recognised trade unions: non-unionised representatives had no legal standing, a serious issue given that it was not until 1974 that union density surpassed 50% of the working population (Bain and Price 1980: 38). RoSPA, for example, noted in 1975 that they wanted everybody to have the possibility of participating and to avoid "*the possible political usage of legislation*".[20] This had implications in terms of the potential impact on democratic legitimacy of this formal means of involving workers in health and safety matters.

Whilst some involved remembered an initial flush of enthusiasm, this was muted by the fact that the Regulations were dependent upon trade union representatives, which created problems where unions were weak: "*[In] '76 safety reps were being trained and there was a huge enthusiasm. I think in the unionised companies that structure worked extremely well, in the non unionised companies sometimes there was very little interest*"

---

[20] RoSPA Industrial Safety Division, Report from Occupational Safety Groups Advisory Council, 14 November 1975, p. 1. RoSPA Archives, D.266/2/24.

(Janet Asherson Interview, para. 45). This also posed problems from the 1980s as trade union membership and traditional industries declined and newer sectors, less likely to be unionised, became a bigger part of the economy. In addition, the response from rank and file union members to the new Regulations was not necessarily keen. As early as 1981, it was noted that "*[t]he post of safety representative is not always a popular one amongst Local Departmental Committee members because of the extra workload it imposes, and at several locations individuals had been persuaded reluctantly to accept the post*".[21] This has remained an issue, as noted by Sarah Lyons of the NUT: "*often [...] in a unionised organisation it's not a question of people clamouring to do the role, it's you get asked to do it, and for me it was good because it led on to, to other things*" (Sarah Lyons Interview, para. 10).

There is some evidence, then, which suggests that the safety representatives and consultative mechanisms suggested by Robens and the HSWA ultimately failed to achieve the levels of uptake, integration, and impact that was hoped. In October 1978, for example, the management side of Southampton Port Accident Prevention Committee was criticised for poor attendance;[22] when the Committee was dissolved in 1982, it was seen as "*a rather ineffective and frustrating 'talking shop'*".[23] This risk was replicated in the recollections of Stan Barnes, former ironworks safety manager and later IOSH President:

> the great problem with safety reps was getting recognition for them in industry. I don't think they ever really did fully get off the ground, except probably in local government. It's a thing which I'm sure was very well intentioned but I never felt [...] it had really succeeded in industry [...] It's much the same with the safety committees. They're easily turned into talking shops. (Stan Barnes Interview, paras. 87–89).

The attention paid to welfare work, which often included health and safety issues, in office environments was also questioned, as in a 1986

[21] BRB and HSE, 'An Outline Report on OSH by the Accident Prevention Advisory Unit' (November 1982): 12. TNA, AN 16/157.

[22] Correspondence of the Southampton Port Accident Prevention Committee, memorandum, 4 October 1978, p. 1. TNA, BK 11/60.

[23] Correspondence of the Southampton Port Accident Prevention Committee, letter, 28 January 1982, p. 1. TNA, BK 11/60.

comment by Geraldine Beech, a part-time departmental welfare officer at the Public Record Office, who noted that *"for none of us is welfare our principle function [...] the welfare service is hamstrung by the fact that its members' first responsibility, except in emergencies, is to their "proper" job [...] there is no Welfare budget"*.[24] When safety committee meetings in the railway industry in the early 1980s failed to prompt subsequent change, *"the system appeared to fall into disrepute with safety representatives taking the view that there was little point in inspecting and reporting if action was not to be forthcoming"*.[25] Similarly, in the Coventry Chain Works Shop Stewards Committee in 1980, it was noted that *"Brother Igoe said that he was disappointed with the Company's response to the safety representatives reports, it seems that requisitions go in but they have no intention of doing many of the jobs unless we draw special attention to a specific hazard"*.[26]

There were also concerns about the operation of arrangements at a higher level: whilst there was no doubt that formal tripartism was progress, Peter Jacques questioned whether it was conceived in such a way as to *"keep the unions in place"* (Peter Jacques Interview, para. 36) rather than as a good in its own right. How much of an impact contact with the inspectorates had was, of course, variable, as Jacques acknowledged:

I don't think we [TUC officials] got on very well [...] with other Inspectorates [beyond the Factory Inspectorate], I don't think they understood who we were and what we were trying to do. For example I didn't get on very well with [the Alkali and Clean Air Inspectorate...]They would never do us anything, wouldn't help [...] The mine working inspectors I was very disappointed with [...] I just thought they were just coasting along nicely to their retirements, I never felt they really wanted to do anything. (Peter Jacques Interview, paras. 40–42)

It is necessary to be cautious, then, about viewing the period before the 1980s as one in which the trade unions had excellent access to and influence upon health and safety policy and practice.

[24] G. Beech, Report 'The Problems of a Part-Time Welfare Officer', May 1986, pp. 5–6. MRC, MSS.381.W.2.1.27.

[25] BRB and HSE, 'An Outline Report on OSH by the Accident Prevention Advisory Unit' (November 1982): 13. TNA, AN 16/157.

[26] Coventry Chain Works Shop Stewards Committee Meeting 5 March 1980, minute 24. MRC, MSS.249/1/24.

For much of the period after 1960, the unions—certainly those in older, gradually declining industries—had to work with employers to balance security of employment with health and safety issues. It is apparent that whilst there was a move away from older ideas of 'danger money' for particularly hazardous roles, in certain sectors, the practice was hard to eradicate. At the 1969 TUC Congress, one speaker noted that "*[w] e refuse completely, in the interests of productivity or anything else to prostitute safety for production*" (Grayson and Goddard 1975: 4); yet, a comment ten years later demonstrated that the issue was still a live one: "*efforts have repeatedly been diverted by offers of compensation and danger money. Health has been sold very cheap*".[27] Handling asbestos in the docks was a further area where danger money was still being applied at least in the 1970s (David Morris Interview, para. 28).[28] At the same time as working around individual members' interests in economic security, many unions also had to work within a context of declining economic viability of particular industries. So, in one 1980 Joint Industrial Committee meeting, the 'reasonably practicable' clause was invoked by employers to argue against the limits on noise levels that the unions were seeking:

> in the wire and wire rope industries there was a great deal of machinery that was somewhat old. The industry could not afford to re-equip in order to meet unrealistic noise levels [...] he felt personally that standards could not be proposed which would shut down a major part of British industry.[29]

The conditions under which many unions were trying to secure health and safety for their members were, therefore, increasingly difficult into the 1980s.

Unionism has also been heavily politicised and contested during this period, bringing issues of 'high' politics further into the workplace (Moran 2003). In particular, the rise of the neo-liberal agenda during

[27] George Woodhead, 'Forge Ahead', *Work Hazards*, 21 (c.1979), p. 9. Samuel Barr Collection, Glasgow Caledonian University [GCU] DC 140/2/1/2.

[28] Owen McIntyre, Interviewed by David Walker (2009). Scottish Oral History Centre, University of Strathclyde Archives and Special Collections, GB249 SOHC18 (2009); Alfred McMillan, interviewed by David Walker (2009). Scottish Oral History Centre, University of Strathclyde Archives and Special Collections, GB249 SOHC18 (2009).

[29] Minutes of Meeting of Joint Industrial Council for the Wire and Wire Rope Industries, 23 January 1980, Minute 100. Glasgow City Archives, Mitchell Collection, TD914.18.1.

the 1980s (Gamble 1994; Harvey 2005) had a severe and lasting impact on the ability of the trade unions to influence policy and practice, in health and safety as in other areas. However, there was an additional problem for health and safety, in that an increasing political antipathy towards health and safety as emblematic of state 'interference' adversely impacted on those areas in which the unions had a more formal involvement. So, the corporatist experiment of the HSC was understood on the political right as a vestige of an increasingly outdated mode of politics. Peter Jacques reflected that:

> it was pretty clear that the trade union position as a whole was going to deteriorate, I don't think there's any doubt about that. And it did [...] I could feel it from the Executive that they were very much more cautious, very much more circumspect about how they were dealing with things and so on. Much more, not respectful but, what's the right word? Attentive to what the employers were saying, a bit loath to bring in regulatory regimes, wittering on about advice and so on. (Peter Jacques Interview, para. 149)

Since approximately 1980, trade unions therefore appear to have seen a reduction in their ability to influence outcomes and a marginalisation in the policy-making process, moving from having had some role (however constrained) to vocal, but external, critics. Even the early promise of the 1997 Labour government and the 'Revitalising' strategy was viewed as coming to little, so far as the unions and health and safety were concerned:

> The expectations were raised with Revitalising that there would be a lot more law, particularly relating to directors and trade unions, and not surprisingly the trade unions got really cheesed off that the expectations that they had built up had failed to materialise [...] the upset was absolutely phenomenal [...] the longer you went into the Blair/Brown governments, in terms of some of the changes that one of the big stakeholders in the Labour Party, the trade union movement, expected to see, they just didn't happen. (Neal Stone Interview, para. 50)

At the same time, the apparent relevance of trade unionism at a national level has also suffered, as a result of changes in the occupational structure and focus of health and safety issues in the UK. With the decline of heavy industries, the older unions have declined in size and power; there was a shift away from large-scale employers towards small and medium-sized enterprises which tended to be less unionised and employed more temporary or casual workers.[30] Newer sectors, such as office and shop work, were those in which the workforce was less likely to be unionised and health and safety issues were perhaps less immediately obvious (Dan Shears Interview, para. 80; Ian Tasker Interview, paras. 10–12). The General Secretary of the National Federation of Professional Workers noted in 1966 that "*the proportion of clerical workers and shop assistants who are organised in trade unions is generally low outside the public services*".[31] The TUC appears to have focused for much of the 1980s and into the 1990s on traditional 'core' areas of concern around the heavy industries, rather than emerging sectors (Peter Jacques Interview, paras. 60, 178, 192); and whilst occupational health was on the agenda, it remained rather divorced from some of the TUC's political lobbying and representation, as work was conducted by the Centenary Institute of Occupational Health at the London School of Hygiene and Tropical Medicine (Peter Jacques Interview paras. 24, 28).

Other voices increasingly entered the health and safety debate, diminishing the extent to which the unions might be heard. The decline in union membership from 1980 meant that policy-makers had to think more widely about whose voices were captured in the decision-making process. It was no longer sufficient to assume that trades unions also represented the public, and from the mid-1980s, public opinion came to matter more. This was related to the decay of the previous vision of tripartism and the emergence of a more pro-business agenda within government, as well as a movement towards the viewing of employment and public rights in individual, rather than collective, terms (Brown et al. 2000). This arguably meant that trades unions, who had traditionally been thought to represent the 'public interest', could not necessarily be taken to do so in totality, thus requiring that new alternative methods be found to represent a broader range of public constituencies. As a result, the older but more limited understanding of governance focused around industrial partners—in particular employers, unions and

[30] M. Leahy 'A Step in the Right Direct' *Focus* (February 1998), p. 2. MRC, MSS.36.2000.72.

[31] Letter from John Fryd to *The Municipal and Public Services Journal*, 25 March 1966. MRC, MSS.239/3/1/95.

regulators—broadened to include a more diverse group of stakeholders, including the public. Beginning in the mid-1980s, this has been consolidated since 2000 (Almond and Esbester 2016).

It is also possible that the extension of health and safety beyond the workplace, as discussed in Chapter 5, started to push more traditional workplace health and safety issues—those with which the unions were more directly concerned—to one side. As Jenny Bacon, then Director General of the HSE, noted in 1998:

> [o]ver the past 10 years, or maybe more, I think we've been drawn into a lot of interesting high-profile cases about the safety of the public affected by work activities. But worker safety and, in particular, health are things that these Ministers are very keen to see us work on [...] it's a slight rebalancing of worker versus public.[32]

It therefore seems that union voices had to fight harder to be heard at a national level from the mid-1980s on. At the same time, health and safety remained a 'day-to-day' feature of business, via representative and consultative measures at the level of individual companies. There was a considerable feeling that the unions *were* having an impact, and that activity in individual workplaces was a 'safe haven' in which trade unionism might remain relevant, even in times of conflict elsewhere (Lawrence Waterman Interview, para. 12). Trade union involvement in safety was viewed as *useful*, as people who knew the area and issues, union reps carried a lot of expertise (often more than employers themselves). This was certainly articulated by the unions themselves: the General Secretary of the Amalgamated Union of Engineering Workers claimed in 1972 that trade unions made an "*extensive contribution*" to accident prevention.[33] But it was also echoed by others outside the union movement, such as Tim Carter, a former HSE official, who recalled:

---

[32] Quoted in H. Fidderman, 'Mood Music with Jenny Bacon', *Focus* (February 1998): 4. MRC, MSS.36.2000.72.

[33] *Amalgamated Union of Engineering Workers Journal*, September 1972, p. 395. TNA, LAB 96/476.

what you see as soon as you're in HSE, actually on casework and working conditions, most union reps and shop stewards do their best to do a good job. (Tim Carter Interview, para. 105)[34]

Across the period, part of the day-to-day work of unions and worker representatives was negotiating around health and safety issues. It is important to note here that some of the evidence gathered suggests that safety was used as a negotiating tool for wider goals of extracting concessions, securing compensation for members, or maintaining pressure on employers (David Maidment Interview, paras. 28–29). Responding to the Robens' Committee claim that "*there is no legitimate scope for 'bargaining' on safety and health issues*" (Robens 1972: para. 66), one TUC article in 1975 noted that that "*most if not all major reforms in the field of safety and health at work have been instituted as a result of industrial and political 'bargaining' by trade unions [...] it is essential in the present situation to intensify trade union bargaining on safety issues*" (Grayson and Goddard 1975: 18). When asked whether he felt health and safety was used as a lever by employers or unions, David Morris, an HSE field inspector in the 1970s and early 1980s, replied:

Oh all the time, oh yeah, absolutely [...] if there is a health and safety issue at the same time as there is an industrial dispute going on, it is bound to be brought into play, but if it's a genuine health and safety issue then it's the job of HSE to help them sort it out [...] invariably, the health and safety issue was just part of the bricks that were being thrown from one side to the other and the broader issue was something completely different, and you just had to bear that in mind. (David Morris Interview, paras. 91–95)

Safety issues thus became part of the 'game playing' that all parties would engage in. According to Peter Jacques, this included high politics, particularly in the late 1960s and early 1970s:

there was the question of what could the relationship between the Labour Government and the trade unions be? [...] if they can't give you the

---

[34] See also Chris Marchese Interview, para. 51 for an example of the positive reception of union involvement by employers, and David Morris Interview, para. 89 on the influence of shop stewards.

money what else can they give you? Those sorts of things may be the reason why health and safety had moved up the agenda [...] the reasons for this tripartism weren't some sort of, 'oh well this is a good idea,' it's knowing what can we do to keep the unions in place? And that was to give them places round the table. (Peter Jacques Interview, paras. 34, 36)

While the motives behind this engagement were not always pure, the outcomes remained, for the most part, productive; the value of consultation was recognised and endorsed (Paul Clyndes Interview, para. 62).

As union membership has declined since the 1980s, so the spaces in which consultation occurs have contracted. Arguably a lot of the impetus behind the promotion of health and safety as an issue has been lost:

in the last ten years there's been a gradual loss in influence, to the detriment of the management side I feel, from the loss of empowerment by trade unions. In the [nuclear] industry fewer people belong to trades unions [...] Because there are fewer trade union members their organisations are much smaller, so they are not in a position to support their own representatives as much as they used to. (Chris Marchese Interview, para. 53)

At the same time, there is evidence that over the last ten years, a more consensual and cooperative working relationship had emerged in particular areas that were either hostile to unionisation or new to the 'health and safety' agenda, such as the rail industry (Paul Clyndes Interview, para. 32), and the retail and online delivery sector (Asda, Amazon and Yodel: Dan Shears Interview, para. 74). Possibly this improved dialogue has resulted from closer working relationships promoted over many years through institutions such as the HSC, and partly from other 'bottom-up' forms of engagement, which are more avowedly pragmatic, and less informed by the high politics of industrial relations, as Dan Shears of the GMB Union suggested:

it's probably more from the ground up than the top down now [...] we [GMB] will work with you to improve your practices to reduce the risk to our members and your employees [...] the approach that we take is, we'll work in partnership where we can, and if we can't work in partnership we will have to go very hard at this. (Dan Shears Interview, para. 78)

Whilst too limited to draw strong conclusions from, this at least suggests that union influence (and corporatism) have had a positive development.

It would also suggest that there is still an important democratic legitimacy role for trade unions in a post-industrial world (James and Walters 2002; Quinlan 1999; Walters 2006), which is particularly pronounced in industries where health and safety is either less immediately relevant (the public sector) or subject to specific risks (construction, extraction).

HEALTH—THE FORGOTTEN ELEMENT?

The issue of health has often been cast as the 'poor relation' in the field of health and safety, afforded a lower profile and less attention from regulators, employers, unions, and the wider health and safety sphere. The reasons for this have usually been linked to the diffuseness, long latency, and indirectness of health-related victimisation, and the subsequent invisibility of the issue when compared to the immediate physical impacts of safety violations. The evidence suggests that a key driver of the relative imbalance has also been the complexity of addressing these issues, and that solutions to health-related problems often tended to be specific and prescriptive, putting them at odds with the overarching regulatory approach. David Eves, for example, suggested that:

> it wasn't so much that health was on the back burner, health was rather more difficult to deal with than safety. You could usually find engineering solutions to mechanical problems [...] other sorts of hazards needed different approaches [...] through the 50s and 60s, there was quite a lot known about industrial health, but the regulatory answer was generally to make the regulations very specific, very prescriptive, and often requiring medical examinations of workers. (David Eves Interview, para. 19)

So, it was not necessarily that there was a lack of awareness of health issues, but that their ability to act was constrained by a variety of economic, social, cultural, and political factors, particularly when compared with what appeared to be more immediately pressing issues like deaths and injuries. Certainly, as Chapter 3 illustrates, the recent history of health and safety has been much more readily and fundamentally influenced by disasters and incidents involving immediate physical harms, rather than by health-related risk issues and events. This was exemplified in a discussion at the Teesside Industrial Accident Prevention Committee in February 1967, which concluded that *"the general feeling of the meeting was that accident injuries were a greater problem than the health hazard [...] Obviously the health hazard should not be forgotten and work in this field should be continued*.[35]

Nevertheless, even if health issues did not receive the same degree of coverage as safety, they were still accorded meaningful attention before the 1980s. Particularly in the earlier part of our period, the prevailing approach might perhaps be characterised as reactive, in that actors responded to known health hazards rather than actively seeking out new issues before they emerged; this reflected a longer standing concern with particular occupational diseases, including lead poisoning and lung diseases such as asbestosis, byssinosis or pneumoconiosis (Bartrip 2002; Johnstone and McIvor 2007; Stark 2012; Wikeley 1992). So, in the late 1950s and early 1960s, a series of state reviews looked at health in particular industries of concern and anthrax (Ministry of Labour and National Service 1958, 1959, 1960),[36] and similar issues were to be found in the reports of the Chief Inspector of Factories on Industrial Health, issued annually between 1957 and 1966, under the Chief Inspectorship of R. K. Christy. It is perhaps indicative of the extent to which health issues were dependent upon the personal drive of those involved that, apart from between 1957 and 1966, health was to be found within the annual report of the Chief Inspector of Factories as a chapter rather than the subject of its own report.

Health issues were also covered by medical agencies and practitioners within and outside the state. Larger firms might employ factory doctors and/or nurses or contribute to Industrial Health Centres, though it was difficult to get smaller firms to subscribe to this arrangement (Johnston and McIvor 2008).[37] The Appointed Factory Doctor Service started in the nineteenth century and involved 'approved' doctors examining workers for fitness for employment. In 1972, this role was merged with the Medical Branch of the Factory Inspectorate, to form the Employment Medical Advisory Service (EMAS), which, although technically distinct from the Factory Inspectorate, worked closely with it. Tim Carter, who occupied several key roles within the HSE including as Head of EMAS

[35] Minutes of Teesside Industrial Accident Prevention Committee, 15 February 1967, Minute 3108. RoSPA Archives, D.266/2/18/10.

[36] Ministry of Labour and National Service, Working Party on Washing Facilities in Docks (1960–1964). TNA, LAB 14/957.

[37] See, for example, the glum assessment of 1956 that 'little progress was made among the smaller factories' (Ministry of Labour and National Service 1957: 88).

and Director of Health Policy, noted that bringing EMAS into the HSE in 1974 brought a legitimacy problem of sorts:

> it had been created with this very broad remit which was almost a pre-tripartite remit, to give advice to everybody, doing everything about occupational health. And when it stood outside HSE there was some chance it could do that, but putting it into HSE immediately put it into the same management structure as enforcement activities and that, not surprisingly, and you can't really document it, but it clearly affected attitudes to it. (Tim Carter Interview, para. 50)

The creation of EMAS did not end calls for an occupational health service in parallel to the NHS, though this was not to materialise (Long 2011).[38]

Although safety was often a more visible problem, there was also trade union input into health issues on some issues, such as byssinosis in the textile industry (Bowden and Tweedale 2003). The TUC established the Centenary Institute of Occupational Health at the London School of Hygiene and Tropical Medicine in 1968, making a financial contribution to its operation for the following twenty years. The unions were increasingly vocal in raising concerns about asbestos.[39] In the 1970s, the tripartite system, including union input, was noted as working effectively to address the dangers posed by vinyl chloride, but in other health-related areas, there was "*a strong feeling that the traditional unions were not looking after their members properly*" (Tim Carter Interview, para. 50). As might be expected, the focus of attention before the 1980s was on health in the more traditional heavy/manufacturing industries. In private correspondence in 1960, an official of the Institute of Directors compared the hazards of office and factory work, concluding that "*[t]here are probably no office conditions which are actually detrimental to health [...] office conditions simply are not a menace to health although they may be some menace to efficiency*".[40] At the same time, from the 1950s,

---

[38] For example, see the 1978 Trade Union Research Unit paper 'The need for an occupational health service', which made its case on the back of a recent EMAS survey of private sector occupational health services. TUC Library, HD 7261.

[39] R. Williams, 'A Key Issue', *Work Hazards* 21 (c.1979), p. 5. Samuel Barr Collection, Glasgow Caledonian University, GCU DC 140/2/1/2.

[40] Letter from Miss. B. O'Donovan (Institute of Directors) to Mr. Farr, 'An Office Workers' Charter', 6 January 1960, pp. 1, 2. Brighton Design Archive, 1685/1.

there was an awareness of changes in occupational health issues, including ionising radiation and new toxic chemicals (Sirrs 2016a). A specific Industrial Hygiene Unit of the Factory Inspectorate was established in 1966, concerned particularly with environmental issues such as noise and dust. The Chief Inspector of Factories assessed the position as he saw it in 1966: "*Today we are left with only the difficult problems; the easy ones have been solved [...] We shall continue to make advances in preventing disease, but each successive advance will be more dearly bought than its predecessor*" (Ministry of Labour 1967: 5).

In terms of the profile of health, the HSWA had marked the beginning of a change (from the ordering of the terms in the title of the statute onwards). While some progress had been made prior to the Act, it did allow for a more proactive approach to be taken around issues like asbestos. Jenny Bacon, who was involved in drafting the HSWA, saw the change:

> The emphasis began to move more to health and away from safety [after the 1970s]. (Jenny Bacon Interview, para. 77)

At the same time, there was still confusion over what health issues were. One former BR manager noted that "*[w]hen we talked about health issues even in the '80s, the sorts of things that came up were injuries, not health*" (David Maidment Interview, para. 309). Evidently, the process was a long one; the 1990–91 HSC Annual Report remarked that "*[w]e signalled in our plan for 1991/92 our intention to increase the emphasis that we and industry give to promoting occupational health*" before going on to note the introduction of the COSHH Regulations (HSC 1991: x). These regulations, along with the Display Screen Equipment Regulations from the six-pack, appear to have had a big impact upon the legitimacy of health and safety, and it is significant that health played a large part in this.

As noted, the reasons why health has received less attention were varied. The long latency and relative invisibility of health issues were noted by one leading health and safety practitioner:

> 2,500 people died of asbestos related diseases this year. But it's a bit like road accidents, they die all over the place and they die invisibly in hospitals and hospices and at home. Whereas Buncefield didn't kill anyone, but was bloody visible, that smoke plume [...] it makes people think about explosive risks. (Lawrence Waterman Interview, para. 18.)

This relative invisibility has meant that occupational health issues have rarely provoked significant public awareness and have received less professional attention than safety. Asbestos was one example of this trend, albeit with a relatively unusual public profile. Since the 1930s, it had received sporadic attention, including regulation (Bartrip 1998, 2001, 2014; Greenburg and Wikeley 1999; Higgison 2005; Tweedale 2000; Tweedale and Hanson 1998; Wikeley 1992) and peaks and troughs of public attention. Rather than a single 'moment' which shone a spotlight on the issue, asbestos garnered occasional intense public and political attention—notably around the Acre Mill factory in Hebden Bridge in the 1970s and early 1980s, including the TV documentary '*Alice – a fight for life*' (1982; McDougall 2012). Given the interest, the work of the state-appointed Advisory Committee on Asbestos (1976–79) was deliberately conducted very publicly, to "*rais[e] the standard of public debate and the level of public understanding*" (Simpson 1977: 1). At the same time, those exposed to asbestos did not necessarily know or understand the dangers. Testimony from people employed from the 1960s into the 1980s suggests that "*no one knew it was dangerous*" (Anonymous Interview 2009) and therefore employees were unable either to protect themselves or to fight to reduce the risks. This is an ongoing issue. Dan Shears of the GMB criticised a recent HSE campaign aimed at getting small building firms to work safely with asbestos on the basis that there is no such thing as safe work with asbestos (an old message that is still contested):

> what you're seeing is a lot of people who do work for larger businesses, 'don't worry so much about this asbestos, just make sure you've got the right gloves and PPE' [...] That's quite contradictory to what we would say. (Dan Shears Interview, para. 94)

Health represented an extremely diffuse set of problems, which were unlikely to be concentrated in a single physical location or within one firm or industry but rather spread through a production process and supply chain, including workers but also sometimes (public) consumers. The difficulty of managing this risk was noted by one agricultural worker in 1980: "*What really worries me is [...] what effect dairy chemicals can have, also animal diseases can have. There are the things the Health and Safety Executive need to be finding out, this is the kind of information I want, not being told how to do my job*.[41] This also highlights the resistance and hostility to regulators that health issues that might be seen as part of 'day-to-day life' were capable of provoking. The diffusion of health hazards posed particular problems for producers and regulators, not least of which was trying to work out what its role *should* be. According to Carter, inspectors knew:

> the dialogue to have if there are faulty electricals or faulty machine guarding, it's much more difficult for them to have a dialogue about bad seating or bad lighting, because they haven't got easy powers if you're in contravention [...] inspectors, like police or anybody else, they're busy people and they want to score quickly and easily and move on, whereas health things are very rarely quick and easy. (Tim Carter Interview, para. 68.)

Trying to convey complicated technical standards and having to use specialised medical terminology also made health difficult to understand, for employers as well as in court cases (Jim Hammer Interview, para. 24). Some problems were common across safety and health, however, including the economic pressures under which all actors were working and the existence of opposition to stricter standards or enforcement from within industries such as asbestos.

The complexity and newness of the science associated with some health conditions (stress, for example: Mackay et al. 2004) meant that intervention was later and more complex than might be the case for other issues. Unlike many safety matters, which involved discrete fixes, health often required ongoing monitoring of individuals and their working environment (Tim Carter Interview, para. 40). One area of change has been the emergence of health issues as a feature of the new workplaces (offices, retail, education, healthcare), particularly since 1980. On the one hand, these new workplaces have developed joined-up ways of addressing health, welfare and well-being in holistic terms, as found in the NHS, for example (Ruth Warden Interview, paras. 24–26); on the other, there remain legitimacy issues about whether these issues are seen as valid, or obscured via their inclusion within general 'wellbeing' approaches, as Kim Sunley, of the Royal College of Nursing, suggested:

[41] J. H. Brown of Saltburn, letter to *The Land Worker*, March 1980, p. 8. MERL.

from a trade union perspective, and the TUC, we've called for employers to take more notice of work related ill-health. I don't know whether that message has got lost somewhere and it's been translated as health and wellbeing but certainly the work related health isn't getting that much emphasis on the whole health and wellbeing agenda. (Kim Sunley Interview, para. 28)

Stress is one health issue to have emerged as a major concern since the 1980s, though the concept (broadly defined) has a long history (Melling 2014). Initially discussed as a part of broad concepts of welfare work, in 1967, it was suggested that "*[i]ndustry may be a psychological hazard [...] more effort should be made to make jobs more satisfying and to make the men happy*".[42] The genesis of concern about stress was noted by Janet Asherson, along with the difficulty of implementing this:

By the time I left HSE [1986...] stress and psychosocial disorders were just beginning to be looked at [...] I spent most of my lifetime trying to say [...] legislation is not the way to go on stress because of its complexity, so codes of practice, etc., and it took until the 1990s to get the management standards on stress. So politically it was very much there [...] stress didn't manifest itself at the level of an inspector going into an enterprise. That would not have been high on the agenda. (Janet Asherson Interview, para. 48)

Stress also links in to some of the most famous work around workplace conditions and morbidity, such as the Whitehall II studies, which suggested that work pressure, and the psychosocial character of work (job control, status, satisfaction), were linked to diffuse negative health outcomes (Marmot et al. 1991). It is also worth raising the question of the extent to which issues such as stress were gendered. With the rising numbers of women in the workforce, and the increasing concentration in office and retail work, it is possible that stress may have disproportionately affected women (particularly given the lower levels of job control they would historically have enjoyed: Marmot et al. 1991), though it might also be suggested that women were more able to discuss/cope with it, not being bound by codes of masculinity (Frankenhaeuser 1996). In 1982, it was observed that

---

[42] Teesside Industrial Accident Prevention Committee, 15 February 1967, Minute 3108. RoSPA Archives, D.266/2/18/10.

[u]ntil recently, occupational stress has not been a major source of health problems for women in comparison with men [...] as an increasing number of women occupy difficult jobs, such as supervisory and managerial positions, stress-related conditions are increasing. (Steventon and Cox 1982: 65)

As this account of stress suggests, office work was one area where health issues have tended to be raised, both before 1960 (particularly in relation to the welfare of female workers), but also via what would become the Offices, Shops and Railway Premises Act 1963. According to the Institute of Directors in 1961, "*[w]e also consider that the setting of reasonably high standards of environmental hygiene, although initially unpopular, are in the long run the only way of achieving an overall standard of which we can be nationally proud*", especially as "*[a]utomation and the use of electrical and other machines is also likely to become much more widespread over the next decade*".[43] More significantly, the debate over visual display units (VDUs) in the 1980s ensured that occupational health in the office 'came of age'. Suggestions that VDUs emitted harmful radiation, including sufficiently to increase the chances of miscarriage in pregnant women (later disproven), were widely reported in the press (Moncur 1986: 4), and the HSE produced guidance, albeit criticised as "*too little, too late and not enforced*" (Pearce 1985: 18). Suspicions lingered, and with the European Display Screen Equipment Regulations (enacted into UK law as part of the 'six-pack' of 1992; discussed further in Chapter 4), the office became a significant point at which many people encountered health and safety issues, and subsequently began to question the proportionality of regulation, as a former HSE inspector suggested:

we've been pushed into regulating some comparatively low risk areas which have high impact on people. I've always had slight worries about the Display Screen Equipment Regulations, for example, there's a good case for saying that prolonged use of VDUs, computers, leads to muscular skeletal disorders [...] But it's a big step from that to say that every individual work station used anywhere in the country must be properly assessed by a competent person [...] that impacts on more or less everyone at work and

---

[43] 'Comments from the Institute of Directors on the Proposed Bill of Health, Welfare and Safety in Shops, Offices and Railway Premises', March 1961, pp. 1, 5. Brighton Design Archive, 1685/1.

people think, 'fuck this health and safety, it's a real pain, I just want to get on with my job.' And once that mindset is there it's very difficult to get out of. (David Morris Interview, para. 111)

As noted earlier, there were unclear boundaries between health and conditions of work more generally, additionally complicated by the bargaining process between employers, unions and workers. So, in the chemical industry in the 1970s, trade union activity on health caused *"quite a lot of stress, not so much on polyvinyl chloride but on other chemicals because this younger generation wanted to effectively use chemical risks as a way to get more worker control"* (Tim Carter Interview, para. 48). This intersection between health and safety, and politics, has also arisen more recently in white-collar work, according to the NUT's Sarah Lyons:

the NUT has got a big campaign on teacher workloads at the moment, and that is really linked to health and safety, because it's stress, it's bullying, and so in some ways health and safety gets tied in with other areas that might be seen more as conditions of service issues, but they do impact on health and safety, so that helps keep it in the forefront. (Sarah Lyons Interview, para. 16)

Health was also an arena in which issues crossed the (loose) boundaries between work and public spaces, particularly in consideration of personal habits that people bring with them from their private lives into the workplace. Smoking would be one key example, noted by Tim Carter about the 1980s:

There was also huge opposition from inspectors to HSE doing anything on smoking, because if they had things they had to enforce on it they knew they'd meet lots of antagonism [...] voluntarism worked and achieved everything remarkably effectively in quite a short period of time without any regulation, or with quite limited regulation [...] probably wisely, HSE avoided regulation, the regulation came in much more about public places of leisure, like pubs and restaurants. (Tim Carter Interview paras. 135–136)

This echoed similar difficulties encountered around, for example, the introduction of ear defenders to protect against hearing loss in factories where workers were loath to surrender the opportunity to chat to colleagues while working (Sir Frank Davies Interview, para. 34). In sum, then, it is clear that the political context around health interventions was a crucial factor—and that without it, difficulties would be incurred:

> sick building syndrome, stress, the effects of smoking at work, all things that were sort of health related issues, but Ministers were very suspicious of these because they thought it was just all a shirkers' charter [...but] these were the things that were actually causing people to lose time at work. (Jenny Bacon Interview, para. 77)

These 'soft issues' have attracted criticism and a lack of political support at times, being seen as less legitimate areas of intervention by government, and as a lesser area of importance by experts, unions, and other actors. One upshot of this was that, in some sectors, the emphasis remained upon short-term, 'injury-type' health issues rather than long-term illnesses and conditions, some of which are still being grappled with at the current time. Health has also long been a frontline in the balancing of productivity and lost time (with pressure to reduce sick leave taking an explicitly economic character). David Maidment recalled of British Rail in the 1980s the attitude towards multiple days of sick leave: "*the sort of assumption was half of them, they've been on the booze over the weekend and it's Monday morning sickness*" (David Maidment Interview, para. 296). All of this has reduced the perceived legitimacy of health as an area—seen as too open-ended, where the needs are not clear enough, and which lacks the immediate, indisputable seriousness of safety and injury.

Conclusion

The three key themes this chapter has explored—the effects of economic and industrial change, the role of trades unions and the nature of worker representation, and the position of health in health and safety—are all concerned with democratic legitimacy: the basis upon which a range of people have been involved in health and safety and represented by the processes involved in its regulation. As the British economy and workforce have changed, so too have ideas about health and safety and its coverage. With diversification of the workforce, the nature and type of occupational health and safety issues have increased, bringing challenges to its legitimacy—perhaps exemplified by the decline of heavy industries and the rise of office work and relatively low-risk workplaces. How people are involved in decisions about health and safety, where those decisions take place, and how those decisions must go about accurately reflecting the realities of the modern workplace, have changed fundamentally since 1960. From a relatively narrow conception of constituencies affected by occupational issues, and even more limited arenas for consultation or input, the growth of tripartism into the 1970s at least accorded trade unions a seat at the table and locations at which health and safety issues might be raised, certainly within the older industries with stronger unionisation. Health and safety matters were seen here as 'natural' homeplaces for union involvement, around which organisation and recruitment could take place, and through which union members might obtain tangible and direct benefits from their union membership. Yet, even at what might be seen as a highpoint of democratic input into health and safety, it is important to be cautious about who was represented: newer workforces and newer workplaces appear not to have been so fully covered. How far women workers, or those of minority ethnic backgrounds, had their health and safety concerns addressed in this period is open to debate. In many respects, the broadening of health and safety—particularly from the 1980s—to include more workers and increasingly the public has been significant in terms of the potential for improving democratic legitimacy. Yet, the extension of occupational health and safety regulation to encompass the public has also proved problematic, something explored in the next chapter.

REFERENCES

PRIMARY SOURCES

Amalgamated Union of Engineering Workers. (1972, September). Journal. The National Archives of the UK, London [TNA], LAB 96/476.

Beech, G. (1986, May). The Problems of a Part-Time Welfare Officer. MRC, MSS.381/W/2/1/27.

BIS. (2014). *Trade Union Membership 2013 Statistical Bulletin*. London: Crown.

British Medical Association. (1963). BMA Conference 'The Immigrant in Industry. Minutes. MRC, MSS.292B/141/3/3.

British Rail Board. (1970, March 9). British Railways Board Accident Prevention Policy. Internal Memorandum. TNA, AN 171/550.

British Rail Board & HSE. (1982). An Outline Report on OSH by the Accident Prevention Advisory Unit. TNA, AN 16/157.

Brown, J. H. (1980, March). Letter to *The Land Worker*. Museum of English Rural Life, University of Reading [MERL].

Cement Makers Federation. (1976, February 16). Safety Representatives and Safety Committees. Modern Records Centre, University of Warwick [MRC], MSS.200/C/3/EMP/4/40.

Coventry Chain Works Shop Stewards Committee. (1980, March 5). Meeting Minutes. MRC, MSS.249/1/24.

Cunningham, M. (1978). Safety Representatives: Shop Floor Organisation for Health and Safety. *Studies for Trade Unionists*, 4/13.

Fidderman, H. (1998). Mood Music with Jenny Bacon. *Focus* OSH Newsletter for Reps of the Iron & Steel Trades Confederation (February 1998): 4–6. MRC, MSS.36.2000.72.

Fryd, J. (1966, March 25). Letter to *The Municipal and Public Services Journal*. MRC, MSS.239/3/1/95.

Gillie, O., & Gillman, P. (1976, July 4). Asbestos Safety Campaign Advert Claims too Much. *Sunday Times*.

Grayson, J., & Goddard, C. (1975). Industrial Safety and the Trade Union Movement. *Studies for Trade Unionists*, 1/4.

HSC. (1991). *Annual Report 1990/91*. London: Crown.

HSE. (1979). Health and Safety, Manufacturing and Service Industries (Report for 1977). London: Crown.

HSE. (1981). Health and Safety, Manufacturing and Service Industries (Report for 1979). London: Crown.

HSE. (1982). Health and Safety, Manufacturing and Service Industries (Report for 1981). London: Crown.

Institute of Directors. (1960, January 6). Letter from Miss. B. O'Donovan to Mr. Farr, 'An Office Workers' Charter. Brighton Design Archive, 1685/1.

Institute of Directors. (1960). Outline for Aide Memoire on Working Conditions in Offices. Brighton Design Archive, 1685/1.

Institute of Directors. (1961, March). Comments from the Institute of Directors on the Proposed Bill of Health, Welfare and Safety in Shops, Offices and Railway Premises. Brighton Design Archive, 1685/1.

Joint Industrial Council for the Wire and Wire Rope Industries. (1980, January 23). Meeting Minutes. Glasgow City Archives, Mitchell Collection, TD914.18.1.

Justice of the Peace and Local Government Review (1966, January 22) MRC, MSS.239/3/1/95.

Laurance, J. (1988, March 9). Paying the Wages of Peril. *The Guardian*, 23.

Leahy, M. (1998). A Step in the Right Direction. *Focus*. OSH Newsletter for Reps of the Iron & Steel Trades Confederation (February 1998): 1–2. MRC, MSS.36.2000.72.

Löfstedt, R. (2011). *Reclaiming Health and Safety for All: An Independent Review of Health and Safety Legislation* (The Löfstedt Review). London: Crown.

McIntyre, O. (2009). Interviewed by David Walker. Glasgow Dock Workers Project, Scottish Oral History Centre, University of Strathclyde Archives and Special Collections, GB249 SOHC18 (2009).

McMillan, A. (2009). Interviewed by David Walker. Glasgow Dock Workers Project, Scottish Oral History Centre, University of Strathclyde Archives and Special Collections, GB249 SOHC18 (2009).

Ministry of Labour. (1962, January 10). Note of Meeting Under the Committee of Inquiry on Anthrax, Between Ministry of Labour and Worker and Trade Union Representatives. TNA, LAB 14/957.

Ministry of Labour. (1967). *Annual Report of H.M. Chief Inspector of Factories on Industrial Health*, 1966. Cmnd. 3359. London: Crown.

Ministry of Labour & National Service. 1957. *Annual Report of the Chief Inspector of Factories for 1956*. Cmnd. 329. London: Crown.

Ministry of Labour & National Service. (1958). *Industrial Health. A Survey in Halifax*. London: Crown.

Ministry of Labour & National Service. (1959). *Industrial Health. A Survey of the Pottery Industry in Stoke-on-Trent*. London: Crown.

Ministry of Labour & National Service. (1960). *Report of the Committee of Inquiry on Anthrax*. London: Crown.

Ministry of Labour & National Service. (1964). Working Party on Washing Facilities in Docks (1960–1964). TNA, LAB 14/957.

Moncur, A. (1986, February 6). 'No VDU Link' with Miscarriage. *The Guardian*, 4.

National Federation of Professional Workers. (c.1913–1963). Paperwork Relating to Office Conditions. MRC, MSS.239/3/1/103.

O'Donovan, B. (1960, January 6). Correspondence to Mr. Farr, 'An Office Workers' Charter. Brighton Design Archive, 1685/1.

O'Gorman, P. (1970, October 8). Written Evidence to Robens Committee. TNA, LAB 96/36.

Observer (1949, December 29), note on office conditions for female workers. Brighton Design Archive, 1685/1.

ONS. (2002). *Annual Abstract of Statistics*. London: HMSO.

ONS. (2010). *Annual Abstract of Statistics*. London: HMSO.

Pearce, B. (1985, December 19). The Sensible Way of Looking at the VDU. *The Guardian*, 18.

Lord Robens. (1972). *Safety and Health at Work: Report of the Committee 1970–72* (The Robens Report). London: HMSO.

Lord Young. (2010). *Common Sense, Common Safety* (The Young Review). London: Crown.

RoSPA. (1960). Industrial Group Advisory Council Meeting Minutes. RoSPA Archives, D.266/2/21/1.

RoSPA. (n.d., c.1963). 'Off the Job Safety Planning' Leaflet. MRC, MSS.292B/146/2/2.

RoSPA. (1975, November 14). Industrial Safety Division, Report from Occupational Safety Groups Advisory Council. RoSPA Archives, D.266/2/24.

RoSPA. (1982). *Annual Report, 1981–82.* RoSPA Archives, D.266/1/2/5/2.

RoSPA. (1983). *Annual Report, 1982–83.* RoSPA Archives, D.266/1/2/5/3.

Simpson, B. (1977, June 27). Opening Remarks, First Open Meeting of the Advisory Committee on Asbestos.

Southampton Port Accident Prevention Committee. (1978, October 4). Memorandum. TNA, BK 11/60.

Southampton Port Accident Prevention Committee. (1982, January 28). correspondence. TNA, BK 11/60.

Steventon, J., & Cox, T. (1982, November). Why Women at Work Get a Raw Deal. *Design,* 65.

Teesside Industrial Accident Prevention Committee. (1966–1967). Minutes. RoSPA Archives, D.266/2/18/10.

Teesside Industrial Accident Prevention Committee. (1967–1968). Minutes. RoSPA Archives, D.266/2/18/11.

The Municipal and Public Services Journal (1966, March 4): 691. MRC, MSS.239/3/1/95.

Thompson, M. (c.2000). Interviewed as Part of University of Aberdeen 'Lives in the Oil Industry' project. British Library, London, C963/13.

Trade Union Research Unit. (1978). The Need for an Occupational Health Service. TUC Library, HD 7261.

WH Smith Retail Group. (1986, January 2). Management Executive Minutes. MERL, WHS 1415.1–4.

Williams, R. (n.d., c.1979). A Key Issue. *Work Hazards* 21 (c.1979). Samuel Barr Collection, Glasgow Caledonian University [GCU] DC 140/2/1/2.

Woodhead, G. (n.d., c.1979). Forge Ahead. *Work Hazards,* 21 (c.1979). Samuel Barr Collection, GCU DC 140/2/1/2.

SECONDARY SOURCES

Allen, S., & Wolkowitz, C. (1987). *Homeworking: Myths and Realities.* Basingstoke: Macmillan.

Almond, P., & M. Esbester, M. (2016). Il/legitimate risks? Occupational Health and Safety and the Public in Britain, c.1960–2015. In T. Crook & M. Esbester (Eds.), *Governing Risks in Modern Britain: Danger, Safety and Accidents, c.1800-2000* (pp. 277–296). Palgrave: London.

Bain, G. S., & Price, R. (1980). *Profiles of Union Growth: A Comparative Statistical Portrait of Eight Countries.* London: Blackwell.

Bartrip, P. (1998). Too Little, Too Late? The Home Office and the Asbestos Industry Regulations, 1931. *Medical History, 42*(4), 421–438.

Bartrip, P. (2001). *The Way from Dusty Death: Turner & Newall and the Regulation of Occupational Health in the British Asbestos Industry 1890s–1970s.* London: Athlone.

Bartrip, P. (2002). *The Home Office and the Dangerous Trades. Regulating Occupational Disease in Victorian and Edwardian Britain.* Amsterdam: Rodopi.

Bartrip, P. (2014). "Enveloped in fog": The Asbestos Problem in Britain's Royal Naval Dockyards, 1949–1999. *International Journal of Maritime History, 26*(4), 685–701.

Beetham, D. (1991). *The Legitimacy of Power.* London: Macmillan.

Black, J. (2008). Constructing and Contesting Legitimacy and Accountability in Polycentric Regulatory Regimes. *Regulation & Governance, 2*(2), 137–164.

Bowden, S., & Tweedale, G. (2003). Mondays Without Dread: The Trade Union Response to Byssinosis in the Lancashire Cotton Industry in the Twentieth Century. *Social History of Medicine, 16*(1), 79–95.

Bradley. (2012). *Occupational Health and Safety in the Scottish Steel Industry, c.1930–1988: The Road to 'Its Own Wee Empire'* (Unpublished PhD thesis). Glasgow Caledonian University.

Brotherstone, T., & Manson, H. (2007). North Sea Oil, Its Narratives and Its History: An Archive of Oral Documentation and the Making of Contemporary Britain. *Northern Scotland, 27,* 15–41.

Brown, W., Deakin, S., Nash, D., & Oxenbridge, S. (2000). The Employment Contract: From Collective Procedures to Individual Rights. *British Journal of Industrial Relations, 38*(4), 611–629.

Carson, W. G. (1982). *The Other Price of Britain's Oil.* Oxford: L.M. Robertson.

Clarke, L., & Wall, C. (2009). "A Woman's Place is Where She Wants to Work": Barriers to the Entry and Retention of Women into the Skilled Building Trades. *Scottish Labour History, 44,* 16–39.

Dawson, S., Willman, P., Clinton, A., & Bamford, M. (1988). *Safety at Work: The Limits of Self-Regulation.* Cambridge: Cambridge University Press.

Ebbinghaus, B., & Visser, J. (1999). When Institutions Matter: Union Growth and Decline in Western Europe, 1950–1995. *European Sociological Review, 15*(2), 135–158.

Frankenhaeuser, M. (1996). Stress and Gender. *European Review, 4*(4), 313–327.

Fudge, J., & Owens, R. (Eds.). (2006). *Precarious Work, Women, and the New Economy: The Challenge to Legal Norms.* London: Bloomsbury.

Gamble, A. (1994). *The Free Economy and the Strong State: The Politics of Thatcherism* (2nd ed.). London: Macmillan.

Genn, H. (1993). Business Responses to the Regulation of Health and Safety in England. *Law and Policy, 15*(3), 219–233.

Greenberg, M., & Wikeley, N. (1999). Too Little, Too Late? The Home Office and the Asbestos Industry Regulations, 1931: A Reply. *Medical History, 43*(4), 508–513.

Harvey, D. (2005). *A Brief History of Neoliberalism*. Oxford: Oxford University Press.

Higgison, A. (2005). Asbestos and British Trade Unions, 1960s and 1970s. *Scottish Labour History, 40,* 70–86.

Hood, C. (1991). A Public Management for All Seasons? *Public Administration, 69*(1), 3–19.

Hutter, B. (2001). *Regulation and Risk: Occupational Health and Safety on the Railways*. Oxford: Oxford University Press.

James, P., Johnstone, R., Quinlan, M., & Walters, D. (2007). Regulating Supply Chains to Improve Health and Safety. *Industrial Law Journal, 36*(2), 163–187.

James, P., & Walters, D. (2002). Worker Representation in Health and Safety: Options for Regulatory Reform. *Industrial Relations Journal, 33*(2), 141–156.

Johnston, R., & McIvor, A. (2000). *Lethal Work: A History of the Asbestos Tragedy in Scotland*. East Linton: Tuckwell Press.

Johnston, R., & McIvor, A. (2007). *Miners' Lung. A History of Dust Disease in British Coal Mining*. Aldershot: Ashgate.

Johnston, R., & McIvor, A. (2008). Marginalising the Body at Work? Employers' Occupational Health Strategies and Occupational Health in Scotland c.1930–1974. *Social History of Medicine, 21*(1), 127–144.

Long, V. (2011). *The Rise and Fall of the Healthy Factory: The Politics of Industrial Health in Britain, 1914–60*. Basingstoke: Palgrave Macmillan.

Mackay, C., Cousins, R., Kelly, P., Lee, S., & McCaig, R. (2004). Management Standards' and Work-Related Stress in the UK: Policy Background and Science. *Work & Stress, 18*(2), 91–112.

Marmot, M., Stansfeld, S., Patel, C., North, F., Head, J., White, I., et al. (1991). Health Inequalities Among British Civil Servants: The Whitehall II Study. *Lancet, 337*(8754), 1387–1393.

Mayhew, C., & Quinlan, M. (2002). Fordism in the Fast Food Industry: Pervasive Management Control and Occupational Health and Safety Risks for Young Temporary Workers. *Sociology of Health & Illness, 24*(3), 261–284.

McDougall, W. (2012). Fighting for Life: Alice, Nancy and the Society for the Prevention of Asbestosis and Industrial Diseases (SPAID), 1976–1990. *Scottish Labour History, 47,* 1–27.

McIvor, A. (2013). *Working Lives: Work in Britain Since 1945*. Basingstoke: Palgrave Macmillan.

Melling, J. (2014). Making Sense of Workplace Fear. The Role of Physicians, Psychiatrists, and Labor in Reframing Occupational Strain in Industrial Britain, c.1850–1970. In D. Cantor & E. Ramsden (Eds.), *Stress, Shock and Adaptation in the Twentieth Century* (pp. 189–221). Rochester, NY: University of Rochester Press.

Mitchell, B. R. (1988). *British Historical Statistics*. Cambridge: Cambridge University Press.

Moran, M. (2003). *The British Regulatory State: High Modernism and Hyper Innovation*. Oxford: Oxford University Press.

Nichols, T. (1999). Death and Injury at Work: A Sociological Approach. In N. Daykin & L. Doyal (Eds.), *Health and Work: Critical Perspectives* (pp. 86–106). London: Macmillan.

Otter, C. (2016). Artificial Britain: Risk, Systems and Synthetics Since 1800. In T. Crook & M. Esbester (Eds.), *Governing Risks in Modern Britain: Danger, Safety and Accidents, c.1800–2000* (pp. 79–104). Palgrave: London.

Paterson, J. (2007). The Evolution of Occupational Health and Safety Law on the UK Continental Shelf, 1964–2006. *Northern Scotland, 27*, 43–67.

Prassl, J., & Risak, M. (2015). Uber, Taskrabbit, and Co.: Platforms as Employers-Rethinking the Legal Analysis of Crowdwork. *Comparative Labor Law & Policy Journal, 37*(3), 619–652.

Quinlan, M. (1999). The Implications of Labour Market Restructuring in Industrialized Societies for Occupational Health and Safety. *Economic and Industrial Democracy, 20*(3), 427–460.

Quinlan, M., Mayhew, C., & Bohle, P. (2001). The Global Expansion of Precarious Employment, Work Disorganization, and Consequences for Occupational Health: A Review of Recent Research. *International Journal of Health Services, 31*(2), 335–414.

Simonton, D. (2002). *A History of European Women's Work: 1700 to the Present*. London: Routledge.

Sirrs, C. (2016a). Risk, Responsibility and Robens: The Transformation of the British System of Occupational Health and Safety Regulation, 1961–1974. In T. Crook & M. Esbester (Eds.), *Governing Risks in Modern Britain: Danger, Safety and Accidents, c.1800–2000* (pp. 249–276). Palgrave: London.

Sirrs, C. (2016b). Accidents and Apathy: The Construction of the "Robens Philosophy" of Occupational Safety and Health Regulation in Britain, 1961–1974. *Social History of Medicine, 29*(1), 66–88.

Stark, J. F. (2012). Bacteriology in the Service of Sanitation: The Factory Environment and the Regulation of Industrial Anthrax in Late-Victorian Britain. *Social History of Medicine, 25*, 343–361.

Suchman, M. (1995). Managing Legitimacy: Strategic and Institutional Approaches. *The Academy of Management Review, 20*(3), 571–610.

Tutt, D., Dainty, A., Gibb, A., & Pink, S. (2011). *Migrant Construction Workers and Health and Safety Communication*. King's Lynn: Construction Industry Training Board).

Tweedale, G. (2000). *Magic Mineral to Killer Dust. Turner & Newall and the Asbestos Hazard*. Oxford: Oxford University Press.

Tweedale, G., & Hanson, P. (1998). Protecting the Workers: The Medical Board and the Asbestos Industry, 1930s–1960s. *Medical History, 42*(4), 439–457.

Viscusi, W. K. (1983). *Risk by Choice: Regulating Health and Safety in the Workplace.* Cambridge: Harvard University Press.

Visser, J. (2012). The Rise and Fall of Industrial Unionism. *Transfer: European Review of Labour and Research, 18*(2), 129–141.

Walker, D. (2007). *Occupational Health and Safety in the British Chemical Industry, 1914–1974* (Unpublished PhD thesis). University of Strathclyde.

Walters, D. (2006). One Step Forward, Two Steps Back: Worker Representation and Health and Safety in the United Kingdom. *International Journal of Health Services, 36*(1), 87–111.

Wikeley, N. (1992). The Asbestos Regulations 1931: A Licence to Kill? *Journal of Law and Society, 19*(3), 365–378.

CHAPTER 6

Health and Safety in Action

INTRODUCTION

Quite apart from any political and conceptual significance it holds, health and safety is primarily a practical issue, concerning questions of action, outcome, and engagement. Health and safety is a major field of practical endeavour, with its own professional bodies and trade associations,[1] conferences and awards,[2] scientific journals, and media outlets.[3] This implementational apparatus plays a key role in sharing information and practice, and in constituting the field of 'health and safety' as a coherent area of activity with its own norms and culture (Okun et al. 2017). As was discussed in Chapter 4, just about every employer and employee in Britain is subject to, or a holder of, the duties imposed by the Health and Safety at Work Act (HSWA) 1974. This means that health and safety

---

[1] Such as the Institution of Occupational Safety and Health (IOSH), the British Safety Council, the International Institute of Risk and Safety Management (IISRM), and the Royal Society for the Prevention of Accidents (RoSPA).

[2] Such as the annual Safety and Health Expo (https://www.safety-health-expo.co.uk/), the IOSH Annual Conference (https://www.iosh.co.uk/Key-IOSH-events/IOSH-2018. aspx), and the British Safety Council's International Safety Awards (https://www.britsafe.org/awards-and-events/isa2018/).

[3] Such as Health and Safety at Work (https://www.healthandsafetyatwork.com/), Safety and Health Practitioner (https://www.shponline.co.uk/), and Safety Management (https://www.britsafe.org/publications/safety-management-magazine/safety-management-magazine/).

© The Author(s) 2019   183
P. Almond and M. Esbester, *Health and Safety in Contemporary Britain*, https://doi.org/10.1007/978-3-030-03970-7_6

intersects with the day-to-day lives of many different stakeholders, who gather some form of direct experience of it as a result. The legitimacy of health and safety is thus heavily influenced by the way it is seen to operate in practice, and the outcomes it produces. These 'functional' elements of 'health and safety', relating to issues of implementation, application, and enforcement, fundamentally shape the perceived effectiveness and value of the undertaking as a whole, as well as evaluations of the people and bodies who do it, and whether they are seen as 'doing the right thing' for the 'right reasons', or not.

Value judgements relating to functional matters, such as efficiency, expertise, and effectiveness, raise questions of *input* legitimacy, in that they provide a basis on which to conclude that a norm or institution has a valid reason for existing (Scharpf 1999) which provides the grounds on which regulators are trusted and given the 'freedom to manage' their functions (Baldwin 1995: 45; Hood 1991: 6; Löfstedt 2005). They are also 'consequential' considerations, which base assessments of worth on the quality of the outcomes that result (Suchman 1995: 580; also Meyer and Rowan 1991). The efficiency and effectiveness of a regulatory 'system' in achieving its goals without waste or unnecessary cost (Hood 1991: 11) underpin assessments of whether it is *deserving* of acceptance. This is a form of 'throughput' legitimacy; it is not only *why* action is undertaken, and *on whose authority*, that matters to the question of whether it is legitimate, but also *how* this exercise of power is conducted. This chapter explores the influence of four specific functional features of health and safety (relating to what it does, and how well it does it) on the perceived legitimacy of that concept.

Firstly, across our period, there has been increasing emphasis on health and safety 'beyond the workplace', in public spaces and areas of life not governed by the employment contract. This has exposed the idea of health and safety to new challenges, some of which may not appear particularly serious or obvious to lay observers. It has also exposed a range of new stakeholders, in settings where the need for action was not historically well established, to the *idea* of health and safety, bringing functional debates about its proportionality and suitability to the fore. Secondly, a 'safety profession' with a distinct identity has emerged in the last fifty years. A process of modernising and consolidating the provision of health and safety has gone hand in hand with the shifting of frontline responsibility for that provision into the hands of a highly influential interest group. This has 'decentred' much of the practice of health and safety from both duty-holders and the public bodies that enforce them, and has raised important questions about the balance of influence between the traditional categories of government, labour, and business. Third, and in conjunction with this process of professionalisation, the nature and availability of expertise in health and safety has also developed since 1960, bringing considerable legitimacy benefits, but also posing new challenges associated with avoiding a shift towards 'technocracy', or 'rule by experts', which may undermine perceptions of health and safety in the long run (Radaelli 1999). Finally, since the 1980s, there has been a movement towards a new model of regulation by the public bodies responsible for health and safety, which has seen less emphasis placed on contact via inspection and enforcement, and which has reinforced the legitimacy impacts, both good and bad, of the process of decentring identified elsewhere in the chapter.

Together, these themes demonstrate the degree to which the functional legitimacy of health and safety has developed through, and been challenged by, an overarching movement of responsibility away from the traditional 'core' of health and safety (government, unions, and industrial employers) and towards a new periphery of non-industrial stakeholders (something explored in Chapter 5). If expertise, efficiency, and effectiveness are seen as public goods which can be delivered better by decentred actors of this sort (a view which mirrors broader government policy over the last 30 years: Hood 1991), then it follows that the ability to legitimate and secure support for health and safety via an appeal to those values will also have been changed by those new delivery arrangements. We have seen elsewhere (Chapter 4) that the new legal framework for health and safety made the constitutional legitimacy of core participants contingent upon the actions of peripheral ones. Here, the motives and accountability attaching to action taken at this periphery by new classes of actor created legitimation needs and reputational risks for 'health and safety' as a whole. At the same time, however, this decentred delivery also provided a means of responding to these challenges, by invoking many of the norms of 'good governance' that underpin arguments about functional legitimacy.

HEALTH AND SAFETY 'BEYOND THE FACTORY FENCE'

One of the key features of the new legal framework outlined in Chapter 4 was that, along with broadening the duties that the law imposed, the HSWA 1974 also brought a wide range of new contexts within the scope of the law. Section 3 of the Act imposed a duty on employers to ensure the health, safety and welfare of people other than their employees (sub-contractors, customers, service users, and members of the public) who may be affected by their business activities. Because 'the effects' of business activity are more diffuse than employment itself, and because the category of 'other persons' was not delimited, this section had the effect of bringing non-workplace settings within the scope of health and safety regulation, for the first time in some cases. So, for instance, the owners of a factory owed an enforceable duty of care not only to the employees who worked there, but also under s.3 to any visitors to that factory, whether they were working there (e.g. a sub-contracted electrician) or not (e.g. customers), as well as to anyone else on whom their business might have had an impact, such as passers-by or neighbours who might be affected by a fire or chemical leak from the factory. Over time, this expanded scope has become taken for granted, at least within the formal sphere of health and safety itself, so that even potentially sceptical Government reviews have accepted it, even when expressing doubt about the inclusion of 'low hazard' workplaces (Young 2010) and self-employed duty-holders (Löfstedt 2011). Politically and socially, however, it has been much more controversial, with much negative media coverage of health and safety since approximately 2000 focusing on the application of the law to areas of civil and social life that have seemed relatively distant from traditional workplace settings (Almond 2009).

Extending the Focus—HSWA Section 3

The changes that s.3 introduced were foreshadowed by previous legislative reforms, not least the Offices, Shops and Railway Premises Act 1963. This Act was intended to establish basic standards for general welfare and working conditions (such as temperature, ventilation, cleanliness, overcrowding, and sanitation) across the working population in response to a perceived need for general health improvements in the working population (Samuels 1963). The previous Factories Acts, such as the 1961 Act, were typically much narrower in their focus, and applied only to specified workplaces, including building sites (s. 127), docks and ships (ss. 125–126), and factories where people engaged in a number of precisely specified manual labour-based processes (s. 175). The 1963 Act shifted towards universalism, bringing the vast majority of workplaces within the scope of systematic legal oversight and extending protection to 8 million workers, though it left a further 5 million unprotected at that time (Robens 1972: para. 35). The Act also gave a more developed role to Local Authorities (LAs) as enforcers of workplace safety conditions (discussed in Chapter 4; Dawson et al. 1988: 214), and exposed a whole new class of persons to the issue of health and safety, and this raised new questions about that new context, as one former Civil Servant (and later Health and Safety Executive [HSE] Director-General) recalled:

> [The Act] brought a whole lot more workplaces in that hadn't been covered before, so some of the early debates about who should be regulated, how heavily should they be regulated, etc., were coming up in a different context, not the sort of heavy industry [...] that it had been in the past. (Jenny Bacon Interview, para. 6)

The inclusion of new workplaces within the scope of the HSWA 1974 was significant, in that it applied the same duties to all workplaces, resolving these tensions and bringing the remainder of the workforce within the scope of the law (establishing a norm of 'universalism' that endured until the 2010s: Almond 2015: 206). More was said about the challenge of the move from traditional industry to 'new entrant' service sector workplaces in Chapter 5; suffice it to say that concerns about new workplaces and new hazards remained pertinent at the time of the Act's passing due to the endurance of old attitudes towards workplace risk, as one veteran safety officer recalled:

> [Health and safety was seen] very often not very seriously, [...] there was a different attitude to risk [...] you accepted a higher level of risk than you would today, by a long chalk. It was your norm. We knew we had to step outside that. (Stan Barnes Interview, para. 13)

The other significant impact of Section 3 was the extension of health and safety into the public sphere, and to the general public. On the one hand, this was a response to the legacies of a number of previous incidents which illustrated the impact that the perils of workplace activities could have on the public. One such incident was the collapse of a mobile crane onto a passing coach at Brent Cross in London in 1964, killing seven and injuring 32 (Fay 1968; Sirrs 2016). The resulting public enquiry went beyond the incident's immediate causes and made recommendations on civil engineering safety more generally, including the suggestion that the public should come under the scope of legislative protection (Ministry of Labour 1966: 30–31).

Another incident was (at the time) the worst public safety incident of the post-war era: the 1966 collapse of a coal spoil tip in Aberfan, South Wales, which engulfed a primary school and killed 144 people, mostly children (McLean and Johnes 2000). Because this incident did not kill any mine employees (although it did cause the deaths of several school employees), it was not formally 'reportable' as a workplace incident under the Mines and Quarries Act 1954. This incident highlighted the inflexibility and reactiveness of the existing regulatory regime, and the increasingly blurred line between workplace 'health and safety', and 'public safety', in the era of the 'industrial disaster' (Sirrs 2016: 265). Still, Aberfan remained in the second category in the eyes of many observers, and of the policy process that followed the disaster (McLean and Johnes 2000; Pantti and Wahl-Jorgensen 2011). As one Health and Safety Commissioner observed: "*Aberfan didn't strike one as being a health and safety issue, it struck you as being a public safety issue*" (Rex Symons Interview, para. 10). The difficulty of responding to public and industrial disasters through the lens of purely *industrial* safety legislation remained an active concern right up until the passing of the HSWA; the Flixborough chemical plant explosion of 1974, which occurred during the legislative process (the Act received royal assent 9 weeks after the incident), killed 28 workers and also had very visible spill-over effects on local populations. The widespread damage it caused to local property, and the potential that had existed for much greater loss of life, prompted significant media coverage and public concern, and led to it being dubbed the "*holocaust at Flixborough*" (*Guardian* 1974: 10). The subsequent Court of Inquiry report framed the issue as a one-off failure of specific components and process elements, and not a sign of wider systemic problems within industry (Department of Employment 1975); it sought to provide reassurance that this was a "*coincidence of a number of unlikely errors*" and not an "*unacceptable risk*", and that it was thus a controllable failure which was "*very unlikely ever to be repeated*" (1975: para. 226).

But a rather different conclusion drawn from Flixborough was that it showed how far society had entered a new era of 'inevitable', latent risk and danger. In a 1977 Interview, Cyril Bell, the then Plant Safety Manager for Flixborough, reflected that: "*we [...] must accept the view that chances are, that somewhere, sometime there will be a major industrial incident. [...] such a thing is inevitable in the sort of times we are living with our needs for a technological society".*[4] These concerns about the potential risks posed by large-scale industrial hazards echoed the Robens Report's *"[a]pprehension at this consequence of modern technology"* (1972: para. 296), particularly in relation to the explosive and flammable material processing and storage industries. There was a sense that public safety was more systematically being threatened by the complex and potentially unknowable dangers inherent in more sophisticated industrial and technological processes, such as nuclear power, chemical processing, and so on.

These concerns perhaps reflect the core ideas of Beck's 'risk society' (1992), which proposes that 'late-modern' society has, in the last sixty years, been reorganised around the control of man-made and quantifiable risks. According to this view, this shift has brought a heightened public risk consciousness in the face of increasing social complexity, a greater tendency towards precaution and an increased focus on risk at all levels, resulting in a heightening of both technical precision and procedural bureaucracy within regulatory systems (Black 2005, 2008; Braithwaite 2008; Levi-Faur 2013; Scott 2004). On some level, of course, this concern was not new, as pollution from the factories of the industrial revolution had, for hundreds of years, negatively impacted on the health of those living in nearby communities, and the myriad risks inherent in human life before the 1945 were no less meaningful or serious for the people they affected because of their 'non-modernity' (Boudia and Jas 2007; Dingwall 1999; Fressoz 2007; Otter 2016). What was arguably new was the explicitness with which the trade-offs between community safety and industrial and economic activity was recognised. From this point on, regulators like HSE would have to be clear about the ways in which they sought to *"preserve the sometimes fragile balance between the interests of economic activity on the one hand and the public welfare on the other"* (Hawkins 1984: 9).

This concern added impetus to what would become s.3 of the HSWA, in that the duty it imposed to ensure the safety of those beyond the workplace was envisaged as a means of bringing these risks within the scope of the new regulatory regime. Some five years after the HSWA came into force, it was observed that it reflected *"a new line and this is [...] basically the acceptance by the establishment of 'major*

---

[4]Interviewed for BBC TV programme 'Red Alert', aired on 5 July 1977. BBC Written Archives, Caversham.

hazards' (Flixborough type) as a threat to the whole population, themselves included".[5] Section 3 was broad enough in scope to cover risks posed to 'any person' not employed by a company—customers, passengers, patients, and service-users; agents, sub-contractors and other non-traditional employees; and members of the general public who came within the scope of an organisation's operations (such as passers-by, road users, or residents). The duty here was similar in breadth to that owed to employees under s.2; as such, it exposed organisations to a very wide range of extended obligations, including not just those around major hazards and disasters, but also relatively mundane risks such as those attaching to the use of car parks. Partly, this focus on public risks was a product of the new Act's focus on a process-oriented and system-focused approach to health and safety, which rendered obsolete the differentiation of attention according to outcome and affected party. It also reflected the emergence of a new range of civil society interest groups and actors, beyond the scope of the formal tripartism that defined the structure and work of the Health and Safety Commission (HSC) and the HSE (Almond and Esbester 2016). The old assumption that 'the public' was inseparable from 'the working population', and hence represented within policy-making by trade unions, was gradually displaced by a recognition that these were two separate constituencies, which placed different legitimacy demands and needs upon regulators (Almond 2007). As the Chair of the HSC noted in 1982: "*I often quarrel with the idea that, because a person contributes to a trade union, he automatically disqualifies himself from being a member of the public*".[6]

What followed from the Act was a "*growing struggle*"[7] to tackle public safety hazards in all their forms, from asbestos exposure, to violence at work, to the management of leisure facilities and public spaces (from the late 1980s), as well as major transport disasters (like the sinking of the *Herald of Free Enterprise* in 1987, and rail crashes like that at Clapham Junction, 1988) and other hazards that went beyond the 'factory fence'. This was prompted in part by the European Seveso directives, which set out to impose standards of governance over major hazard industries

[5] B. England, "Discussion", *Work Hazards* 21 (c. 1979), p. 18. Samuel Barr Collection, Glasgow Caledonian University [GCU] DC 140/2/1/2.

[6] *The Working of the HSC and E* (1981/82 HC-400iii), Minutes of Evidence 23 June 1982, p. 25.

[7] R. Williams, 'A Key Issue', *Work Hazards* 21 (c. 1979), p. 3. Samuel Barr Collection, GCU DC 140/2/1/2.

(Baldwin 1996), and the spectre of disaster served to drive 'workplace' health and safety regulation into new areas. So, for example, when in January 1985 an explosion (thought to have been caused by the faulty installation of gas appliances) at the Newnham House flats in Putney, London, killed eight people, the HSE was called into investigate, the first time the body had dealt with the problem of domestic gas safety (Cook et al. 1985: 1; Times 1985: 1). The then-Chief Inspector of Factories, David Eves recalled that this:

> elevated gas safety as an issue [...] everybody became aware that about 30 people a year were being gassed in their homes [...] so we became involved with domestic gas issues. Now I think I'm right in saying this was the first time [...] that we had powers, or even the willingness, to enter domestic premises. (David Eves Interview, para. 57)

As the *Guardian* observed in 1988, the ongoing *"series of disasters [of which Putney was part...] put worker and public safety high on the agenda and the role and focus of the HSE is being closely examined"* (Brown 1988: 1). The expectations incumbent upon public regulators were expanding to incorporate a broader range of social, public, and domestic responsibilities, and the associated legitimation challenges posed by the work of ensuring public safety were becoming accordingly more difficult to fulfil. As the Chair of the HSC observed in 1988, *"technological change is now probably swifter [...] the public has become increasingly conscious of and knowledgeable about its implications. The reassurance provided by a fully effective and respected state regulatory body is more and more important"* (Employment Committee 1988: 2).

Risk, Tolerance, and the Public

One means by which public regulators, such as the HSE, sought to address this kind of legitimation challenge was via the development of decision-making models that explicitly set out the ways in which public safety considerations, and public risk perceptions in particular, ought to be weighted and incorporated into regulatory decision-making processes. The Sizewell B nuclear power station planning inquiry of 1982–1987, considering the siting of a new nuclear facility, prompted the HSE to explicitly measure and assess the *"tolerable levels of individual and societal risk to workers and the public"* in relation to nuclear power (Layfield 1987). Within this Inquiry, it was recommended that public concerns ought to be factored into the otherwise quite technocratic probability-based assessments that regulators were used to relying upon within decision-making (Pidgeon 1998: 7). The 'Tolerability of Risk Framework' (HSE 1988) was developed to provide a means for weighing assessments of risk and the costs of prevention alongside a sense of what was deemed socially appropriate (McQuaid 2007; Pidgeon 1998). It explicitly included justice-based societal values about the acceptability of hazards as part of its decision-making matrix, to be balanced against scientific calculations of probabilities and costs of prevention (Bandle 2007; McQuaid 2007). One senior scientific advisor working for the HSE at the time described this as a "political-scientific" threshold:

> There is a scientific process of assessing how big a risk is, but in the end it's a political judgement, you draw the line between what level of risk is tolerable and what is intolerable. That HSE document was a fascinating exercise, we all ended up sitting round late at night as John [Rimington] was composing away, producing these guidelines...you're not saying risk is acceptable, you're saying you'll tolerate some small areas of risk. (Tim Carter Interview, para. 132)

The quantified risks were thus deemed either 'minimal' (could be taken without precaution), 'tolerable' (could be taken subject to reasonable practicability and cost–benefit calculation), or 'intolerable' (not to be taken under any circumstances), depending on the underlying level of risk (how serious, how likely), what was possible in terms of risk control (feasibility, costs and benefits), whom that risk affected and benefited, and the broader societal acceptability of those risks (from 'broadly acceptable', through 'tolerable', to 'unacceptable') (Bandle 2007; Pidgeon 1998).

The Framework allowed for principled assessments of risk control to take place in high-hazard areas like nuclear power, allowing for more objective framing of contested policy issues and the trading-off of cost and benefit in a way which could be justified to sceptical public, social, and scientific audiences (Hutter 2005: 6). Having allowed HSE to respond to high-hazard settings like the nuclear industry, the 'Tolerability of Risk Framework' was revised in 1992 so as to extend and apply the approach developed in the nuclear industry to lower hazard sectors (Bandle 2007). The Framework's synthesis of 'technical' and 'socio-political' approaches to risk would prove highly influential on future risk theorists and policy actors:

In terms of theory, I was always fascinated by the Tolerability of Risk Framework and how HSE really put a strong evidence based focus on that; I think that's an excellent document. (Ragnar Löfstedt Interview, para. 5)

This influence, and the framework itself, reflected the growing importance (in terms of democratic legitimacy) of public attitudes towards health and safety, as well as an increasing recognition of the duties towards the public that existed under HSWA Section 3. It also reflected a new way of thinking about risk in general, as one nuclear industry executive recalled:

Protection of [workers and...] the public, the culture of that hasn't changed. What has changed is the quantification of it. I would describe the attitude towards safety cases in the '70s as being that of a good, scientific, enquiring mind...you would say 'what is the protective standard I'm trying to achieve [...?]' and you'd put a number to it. The change comes halfway through my career. Up until the Sizewell B Inquiry, things were almost non-numerical...By the mid-80s...the methods get enshrined and...you start to get numerical limits on those safety calculations. (Chris Marchese Interview, paras. 13–14)

The quantification of risk was a product and reflection of the wider culture within government and regulatory circles. The 'New Public Management' era (Hood 1991) of performance targets, fiscal discipline, and managerialism placed a particular emphasis upon the measurement and assessment of risk as a basis for resource allocation and targeting, as well as the prioritisation of areas of concern depending on what outcomes matter to the regulator (Sparrow 2000; Baldwin and Black 2016). The 'Tolerability of Risk Framework' (HSE 1988) was thus representative of a movement towards a more quantitative approach towards regulation and public policy in general, and shows the difficulty involved in translating complex public risks into a more easily legitimated, numerical form (Almond and Esbester 2018).

The Culture of Fear?

By the early 1990s, 'health and safety' was increasingly understood as a formation of policy and practice which included issues of public, as well as workplace, safety. In part, this was a process being driven by the contextual changes discussed in the preceding chapters, such as the move away from manufacturing within the British economy, the open-ended duties involved, the growing role of LAs, and a movement towards seeking redress following public disasters, via, for example, corporate manslaughter prosecutions (Almond 2013; Wells 2001). The change of focus at this time was captured in this exchange with the Director-General of the HSE, as recalled by a former Trades Union Congress (TUC) health and safety policy-maker:

> I said, 'you need to look again at worker safety' [...] he put his hand up and said, 'no, no, no, stop my boy [...] that's worker safety. That's a dead volcano', he said. 'The live volcano is public safety. That is what's going to energise everyone'. (Roger Bibbings Interview, para. 19)

The movement towards a focus on public safety was driven and facilitated largely by the scope of HSWA Section 3, which has subsequently been used in a wide range of settings, from the deaths of passengers on the rail network, to the shooting of Jean-Charles de Menezes people during an anti-terror police operation, to deaths and injuries sustained during the Hillsborough football stadium disaster. Much of this seemed to go beyond what had originally been envisaged as the proper scope of the HSWA; as one former Director-General of HSE (who was also involved as a civil servant during the passing of the 1974 Act) observed:

> Robens did envisage that there would be some limited duties towards the public. But I think it's fair to say that it got extended. Not just in terms of the major industrial hazards and things affecting the public like [Flixborough...] when the Act was going through, I don't think that the very extensive ways in which public health and safety have come into play was envisaged. (Jenny Bacon Interview, para. 23)

The fundamental definitional problem being grappled with at this point in time was that the duties to ensure safety within the HSWA are applicable to any affected person within almost any workplace; this did not leave much scope for defining any areas of human life or activity as outside the scope of the legislation. Every leisure facility or public space constituted a workplace, as lifeguards, gardeners, security guards, and wardens undertook their working activities there (Helen Leiser Interview, para. 18; Lawrence Waterman Interview, para. 59); the same was true of public transport, military establishments, and other spaces which did not obviously constitute 'workplaces' in the traditional sense. The scale of this challenge has only been heightened in recent years by trends such as the increasing privatisation of public space, which increases the level of disciplinary and contractual control exercised over such spaces by a wide array of organisational actors (Németh and Schmidt 2011; de Magalhães 2010), and the increasing demands placed upon public actors, and LAs in particular, to act as 'risk managers' for their very diverse portfolio of risk liability activities, from the provision of health and social care services to the maintenance of public highways (Halliday et al. 2011).

Two particular legitimacy challenges that emerge from this extension of coverage can be identified. First, it created new areas and sectors where the management of health and safety was required by law, in many of which those responsible for enacting health and safety were less experienced or skilled at this kind of work. As a result, these new demands were often either viewed as excessive, or prompted sub-optimal responses from the 'new-entrant' organisational actors subject to them, as a former HSE policy official recalled:

> 'white-collar'-type employers had new duties of care towards the safety of members of the public who might be affected while visiting shops, offices, hospitals and so on. The hazards and risks for this newly-protected category were generally much less self-evident to those running these types of organisation […] but capable of [prompting…] disproportionate public and media reaction. (Helen Leiser Interview, paras. 67–9)

This seems to have resulted in some incidences of the application of 'health and safety' in ways that were either disproportionate, partial, or which lack an immediate public justification. Public attitudes research (see Chapter 2) has demonstrated that LAs, the principal regulators of low-risk workplaces, have struggled to combat perceptions of regulatory 'spill-over' and suffered a significantly damaged public trust profile as a result (Halliday et al. 2011; Walls et al. 2004; Young 2010).

The second legitimacy challenge associated with this extension into the public sphere is more conceptual, relating to a sense that there are areas of social life in which regulators, and the state, have no business interfering. The idea of 'over-reaching' by regulators is not necessarily a new concern (Wilson 1980; *Lifting the Burden* 1985), but it was not necessarily a live political concern in relation to health and safety outside the traditional workplace until the mid-1990s, when Select Committee time was increasingly devoted to issues like domestic gas safety and safety in schools (Employment Committee 1993: paras. 15, 22). The public disasters at the end of the 1980s (such as those at King's Cross and Hillsborough) paved the way for an increased focus on safety in non-employment settings. The 1994 Lyme Bay canoeing tragedy, where four children died on a school outing, led to the introduction of a more stringent regulatory regime for outdoor activities and for school trips and events (Ball and Ball-King 2013), as a largely politically driven response to a perceived public concern about safety in these settings:

> After those canoeing deaths in Lyme Bay HSE got pushed into a lot of the regulation of leisure, adventure-type activities. And they knew that was an absolute loser and it wasn't wanted, but it got dumped on HSE [...] the driver is [civil liability...], not health and safety. (Tim Carter Interview, para. 143)

This laid the foundation for a subsequent backlash against 'public' health and safety. The HSWA was thought to have significantly exceeded the reach and scope that was envisaged by Robens and by Parliament, and to have encroached into the lives of private citizens, where individual agency and 'common sense' might be a more suitable basis for decision-making than regulatory precaution (Ball and Ball-King 2013; Furedi 1997; Young 2010).

Alongside this concern can be discerned a broader ideological opposition to the perceived encroachment of 'rational', procedural regulatory expertise into areas of moral and individual freedom, something perceived to pose an existential threat to traditional ways of life (Almond 2009). This view was reinforced during the 2000s via a significant amount of negative media coverage of 'excessive' health and safety interference in the private lives of citizens (Almond 2009, 2015; Dunlop 2014). One influential example, the idea that schoolchildren were required by law to wear goggles to play conkers, was actually a publicity stunt designed to critique the contextual changes introduced after Lyme Bay (Almond 2009). Most of these stories about 'health and safety gone mad' were either apocryphal or anecdotal in nature, but they did prove influential enough to shape the policy agenda, challenging the HSWA system's foundational underpinnings of generality, universalism, and inclusiveness (Almond 2015). Regulators came under increasing political pressure to limit their involvement in non-industrial areas, as a former HSC Chair testified:

HSE's resources were being spread too widely by a concentration on Section 3 issues [...] We were getting drawn into debates about C.Diff and MRSA [...] We tried many times [to] put some walls around s.3, but we couldn't. (Bill Callaghan Interview, paras. 105–109)

At the same time, however, there are clear arguments in favour of the extended scope that Section 3 gave to health and safety. First, the involvement of the public within debates about safety brings significant democratic legitimacy benefits (Black 2008; Dingwerth 2007), broadening the range of stakeholders and issues considered within the policy-making process. It might also allow for the development of a more 'risk-literate' society, where health and safety is better understood and more fully engaged with at all levels of society (Löfstedt 2011: 94). Second, it is clear that the notion of organisational responsibility for public safety is not new, having been embedded in civil law for many years, and that these ideas are also widely accepted within society (see Chapter 2). Effective accountability for public safety failures has been secured only via hard-fought campaigns in the aftermath of events like Aberfan, Flixborough, Hillsborough, and others, and serves a greater public purpose than simple 'blaming' (Almond and Colover 2010). And it is only via the application of Section 3 and 'public'-facing safety regulation that accountability for an event like the 2017 Grenfell Tower fire might be secured in the future. Finally, providing a 'public' element to health and safety coverage has allowed the regulatory system to adapt to a new era of '*Uberised*' quasi-employment relationships between contracting firms and individual contractors who may not otherwise meet the definition of 'employees' (Prassl and Risak 2015). Finally, it should also be recognised that the extension of 'health and safety' into public life was not purposely engineered by regulators or law-makers; rather, it emerged in response to broader social demands around the evolving frontiers of work and changing conceptions of citizenship and social obligation; regulatory intervention was one way of addressing these trends, but has brought considerable legitimacy challenges for the actors and institutions involved.

THE SAFETY PROFESSION AND THE FORMALISATION OF EXPERTISE

This section will explore the role and influence of two interrelated themes, both of which relate to the functional legitimacy criterion of 'expertise', and the need for actors and decision-makers to demonstrate their credentials and competence, in order to justify their decisions and secure public trust (Baldwin 1995: 45–46; Löfstedt 2005). Professional health and safety organisations are today one of the most influential means of transmitting and developing workplace health and safety information and expertise (Okun et al. 2017). They also represent a relatively distinctive feature of the modern health and safety system, which encompasses a wide range of different actors, organisations, and stakeholders who are involved in the delivery of health and safety. While Robens and the 1974 Act envisaged a three-legged stool (state regulators, trades unions, and employers) as the basis of safety regulation and control, the safety profession has emerged as a central player in the contemporary health and safety sphere. Functionally, this might be understood as a form of policy network, a collective of mutually dependent actors clustered around a problem area (Klijn and Koppenjan 2016), who each have *"an interest in a policy sector, the resources to effect outcomes, and a need for other resources (which it does not possess) to pursue its policy objectives"* (Rhodes 2001: 19). These networks provide the context within which health and safety is implemented, and are increasingly the key means via which individual workers, citizens, and stakeholders interact with this issue; as such, they have important implications for the popular legitimacy of health and safety as a whole, not least in providing a basis of trust and confidence in the institutions to which they apply (Dingwall and Fenn 1987).

Second, the emergence of a recognised body of formal health and safety expertise, based on technical knowledge and qualifications, provides a basis on which regulators and others can be judged as legitimate. Health and safety intervention is respected and seen as necessary or desirable when it is linked to a strong basis in expertise; evidence on public attitudes highlights the centrality of expertise to these legitimacy assessments (Baldwin 1995; Black 2008; Tyler 2011; Pidgeon et al. 2003; Walls et al. 2004). In general, as Chapter 2 showed, health and safety is an area where the expertise of participants is positively received and widely recognised; the HSE, and the safety profession, are seen as competent, and this translates into a positive perception of their legitimacy in general. But when expertise is seen as being misused, either because decisions lack the quality of 'common sense'/fail to fit with face-value, practical knowledge (Geertz 1975), or because it is used to justify 'gold-plating' (Young 2010), or because the qualifications and standards on which it is based are unreliable or untrusted, then it can also act as a barrier between public audiences and health and safety, and do more damage than good. So the core question to be addressed is not so much '*is health and safety based on expertise?*', but rather, '*how and where is that expertise used*'? Together, the issues of professionalism and expertise both speak to the claims that can be made about whether health and safety is *deserving* of acceptance or not.

'Well-Intentioned Amateurs': The Emergent Safety Profession

The first organisations to come into existence in order to promote or implement issues of health and safety, aside from the Factory Inspectorate, emerged in the early part of the twentieth century, with a mandate to develop formalised skill in safety management. The Royal Society for the Prevention of Accidents (RoSPA) was formed in 1917 as the 'British "Safety First" Association' (it changed its name in 1941), drawing on the USA-led 'Safety First' movement, as well as other international examples, (Aldrich 1997; Swuste et al. 2010). This was mirrored by the emergence of specific industry-wide safety campaigns, such as that in the rail industry (Esbester, forthcoming). Other bodies, such as the Institution of Industrial Safety Officers (IISO, 1945, later to become the Institution of Occupational Safety and Health, or IOSH, in 1953), and the British Safety Council (1957), emerged in the post-war period to develop and share knowledge and understandings of safety management in industry and beyond. All of these organisations were focused primarily on the generation of expertise among safety committee members, which were otherwise "*mainly comprised of well-intentioned amateurs [...] accident prevention problems were being dealt with in an amateur way over probably 10 per cent – 20 per cent of the potential accident field*".[8] One future IOSH President's experiences as a safety officer in the iron and steel industry in the early 1950s also underline the amateurism of this early undertaking:

---

[8] R. Barry, 'Accident Prevention in the Iron and Steel Industry', talk to the Central Metropolitan Group, London Industrial Accident Prevention Committee, 13 March 1967, noted in the minutes of the meeting, p. 3. MRC, MSS.292B/146/2/4.

There was a vacancy for an assistant safety officer in the ironworks. I applied and to my amazement, with absolutely no knowledge of industry at all, I got the job. That will give you an idea of what the understanding of occupational safety was at that time, almost nil. (Stan Barnes Interview, para. 3)

This essential informality endured until as late as the early 1970s, according to Andrew Hale, Professor of Safety Science:

when I first started, the face of safety was very much RoSPA in its guise of voluntarism, and safety awareness was a word which was bandied about that they were trying to promote, and it was very amateurish. It tended to be run by people who had a social conscience, a great personal concern, but not any in-depth knowledge. (Andrew Hale Interview, para. 51)

Early efforts at developing the capacity and knowledge base of these safety officers also invoked quasi-Victorian values, in this case, of didactic formal schooling:

We organised tuition mornings in one of the Cutler's halls, and all the firms sent apprentices in for health and safety instruction, rows and rows of apprentices in front of you. (Stan Barnes Interview, para. 67)

The Robens Committee recognised the potential of the emergent safety profession in terms of its capacity-building and training functions (1972: para. 387), advocating a greater coordination via what would become HSE, but it was also highly critical of the quality and nature of the existing safety officers, who it considered to be "*undervalued*", often not "*of the right calibre*", and very variable in terms of their capacity and status (1972: paras. 54–56). Notably, however, the role played by bodies like IOSH, RoSPA, and the British Safety Council within the Robens process was quite minimal (all of these bodies gave evidence to the Committee, but did not feature heavily in its report), and seemed to reflect a broader perception that their general role and influence were quite limited. This was a view shared by the bodies themselves; testimony provided to the Robens Committee by a RoSPA representative recognised that safety professionals had "*to earn people's respect before they will expect much from us*" but at the same time groups like RoSPA had, over the preceding few years "*introduced professionalism into the work [and] people are listening to us*".[9] Accident Prevention Groups affiliated to these bodies were seen as "*useful*" (1972: para. 91), but Robens' view of the future of industry-level activity did not mention the professional bodies at all (1972: paras. 94–96). Perhaps in response, the Chief Inspector of Factories, Bryan Harvey, in 1973 stated that "*I have long held a strong personal belief that there should be close ties between the Inspectorate and bodies outside the Government service involved with the same problems*" (Department of Employment 1974: xvii).

The Robens Committee, and the subsequent HSWA 1974, were predicated upon the notion of a tripartite structure of authority and influence, with the state (HSE), organised labour (TUC), and industry (Confederation of British Industry) as the three dominant constituencies. Safety professionals, insofar as they had a role, were positioned very much within the latter constituency, as an aid to management, rather than being seen as an independent professional entity[10]; this perception was reinforced by the lack of clarity about the constituencies and subject matter involved, as well as by the politicisation of the underlying industrial relations context. As Atherley and Hale observed, "*[u]ntil the health and safety specialist makes an unequivocal stand on who his client is, he will always be open to accusation and counteraccusation; from trade unions of being a bosses'-man, and from employers of being an unrealistic humanitarian*" (1975: 331). In reality, the safety profession occupied a space somewhere between the three constituencies of tripartism; managing risks for employers, holding a union-like commitment to worker welfare, and occupying a quasi-regulatory role within a self-regulatory system (Dawson et al. 1988: 47). This divided identity remained a barrier until the 1980s, when the contested political climate of the time open the door for a safety profession that could play a more mediating,

---

[9] Verbal evidence on behalf of RoSPA to the Robens Committee, 14 June 1971, transcript p. 5. TNA, LAB 96/75.

[10] Discussion document 'The Role and Function of the Safety Officer', RoSPA National Occupational Health and Safety Committee, 25 November 1976, point 4.6. RoSPA Archives, D.266/2/24.

less partisan role. The Safety Representatives and Safety Committees Regulations 1977 gave trade unions significant power to appoint safety representatives, who were then the focal point for employee consultation processes (Dawson et al. 1988: 19). The political climate of the 1980s would make it more difficult to achieve anything via this route, however:

> it looked as though the country was going down the road of training safety representatives, and giving them a much more active role [...] the clashes between the trade unions and employers and government poisoned that route, and meant that safety representatives, who had a lot more power and influence in the late '70s and early '80s, lost that power, and the professionals somehow picked it up. They [...] didn't have that antagonistic aspect. (Andrew Hale Interview, para. 62)

From this point on, arguably, the safety profession established itself as a preferential bargaining partner for employers, as although they possessed a broad similarity of interest with trade unions, they were also an internal constituency that was broadly responsive to business imperatives (Dawson et al. 1988: 49).

From Officer to Advisor

By the mid-1980s, a distinctive role for safety professionals had thus been carved out, embedding self-regulatory processes of auditing and assurance at a business level. The 1974 Act placed a much greater emphasis on the systematic management of health and safety, but, at least at first, this did not filter down into the approach taken in appointing safety officers, as one leading health and safety practitioner recalled:

> The standard thing was the Chief Executive [...] calling in the security officer in the autumn of 1974/spring of 1975, and saying, 'I've just read in the Sunday Telegraph that there's this Health and Safety at Work Act. Here's an application for ten bob a year, I think you ought to be our Security and Safety Officer'. So I think a lot of people got involved in safety at that time because management teams wanted safety to be addressed. (Lawrence Waterman Interview, para. 6)

As the self-regulating responsibility of business became more embedded, and the need for more sophisticated management of health and safety grew, so did the demands for qualified, 'respectable' professionals to take on this role (Richard Jones Interview, para. 40).[11] The emergence of new, technical, pervasive requirements that needed managing, like the six-pack (and particularly the written risk assessments introduced by the Management of Health and Safety at Work Regulations 1992, noted in Chapter 4), meant that safety was not a job that anyone could do any more, at least not in large organisations; the associated professional status that came with the demarcation of this specialism also established this as a legitimate arena of practice. The reduction of the capacity of public regulators to perform an all-encompassing advisory and supervisory function (noted by the Chief Inspector of Factories as early as 1978) also gave rise to a need for this 'localised' means of exerting control: "*Since inspectors cannot be expected to act as consultants to all industry and all public authorities such bodies have increasingly to establish or find their own sources of expertise*" (HSE 1980: vii).

One aspect of this shift was the development of more academic, university-based, professional training in health and safety. Section 11 of the HSWA imposed a requirement that coherent 'information services' be provided by HSE to the regulated population, something that was undertaken via innovative technological means. The HSELINE database, created in the late 1970s, actually made use of the European Space Agency's international computing platforms (a precursor to the internet), the first such body to do so (Sheila Pantry Interview, para. 50). Alongside this, the 1974 Act broadened the law's coverage, creating a need for accessible, non-specialist materials, such as HSE's 'DO AND DON'T' leaflets, written in accessible language, and the creation of telephone enquiry lines, among other things. Each of these steps decentred health and safety and shared expertise outwards, building capacity within the 'periphery' of the new system and empowering local actors to perform as risk-managers and decision-makers. In this way, the regulatory system focused on creating '*sociological citizens*', individuals capable of implementing regulatory requirements within their organisations (Haines 2011; Huising and Silbey 2011). These sociological citizens were actors who possessed the individual agency, status, and expertise required to discharge their obligations as risk managers via processes of "*relational regulation*", by developing, experimenting, collaborating, and adapting in a practical way, so as to "*govern the gap between regulatory*

---

[11]On the need for training and qualifications, see verbal evidence on behalf of RoSPA to the Robens Committee, 14 June 1971, transcript p. 11. TNA, LAB 96/75.

expectations and performances with an appreciation of the ongoing production of organizational and material life through human transactions" (Huising and Silbey 2011: 17).

This role was a necessary one within a decentred, self-regulatory system, and was presupposed on two key features: a requisite level of knowledge and expertise to allow them to demonstrate "*high levels of reflexivity and goal-directed action*" (Haines 2011: 139); and a capacity to communicate and build relationships effectively with others. Safety professionals, as sociological citizens, were thus expected to operate at a level of abstraction, working collaboratively, developing innovative strategies, and solving complex regulatory problems (Pires 2011). This is a more complex task than simply overseeing mechanical processes, and so the nature of the role has changed:

> the move from officer to advisor has been almost ubiquitous. So the idea that managers manage, but advisors advise, I think is very common. And in all but the smallest of workplaces is the way that it is now [...] you might still get security officers but you don't usually get safety officers. (Lawrence Waterman Interview, para. 64)

Similarly, the discussions in which safety professionals are now expected to participate are more strategic and managerial, requiring them to show leadership on the issue:

> One of the roles of the health and safety professional is to work with businesses to try and keep the foot on the pedal [...] the job of a health and safety professional is to make sure that you're scrutinising health and safety appropriately. (Graeme Collinson Interview, para. 62)

Outside of the parameters of this profession, however, there are now concerns that expertise about health and safety is scarce. The further away from the core of the system one goes, the more a lack of expertise seem to become an issue, until you reach the general public, who are often characterised as lacking in basic risk competence (House of Lords 2000; Löfstedt 2011):

> there is a crying need to create a basic ability in people to be able to [understand risk...] the failure to educate key people in and around the concept of risk has been one of the weaknesses which has led to the

present cultural impasse between health and safety on one side, and politicians on the other. (Roger Bibbings Interview, para. 43)

The risk associated with professionalisation is that it renders the issue involved the sole concern of those within that profession, and not something that anyone else has to engage with; professions can sometimes exacerbate this by rigidly policing the boundaries of the professional sphere so as to protect their status (Dingwall and Fenn 1987). An increasingly common result of this has been a tendency to 'outsource' health and safety to external professionals, undermining the participatory aspirations of the 1974 Act, as this insurance industry actor observed:

The health and safety manager in the 80s, it was his problem. In the 90s it was his problem. Probably in 2000 it was his problem. Health and safety consultancy has evolved significantly and this concept of packaging up health and safety services, I could see why people might say 'well, we're paying someone to do that, it's not our problem'. (Huw Andrews Interview, para. 123)

Building an Expert Regulatory Agency

The other area where expertise and professionalism had to be developed was within the state regulator itself. The Robens Committee recognised the high levels of technical expertise and knowledge required of Factory Inspectors (1972: paras. 226–230), and advocated a greater degree of specialism within what would become the HSE. Expertise at this time was framed mostly in terms of formal academic and technical qualifications, and the provision and validation of this expertise were a problem that the regulatory leadership sought to resolve via partnership with external bodies. Aston University's Department of Safety and Hygiene, where new training courses were being created, played a major part in this (Robens 1972, para. 230), as one Department member recalled:

in 1973/4, we set up the first Masters course in safety and hygiene. And our main customer was the Factory Inspectorate. So right from the start of setting up that group in Aston, we were in contact with the regulator, Bryan Harvey, Chief Inspector of Factories [...who] was unhappy at the time with the internal training of factory inspectors, and wanted to open it up and go outside the inspectorate. (Andrew Hale Interview, para. 10)

Technical training was a natural response to the safety science that under-pinned the risk-based Robens approach to health and safety. The profes-sionalisation of the HSE's expertise was key to the legitimation of the new regulator in the immediate post-1974 period, particularly in the eyes of new entrant sectors and those in specialised industries. Even then, the status of this expertise was challenged, as in an article in a Communist-aligned publication in 1979, which questioned the neutrality of scientific advice and suggested that "*[e]xpert opinion (through committees) plays an increasingly important role*" in shaping political outcomes.[12]

By the 1980s, the HSE had established itself as a technically adept regulator. It had a strong research base, and was able to contribute this expertise effectively to the process of developing regulations. The HSE was seen as competent in what it did, but perhaps as less politi-cally legitimate as a result; appearing technocratic and lacking a certain political awareness did not help in building good relations with other Departments and with central Government:

> [The HSE] is one of the most highly technically qualified organisations in government. Many hundreds of people with second degrees, technical degrees, doctorates and so on, very highly regarded in their own technical fields. The Civil Service is often not at its best in dealing with something that is as technical as all that. (John Rimington Interview, para. 33)

At the same time, however, when it came to concrete issues, expertise could also be a useful weapon to deploy strategically, as a former TUC policy officer recalled:

> [Rimington's] strategy was that whenever there was a crisis in Government about an issue and a high level meeting of officials was called [...] he always sent an expert, someone technically knowledgeable [...with order to] sit there, listen and when they all start giving opinions about what should be done, step in and say, 'well you need to take account of X and Y and so on' [...] you then get control of the subject. (Roger Bibbings Interview, para. 19)

As well as the governmental audience, expertise and legitimacy must also be demonstrated to public and stakeholder audiences as well (Almond

---

[12] R. Williams, 'A Key Issue', *Work Hazards* 21 (c.1979), p. 4. Samuel Barr Collection, GCU DC 140/2/1/2.

2007). Displaying expertise shows that the decision-maker in question has a sound and valid basis for the other choices and actions it undertakes, having a cross-over legitimating effect on other areas of their work (Baldwin 1995: 45). So, for example, offshore inspectors of the 1980s and 1990s had good working relationships with offshore workers and managers because they *"knew exactly what our job was, they knew that we were professional people [...] with engineering backgrounds, we understood what they were trying to do"*.[13] Expertise has an enduring, robust, and pervasive effect on stakeholder perceptions; it establishes credentials for intervening, as one nuclear and rail industry actor pointed out:

> If your site inspector is somebody that you believe is credible and brings experience and pragmatism to the piece then you'll probably get on quite well. If the site inspector is known for wanting to make a point [not based in proper expert judgement...] it might make things much more difficult. (Paul Thomas Interview, para. 70)

It is clear that the expertise of the HSE and of regulators in the field of health and safety is now largely accepted, particularly in areas of high-hazard and technical challenge, such as offshore (Rex Symons Interview, para. 64) and the nuclear industry (Helen Leiser Interview, para. 63). For the most part, that expertise is recognised and accepted as a taken-for-granted feature of these bodies by duty-holders and other observers. Issues can arise when dealing with areas of new risk, emergent risk, or new regulatory contexts, however, where regulators can seem to lag behind the industry they regulate. For example, when the HSE and the Railway Inspectorate were getting to grips with the regulation of a newly privatised railway system in the 1990s, with a new approach to safety, observers (such as this British Rail safety manager) felt that the regulator's expertise had been eclipsed by the rest of the industry:

> With privatisation, we went to the safety case process [...after] King's Cross and Clapham. We learnt about culture, we learnt about changing the reactive to proactive and we found that we were ahead of any other railway [...] we were also ahead of the Railway Inspectorate. (David Maidment Interview, paras. 119–121)

---

[13] Martin Thompson, University of Aberdeen 'Lives in the Oil Industry' project, tape 2 side b. British Library, London, C963/13.

Similar dynamics were observed by interviewees in relation to the nuclear and biotechnology industries. But gaps in expertise also emerged in less cutting-edge settings too. In particular, some of the 'new entrant' workplaces, where health and safety is being adapted to fit a new world of clerical, non-manual work, also encountered regulators who seemed to be a step behind, as a former head of health and safety for financial and legal institutions recalled:

> I've always been surprised at some of the competence levels that I've been faced with, they've never been quite where I'd hope them to be [...] I would expect everybody that faces me to be an expert [...] very mixed experiences with the regulator, particularly with the banks. (Peter Kinselley Interview, para. 32)

The effective sharing of expertise remains a challenge to this day and there is still more work to be done, both in terms of making the health and safety system work better, but also in terms of improving the perceived legitimacy of this area.

The Pros and Cons of Expertise

What overall impact has the emergence of the new safety profession, and the formalisation of relevant health and safety expertise, had upon the perceived legitimacy of this issue? The former Head of health and safety at the Confederation of British Industries shared many of the concerns that were voiced in the Young Review (2010) and elsewhere in recent times:

> Frankly, it's a consultant's paradise because it's got such low barriers to entry [...] It's a very brave consultant who would ever say you're managing that OK. You can always find something that has to be done [...] it's frustrating to professionals but has also allowed the professional bubble to burst into society and be viewed [...as] making things too bureaucratic. (Janet Asherson Interview, para. 52)

On the one hand, the professional bodies, like IOSH, RoSPA, and the British Safety Council, have established themselves as bodies that not only accredit and represent those who work in health and safety as 'risk handlers', but also as representatives of a broader agenda around health and safety. They are large, visible, and their functional role as participatory fora means they are able to speak with authority in policy-making discussions, as the then-policy lead for one such body observed:

it has been made clear to us by HSE officials, the reason why HSE talks to us [...] is because of our membership base, because we represent businesses who are vital to the UK economy [...] if we didn't have that membership base, I think that our ability to be a voice influencing health and safety would be negligible. (Neal Stone Interview, para. 64)

On the other, however, the safety professions have been perceived as a central component in the bureaucratisation of health and safety, at least in the eyes of the public, and of key actors, including the HSE, whose Acting Chief Executive felt that:

the reputational stuff about health and safety lies at the feet of the health and safety profession rather than the regulator [...] the growth in negative health and safety stories and the growth in health and safety professionals follow a similar trajectory! (Kevin Myers Interview, para. 15)

Safety professionals are still seen as poorly regulated (Young 2010) and sometimes a barrier to good practice, but their positive impact in raising overall standards, and in filling the gaps left as union influence declines, is also acknowledged. Many interviewees saw a greater need for leadership by professional bodies in securing the social benefits of health and safety, via robust accountability and governance:

IOSH is weak at trying to manage out poor health and safety advice. It acts more like the trades union for professional members than talking about professional standards. And as a chartered body, and a registered charity, it needs to strike a better balance. (Lawrence Waterman Interview, para. 59)

Expertise is thus a powerful tool in building a positive public profile for health and safety in general. As discussed in Chapter 2, the public tend to view the 'safety guys' as boring, jobsworths, or meddlers (Pidgeon et al. 2003), and as early as 1971 RoSPA observed that there was a public image problem around health and safety, an issue seen as '*stuffy*'.[14] A second key weakness in terms of public perceptions here is that the professionalisation of health and safety starts to bring into question the

---

[14]RoSPA National Publicity Committee minutes, 22 September 1971. RoSPA Archives, D.266/2/4/3, p. 5.

motives and altruism with which they are seen to act, leading to a suspicion of self-interest. Roles of this sort tend to expand in scope over time, and the drivers of this growth are often associated with the self-interest of those who perform the roles (van der Heijden 2017). Finally, there is often a downside to expertise, in that it leads those who have it to struggle in communicating with those who do not; the challenge here is well summarised by a NHS employer representative:

> health and safety people [...] need to be able to speak the language of the organisation, not just health and safety language [...] because they can be quite niche, and can struggle to explain and persuade others of the relevance, and consequently are sidelined. (Ruth Warden Interview, para. 58)

In terms of stakeholder perceptions, a further negative association of expertise is that it can lead to bureaucratisation and an undue formalism, as real-world issues are 'juridified' into matters of procedure, rules, and paperwork (Teubner 1987). For example, in the early 1980s, an investigation into British Rail's implementation of the HSWA noted that some managers were *"dismissive about the [local OHS] statements regarding them as a paperwork exercise which add little to BR's existing safety effort"*.[15] Similarly, a 1977 British Rail memo noted enthusiasm for health and safety generated by the HSWA, but followed it with a warning: *"It would be a tragedy if, instead, implementation of the Act became a battlefield of disagreement on [...] unreasonable safety standards"*.[16] Many interviewees reflected disapprovingly upon the emergence of a tendency towards the use of 'off-the-shelf', manualised, or standardised documentation as a means of responding to regulatory requests (Andrew Hale Interview, para. 65; Helen Leiser Interview, para. 69; Kevin Myers Interview, para. 48; John Rimington Interview, para. 82; Dan Shears Interview, para. 25; Jim Wilkes Interview, para. 26). Such approaches constitute 'performative' compliance measures (Power 1997, 2007), or 'Potemkin villages' (Gray 2006: 885) designed to give the appearance of 'skin-deep' legal compliance which does not actually reflect any internalisation of safety norms, to the longer term detriment of the regulatory regime (see also Chapter 4).

---

[15] BRB and HSE, 'An outline report on OSH by the Accident Prevention Advisory Unit', p. 1. TNA, AN 16/157.

[16] 'Fatal Accidents to B.R.B. Employees', 16 August 1977, p. 10. TNA, AN 156/936.

It appears that the pooling and valorisation of expertise have the potential to create systematic, instrumentalised approaches to safety on the part of regulated populations, based in technical mindsets that are hard for lay people to understand (Radaelli 1999), and which create a distance between regulators and the regulated. This way of seeing is also identified at a regulatory level; from the 1980s, the trends towards targets, probabilities, and quantification (Black 2000, 2008), while necessary from the point of view of efficiency and effectiveness, have had costs in terms of the accessibility of the systems that they establish, and thus on degrees of social acceptance. It may well be that while changes in the functional reach of health and safety beyond the workplace have exposed an ever-wider range of public and social actors to regulatory requirements and processes, this has tended to be in a relatively passive or background way, as *objects* of regulation; at the same time, however, the agency, knowledge, and capacity required in order to participate as active *subjects* within the health and safety system have arguably not been shared downwards in the same way. While a broader range of actors now constitute part of the regulatory system, the capacity to actually act as 'sociological citizens' is increasingly restricted so as to include many workers and private citizens at the 'frontline' of health and safety (Almond and Gray 2017). So while expertise and professionalisation may give rise to functional legitimacy benefits, they may also exacerbate democratic legitimacy deficits (see Chapter 5) by erecting barriers to more widespread participation. There is thus a need to institute methods and mechanisms capable of eliciting public involvement in health and safety, if health and safety is to be properly embedded and accepted as a social responsibility (Löfstedt 2011: 93–94).

ENFORCEMENT, INSPECTION, AND THE REGULATORY ROLE

The ways in which regulatory tools are used by public authorities and regulators constitute one of the key components of functional legitimacy. Baldwin suggests (1995: 41–46) that a regulator must, via its actions, demonstrate core values of competence, expertise, and efficiency, and so demonstrate a procedural and functional validity that justifies the decisions and actions it takes. Efficiency and effectiveness, in terms of fiscal responsibility and the avoidance of wasted effort, are core values that all public bodies are now required to pursue (Hood 1991: 11), but these values can be implemented in rather different ways at different times according to the resources, time, and political pressures that apply (Mascini and van Wijk 2009; McAllister 2010). Further, the use of tools of regulatory intervention has a functional value for regulators, in that it symbolically demonstrates their responsiveness and reach, and helps to achieve the functional outcome of increased compliance with the law (Hawkins 2002; Hutter 1997; Mendeloff and Gray 2005; Thornton et al. 2005). At the same time, however, the dominant model of regulation has increasingly shifted away from state-led 'command-and-control' approaches, and towards more responsive (Ayres and Braithwaite 1992) and self-regulatory formations which emphasise a more limited role for prosecution and enforcement, which are seen as ways of underwriting processes of regulation 'beyond the state', rather than as a direct means of exercising control over business behaviour. The extent and meaning of these changes are contested, however, and it is far from clear that the story has been one of simple retrenchment. It is also not necessarily the case that making greater or lesser use of these tools is straightforwardly related to the legitimacy of those who use them; it is the nature and quality of the usage, rather than its quantitative amount, that have the greatest impact on these perceptions.

This section examines how two key regulatory tools have been deployed by public regulators: the use of prosecution and enforcement action following breaches of the law, and the proactive inspection of premises. These are the two main 'invasive' tools used by regulators, and capture much of the traditional, 'policing' component of the public regulators' role (Gill 2002; Hawkins 1984; Kagan 1984; May and Burby 1998). They are also the most high-profile, and hence most *politicised*, element of the regulatory role, reflecting the ideological context, operational constraints, and legitimacy needs that exist at any particular point in time (Baldwin and Black 2016: 566). As such, the way they have been deployed impacts very significantly on acceptance of regulation and regulators among both the stakeholder population subject to regulation, and also broader social audiences (politicians, policy-makers, and the public: Almond 2007). These regulatory measures have been used in different ways, and to mean different things, at different times. They have also had a very significant impact on the legitimacy of health and safety as a whole, exerting an influence that trickles down the regulatory pyramid (Ayres and Braithwaite 1992) to both reinforce, and raise questions about, health and safety as a whole.

Prosecution

The passing into law of the HSWA in 1974, and the formation of the new HSE, was not only a matter of creating a new organisational framework for regulation; it also brought about a modernisation of the modes and methods used in relation to prosecution activity. Prior to the 1974 Act, Factory Inspectors pursued a conciliatory approach to enforcement that owed a debt to the first Factory Inspectors. Little meaningful enforcement action took place in relation to factory legislation before the establishment of the Factory Inspectorate in 1833, which created an administrative enforcement infrastructure, and even after this time-point, the frequency of prosecutions remained low and under-enforcement was the norm (Bartrip and Fenn 1983; Bartrip and Hartwell 1997; Carson 1979; Field 1990). Carson characterises this post-1833 model as one of '*conventionalisation*', with breaches of the law "*only rarely subjected to criminal prosecution and [...] often not regarded as really constituting crimes at all*" (1979: 38). Inspectors held ambivalent views about the moral culpability of the factory owners they investigated, and were reluctant to treat or categorise breaches of the Factories Acts as matters of criminal law, preferring instead to tolerate certain levels of violation (Carson 1979: 48–49); as an example, in 1835, just 177 convictions were obtained by Factory Inspectors (Kydd 1857: vol. 2, 83). This was reinforced via the lack of formal *mens rea* components in the offences used (which downplayed their status), and the reliance on informal *mens rea* requirements (only taking action where wilful culpability was present) within prosecution decision-making (Carson 1979; Norrie 2001). Prosecution remained a tool utilised only in the most egregious of cases, primarily so as to avoid alienating the large number of employers who might otherwise be prosecuted under a more adversarial system, something that might damage the longer term legitimacy of the developing regulatory infrastructure as a whole (Bartrip and Fenn 1983: 218).

By the 1960s, and the start of this book's time span, the rate of prosecutions had increased to 2001 *per annum* (in 1960–1961), resulting in an average fine per conviction of £20 (Ministry of Labour 1962: 43); given that the Inspectorate was, by this time, responsible for some 230,000 workplaces across a wide range of industries (1962: 102), it hardly constituted a regular response to workplace safety violations. Similarly, at the end of the 1960s, the 2657 prosecutions brought each year compared relatively unfavourably with a fatality rate of 649 *per annum* (Department of Employment and Productivity 1970: 122) and a high rate of recorded violations of the law, of which approximately 1.5% led to prosecution action (Carson 1970: 391). The Robens Report did recognise that enforcement action was rare, and penalties were low, but did not necessarily regard these outcomes as problematic in themselves, except that they led the Committee not to *"believe that the traditional sanction commands any very widespread degree of respect or confidence in this field"* (1972: para. 260). Prosecution was seen as a blunt and unhelpful tool that was *"laborious"* to use, *"not readily applicable"* to health and safety cases, and *"largely an irrelevancy"* that *"made little direct contribution towards th[e] end"* of accident prevention and improving safety standards (1972: para. 261). They envisaged a broadly symbolic role for prosecution, as a means of applying *"exemplary punishment"* in a case of flagrant, wilful, or reckless wrongdoing that created a public demand for action (1972: para. 263).

For many critics, however, this approach represented an example of steering a *"middle if not altogether ambivalent"* course (Carson 1970: 396) between disrupting relationships and maintaining pressure on employers, on the one hand, and maintaining deference to the limitations on regulatory action imposed by existing power relations in the area (Wilson 1983: 131; David Morris Interview, para. 34). From the outset, concerns were expressed by some about the approach to enforcement that was liable to be taken under the 1974 Act. The socialist *Morning Star* reported that it was necessary for the unions to *"ensure that the new regulations and codes of practice [...] contain penalties and sanctions backed by an adequate inspectorate"* (Paterson 1975: 4). Fears were also expressed in parliamentary debates:

> the [Robens] report seeks to cajole rather than coerce [...it] makes no recommendations about the size of safety and health inspectorates. It appears to accept without reluctance that extra financial resources from industry and Government will simply not be made available.[17]

The approach that Robens envisaged is almost exactly the same as the one subsequently utilised by the HSE since 1974. The 1974 Act made

---

[17] Neil Kinnock, 'Workers' Safety and Health', HC Deb 21 May 1973, Hansard vol.857 cc62-117, 66. See also *Times*, 31 January 1973, p. 7; *Times*, 4 April 1974, p. 16.

the new Executive responsible for the enforcement of the offences contained within the Act (Dawson et al. 1988), and inspectors were empowered under HSWA ss.38–39 to conduct prosecutions in the Magistrates' court, which they did alongside LAs, who also had enforcement powers under the same legislation (see Chapter 4 for more detail). As well as prosecuting, the HSE can also issue improvement notices (under HSWA s.21) and prohibition notices (under HSWA 1974 s.22) to compel a duty-holder to either make specified changes to their processes and activities within a set time frame, or to suspend a process or activity until certain conditions have been met. A significant literature has analysed the enforcement practices employed by the HSE and, while it is not necessarily useful to reproduce the detail of that literature here, some core themes can be summarised, and tied into our investigation. First, prosecution occurs both reactively (in the aftermath of an incident), and proactively (following an inspection), but the majority of reported incidents (and as many as 80% of major incidents) are not investigated (Hawkins 2002; Hutter and Lloyd-Bostock 1990). For critics, this reflects both the relative powerlessness and lack of resource of the HSE, and the power that business interests wield in order to secure a 'collusive' approach from regulators (Dalton 1998; Dawson et al. 1988: 241; Pearce and Tombs 1990; Tombs and Whyte 2007: 102; 2010), something that regulators themselves (such as the former Deputy Director-General of HSE) refute: *"I think we prosecuted too many people for that impression to arise…I don't think we made many friends"* (Jim Hammer Interview, para. 75).

Second, the use of prosecution more generally remains a measure of 'last resort', deployed only when other, more self-regulatory, measures are deemed unsuitable (Hawkins 2002; Hutter 1997; Pearce and Tombs 1990; Tombs and Whyte 2010). Prosecution has always been framed by the HSE as subordinate to its broader compliance-seeking role, and reflects a relatively non-adversarial view of the regulated population, as indicated by this former Chief Inspector of Factories:

> [Inspection] was a source of free technical advice, and provided they didn't get prosecuted at the same time, businesses were quite happy to see the inspector…You have arguments and sometimes you have to flex your muscles and use your enforcement powers, but there is a general willingness, maybe tempered by ignorance of what needs to be done, to comply with health and safety law. (David Eves Interview, para. 65)

Third, the decision to prosecute is one that reflects a primarily instrumental set of regulatory considerations, as reflected in the terms of the HSE's current Enforcement Policy Statement, which states that enforcement action is used:

> to prevent harm by requiring duty holders to manage and control risks effectively [by]…ensuring action is taken immediately to deal with serious risks; promoting and maintaining sustained compliance with the law; and ensuring that those who breach the law…may be held to account. (HSE 2015: para. 3.01)

Prosecution is therefore envisaged as a method of achieving future public policy goals; as Hawkins puts it, a *"utilitarian…conception of prosecution in the occupational health and safety arena as a way of reducing the extent of death, injury, and disease in the workplace is… dominant"* (2002: 4; also Almond 2006; Hutter 1997: 220–222). It follows, therefore, that the HSE will always be aiming, on some level, to move away from the use of prosecution and render it obsolete via the achievement of sufficiently improved levels of compliance. This appears long-standing, at least in the eyes of some of those involved at a senior level of the HSE in the 1970s and 80s:

> You've got to separate out prevention and retribution, and obviously if you have good prevention, there's less scope for retribution…It's a tricky balance…HSE was really concerned about prevention, and I think fairly firmly stuck to that. (Tim Carter Interview, paras. 14–17)

Fourth, prosecution was often deployed as a tool indirectly, in that the threat of prosecution, and the capacity of the Inspectorate to utilise it, was used as part of an 'insistent' strategy to prompt compliance via bluff, bargaining, and exhortation (Hawkins 2002: 43–47; Hutter 1989). This approach is captured in the recollections of this former HSE inspector:

> One of the privileges of being an inspector was the fact that people knew you had enforcement powers, which made the task of persuasion a great deal easier because they never quite knew when you were going to reach into your back pocket for your notices and prosecutions, so you were in a powerful negotiating position to influence change without taking formal enforcement action…it was actually one of the strengths of the Inspectorate that we did both, we were neither the heavy handed enforcer nor the light touch regulator. (David Morris Interview, para. 46)

Fifth, the decision to prosecute was shaped by, and reflective of, a wide range of strategic and cultural considerations that HSE decision-makers refer to, including the formal tenets of enforcement policy and the legal risks involved in prosecuting (Hawkins 2002: Ch. 12), relational factors around the demeanour, track record, and of the offending company (Hutter 1989), the broader political and social context and climate at the time (Hawkins 2002: Ch. 4; Hutter 1989; Hutter and Manning 1990), and the potential for prosecution to communicate a symbolic message to others about the value and importance of compliance with the law (Almond 2007; Almond and Colover 2012; Hawkins 2002: Chs. 7 and 11).

Overall, the use of prosecution has remained heavily contested, with two dominant visions of the role of prosecution, as either a last-resort adjunct to compliance-seeking, or as a powerful and necessary marker of regulatory independence, remaining in fundamental tension with one another. Three particular functional legitimacy challenges arise around the use of prosecution today. First, the available evidence all points to a decline in the rates of prosecution over time, from 2091 per year (plus 123 by LAs) when the 1974 Act came into force, to 1265 per year (plus 257 by LAs) in the mid-1980s (Dawson et al. 1988: 228), to 604 per year (plus 92 by LAs) in 2013–2014, and 593 in 2016/17.[18] This gives rise to questions about the capacity of the HSE to present itself as an active public regulator; prosecution is an important declaratory and communicative process, by which the core values and work of the agency are displayed and validated (Almond 2007, 2013), and it is arguable that the increasing absence of prosecution may have contributed to a decline in the profile of the HSE overall. Secondly, prosecution decision-making by a regulator must be seen to be consistent and principled in nature if it is to be perceived as legitimate (Baldwin 1995); concerns about the detrimental impact of 'regulatory unreasonableness' are longstanding (Bardach and Kagan 1982) and so it has been incumbent on an agency like the HSE to demonstrate that its powers are used consistently. This was clearly a concern that had weighed heavily on the minds of senior HSE figures during their time in post:

> My job was very much the 'management of discretion'…if you give inspectors considerable scope to issue Enforcement Notices, to shut down a factory or to embark on the process of prosecution…then you must set

[18]http://www.hse.gov.uk/statistics/enforcement.htm, also Tombs and Whyte (2010, 2013).

clear boundaries to that discretion...I personally only made two prosecution decisions in 10 years as chief inspector. (Jim Hammer Interview, paras. 85–86)

That said, the modest rates of prosecution undertaken (above) make it relatively hard to argue convincingly that the HSE has been over-zealous in its use of this tool.

Finally, there are also concerns that the threat of prosecution by the HSE wields a disproportionately large influence over stakeholder and public perceptions of health and safety. The nature and form of the 1974 Act allowed for a relatively broad scope of application of the offences it contains (see Chapter 4), and figures within the HSE acknowledged that this gave them a tool that could be used extremely broadly:

> If you want to prosecute people you're infinitely better off with principles-based regulation. You can get anybody under principles-based regulation, for anything. (Senior HSE Source Interview, para. 207 [emphasis added])

While concrete rates of prosecution may now be modest, and the symbolic threat of prosecution may be useful in prompting widespread change and compliance, there are risks associated with presenting a 'powerful' prosecutorial image, namely, that stakeholders and others perceive the regulator to be over-reaching, and the risk of prosecution so great, that they begin to view the regulatory system as a whole as illegitimate. Such a perception may underpin criticisms of a modern culture of 'regulatory punitiveness' (Baldwin 2004), and the spectre of the over-reaching health and safety regulator seems to feature widely in media coverage (Almond 2009), public perception (see Chapter 2; also Almond and Colover 2012), and governmental criticism (Young 2010) of health and safety. As such, prosecution must be recognised as a regulatory tool that has brought almost as many institutional risks for regulators as it resolved.

Inspection

Like prosecution, the use of inspection by health and safety inspectors also has a long history. Inspection of factories was first introduced by the Health and Morals of Apprentices Act 1802, under which external visitors (usually Justices of the Peace and clergymen) assessed compliance with the terms of the Act. Failure to fulfil the terms of the Act could result in a fine of up to £10, and employers were entreated to '*act in strict conformity to the said rules and regulations*' (Bartrip and Fenn 1983: 203; Thomas 1948: 11). A more modern, formal, state regulator was subsequently introduced by the Factory Act 1833 (Almond 2013: 99; Bartrip and Fenn 1983; Carson 1979; Rhodes 1981), and adopted a conciliatory approach to inspection, whereby conformity with the terms of the Act was promoted via negotiation and the provision of advice by Inspectors who avoided '*a rigorous execution of conformity to all [the law's] details*', in the words of Robert Rickards, the Midlands inspector.[19] By 1881, the first year that formal statistics were presented, inspection was an established regulatory tool, with Inspectors visiting 69,545 factories and workshops that year (Home Office 1933: 132). By the start of the 1960s, the number of inspections undertaken had risen to nearly 283,000 per year, across a wide range (230,000) of industries and workplaces (Ministry of Labour 1962: 102), a rate which remained relatively steady across the subsequent decade (Department of Employment and Productivity 1970: 122).

This pre-1972 regulatory system was largely *relational* and knowledge-centric in nature, focused on the transmission and sharing of expertise within the parameters and structure provided by stable, ongoing relationships which were often based on goodwill. Inspectors got to know the premises on their 'beat', and the risks that were present within them, and used inspection as a tool of regulatory 'craft' (Sparrow 2000), based on expertise derived from their frontline experience, as the former HSE Chief Inspector of Factories recalled:

> There were 400 [power] press accidents a year [...] We kept records of every machine in Birmingham and we checked them and visited by surprise and even if they'd stopped the machine you could feel that it was hot and we could get evidence from the operator. (Jim Hammer Interview, para. 8)

Coverage was selective, due to the expansion of the regulated sphere after the Offices, Shops and Railway Premises Act 1963 (Department of Employment and Productivity 1968: xi), but a broadly standardised inspection programme was used. The Robens Report found that

---

[19] Parliamentary Papers (1835) XL, p. 698 (cited in Thomas 1948: 76).

inspection was the "*main day-to-day activity of the majority of inspectors*" (1972: para. 201), but was often "*brief, superficial, and...unproductive*" in practice (1972: para. 218), and argued that increasing rates of inspection was "*manifestly impracticable*" (1972: para. 28). A new 'ideal' of regulatory inspection was proposed, whereby "*[t]he primary objectives... should be first, to support [the] development [of self-regulatory capacity]; and secondly, to concentrate regulatory resources more selectively on serious problems*" (1972: para. 216). Inspections should be targeted at issues "*identified through the systematic assessment of all the available data*" (1972: paras. 217–218).

This can be broadly understood as a form of 'risk-based' practice, whereby "*decision-making frameworks and procedures...prioritise regulatory activity and the deployment of resources...[via] an assessment of the risks that regulated firms pose to the regulator's objectives*" (Black 2005: 516). On this view, inspection was one option among a range of tools that might be used, including as time went on many (self-auditing, third-party accreditation, association memberships, voluntary agreements) that went beyond the traditional, state-centred approach; the choice and use of inspection (or any other tool) were made according to its value in achieving regulatory outcomes, and a consideration of whether the risks involved could be controlled acceptably via less intrusive means than state-led inspection (Baldwin and Black 2016; Black 2005; Hampton 2005; Power 2004; Sparrow 2000). This approach developed in response to political demands for greater accountability and efficiency in the use of public resources, but also as a corollary of the spread of quantitative risk management methodologies as a means of ensuring sustainable 'good' governance more generally (Hood 1991; Power 2007). In the early 1970s, however, 'risk targeting' was to be based on all "*available data*", including "*local knowledge*" (Robens 1972: para. 218) and professional expertise (Hutter 1986: 117), thus "*placing a high value on the judgement of inspectors*" (Warburton 1980: 8). 'Risk-based' inspection was an approach that the HSE developed itself, in response to its own legitimacy requirements (Baldwin and Black 2016), refocusing the work of skilled inspectors and supporting their professional status:

> The reorganization of the inspectorate...was to meet Robens' criticism that there was insufficient informed inspection. Inspectors had to be more expert and this specialisation enabled inspectors to concentrate on what they really knew about. (Jim Hammer Interview, para. 37)

Inspection thus expressed an image of expertise and competence which helped to establish the HSE's organisational legitimacy, and the new social norms around safety that it sought to uphold. As Hutter observed, "*The very physical presence of a regulatory official serves...to emphasise that official's authority of office and remind the regulated of the existence of regulatory laws*" (1986: 122).

Subsequent social science research has outlined the strategic decision-making processes that underpinned the HSE's use of inspection (Hawkins 2002; Hawkins and Hutter 1993; Hutter 1986, 1989, 1997; Hutter and Lloyd-Bostock 1990). While many of these were relational, relating to an inspector's own judgements and interactions with the regulated firm, they have also increasingly come to reflect more systemic organisational considerations, which themselves reflect the prevailing policy 'surround' of the day (Hawkins 2002: Ch. 4). During the 1980s, this surround became both more hostile towards regulation, viewing it as clashing with a prevailing liberal-market political climate which tended to advocate for a rolling back of state interventions (Dawson et al. 1988; Hutter and Manning 1990: 107; Moran 2003). This policy movement became known as the '*New Public Management*' (or NPM: Hood 1991), and sought to introduce a culture of 'managerialism', the setting of performance targets, the introduction of private sector involvement, and greater fiscal discipline, into the public sector; these values of accountability and efficiency provided the functional legitimacy basis for regulatory action at this time. This shifted the way that government operated, with reduced state provision being accompanied by a proliferation of arm's-length regulatory controls focused on secondary 'framework' oversight rather than primary delivery (Hood et al. 2001; Moran 2003: 5; Pires 2011). Two Government White Papers, *Lifting the Burden* (HM Government 1985) and *Building Businesses... Not Barriers* (HM Government 1986), highlighted the perceived disadvantages of regulation, reinforcing the risk-based approach taken by the HSE (Dawson et al. 1988: 234; Hutter 1986: 116; 1997: 110), whereby Inspectors used a computerised Inspection Rating System to determine, for each business. This was explained by HSE's former Chief Inspector of Factories (1975–1984) as involving assessment of:

> what the standard of health was, what the standard of safety was and what the standard of welfare was, and they gave values from 1 to 5...Then they would judge, if the worst happened at the factory, would nobody be killed,

would one person be killed, or would more than one person be killed?...
The last question was 'what confidence have I got in management's ability
to maintain standards?', that was given a figure. When you multiplied these
figures they gave a value between 1 and 100...add up the rating values for
every factory across the country and divide that total by the available number of 'Inspector half-days' to give a value above which we should aim to
inspect...That came out initially at 42. (Jim Hammer Interview, para. 31)

The Rating System systematised and quantified the process of inspection
decision-making, basing it upon a 'responsive' conception of risk which
takes into account elements of behavioural, institutional, and cultural
context (Black and Baldwin 2010: 186), as well as other contextual factors, such as social utility, public concern, and cost. This mirrored the
similarly holistic method of systematic risk calculation contained within
the 'Tolerability of Risk Framework' (HSE 1988, discussed above),
which also involved determining which risks were going to be tolerated,
or permitted to arise, and which were not:

There is a scientific professional process of assessing how big a risk it is, but
in the end it's a political judgement where you draw the line between what
level of risk is tolerable and what is intolerable...you're not saying risk is
acceptable, you're saying you'll tolerate some small areas of risk. (Tim
Carter Interview, para. 132)

As well as fitting well with the broader policy context of the NPM,
this risk-based approach also reflected the core operational constraint
of budgetary restraint (Baldwin and Black 2016: 573; Almond and
Esbester 2018: 52). HSE's Chief Inspector of Factories had felt it necessary to acknowledge that, following a 6% budget cut in the early 1980s,
"*what the Factory Inspectorate can achieve is obviously constrained by the
resources at its disposal*", necessitating cost–benefit analysis to identify
where inspection *could* be applied (HSE 1981: v–vii). But assessing the
value of inspection is difficult (Hawkins 2002: 284–289; Sparrow 2000:
113–115), and throughout the 1980s, HSE repeatedly debated the trading-off of inspection in pursuit of efficiency (Dawson et al. 1988: 231–
233). By 1988, the HSE's Director-General was telling Parliament that
"*[o]ne really has got to decide if one can afford [the] scale of diversion of
one's resources*" that more inspections would entail, and that "*[w]e would
not think it necessary to waste a lot of time on a great many of the premises*"

this would involve (Employment Committee 1988: paras. 18 and 23). As such, the guiding reference point for the use of inspection had shifted from expertise to capacity; the functional legitimacy of the regulator was now assessed not according to the knowledge that it distributed so much as the degree of efficiency and cost-effectiveness with which it did so. This was an adaptation to Government policy and the need to maintain an institutional 'license to operate'; risk-based targeting became a means for the agency to communicate about its compatibility with the broader political agenda of the time.

The HSE has subsequently been subject to numerous policy reviews and resource cuts (Almond 2015; Walters et al. 2011: Ch. 8), and so rates of inspection have fallen, halving in the first ten years of the HSE's existence (from 481,000 in 1975 to 246,000 in 1985: Dawson et al. 1988: 225), and continuing to fall since. One conclusion that might be drawn from this trend is that 'risk-based' strategies are inevitably deregulatory. But we must also consider the ways in which these approaches focus attention *onto* specific problems, as well as *away* from others, by focusing resources onto issues of particular concern to the regulator in question (Almond and Esbester 2018: 54; Baldwin and Black 2016; Black and Baldwin 2010: 184; Rothstein et al. 2013: 217; Sparrow 2000). When the late-1990s Labour Government took a proactive interest health and safety regulation, it spoke to their traditional interests and stakeholders in the union movement, but also brought challenges, according to the HSE Director-General of the period 1995–2000:

> The Deputy Prime Minister…wanted far more inspections…a far stricter regime, and…to be in charge of everything. That was a different kind of challenge, it wasn't one of, 'let's cut the resources, let's undermine the regulation', it was the opposite. But in many ways just as dangerous. (Jenny Bacon Interview, para. 45)

This led to some increase in resourcing (Walters et al. 2011: 184), the introduction of a new corporate manslaughter offence (Almond 2013), and the launch of a 'Revitalising Health and Safety' strategy (DETR 2000), which established that 'problem-oriented' (Sparrow 2000: 123–136) inspection would be used to "*promot[e] better working environments*" (DETR 2000: 18) by focusing attention upon the most commonplace risks (such as 'slips, trips and falls' and 'lifting accidents').

This also allowed for greater organisational control to be exercised over inspectors' choices, and for attention to be placed upon particular industries at particular times.

Inspection was utilised during this era in what can be broadly understood as an 'activist' manner; it was undertaken in a limited number of cases, both as a 'trickle-down' back-up to 'softer' regulatory mechanisms, and as part of a more proactive 'blitz' strategy (House of Commons 2004: 7). So, for example, construction was singled out for both these sorts of attention because it was a highly visible, growing industry, with a poor safety record, and one where strategic targeting of interventions against major firms could have a broader impact, as one Chairman/Director of a major construction firm observed:

> [The HSE] haven't got very many inspectors and so their attitude, quite openly, is we're going to focus on the big boys because we see them as being the ones who've got to set the standard, and indeed, if we get a prosecution against a big boy then it'll get much more publicity and hopefully this will trickle down [...] which to the big boys by and large always seems extremely unfair. (Sir John Armitt Interview, para. 24)

This 'activism' also involved regulators taking 'ownership' of a wider range of issues (in areas such as public safety, workplace health, and well-being), some of which went beyond the scope of established risk-based principles:

> [HSE] rather lost the ability to understand that it would do better to concentrate on a small number of things that mattered in any given area as opposed to telling everybody how to do everything. It moved into that mode in the 1990s and the early 2000s. (Senior HSE Source, 2000s–2010s: para. 31)

The target-based *Revitalising* approach was implemented via the 'Fit3'[20] strategy (Prosser 2010: 96), which continued the principle of *Revitalising* by focusing risk-based inspection practices onto low-hazard, high-frequency issues (Walters et al. 2011: 200–202). But this approach had two legitimacy flaws that highlighted the limits of what can be

---

[20]'*Fit for Work, Fit for Life, Fit for Tomorrow*': http://www.hse.gov.uk/aboutus/strategiesandplans/hscplans/businessplans/0506/fitfor.htm

achieved with a tool like inspection. First, it clashed with the professional culture of HSE inspectors, which tended to prioritise 'high-hazard' issues as the most important ones to focus on (Black and Baldwin 2012: 14; Walters et al. 2011: 208). Second, Fit3's activism remained subservient to economic thinking; the HSE was required to show that inspection was having a positive effect on the 'outcome measures' it was using (Walters et al. 2011: 205), but could not do so:

> [HSE] never tested their interventions by saying, 'if we did this, would it actually work?' They simply assumed if you put more resources into more inspections then there would be improvement...the figures didn't move at all...[Inspectors] thought it was rubbish, and they were right. (Senior HSE Source, 2000s–2010s: para. 70)

Inspection was deployed to achieve *symbolic* changes in response to political concerns, but was measured in terms of the economic assessment of its *instrumental* impact. This meant that the communicative value of inspection was not measured; its functional legitimacy was limited to the core NPM values of efficiency and effectiveness (Baldwin 1995; Hood 1991), and on these measures, was seen to have only limited value. This framing endured through subsequent governmental policy initiatives. For example, the Treasury's Hampton Review framed risk-based inspection as a matter of common-sense cost reduction (Hampton 2005: 1.76–1.8), and pushed the HSE to reduce its inspection activity further, leading to a drop of 50% in inspection rates from 2004 to 2007 (BRE 2008: 22). The 2010–2015 Coalition Government further reduced HSE's rates of scheduled inspection by 33%, seeking "*better targeting based on hard evidence of effectiveness*" (Department of Work and Pensions 2011: 9) and of firms' past performance and safety management systems (BRE 2008: 7). The result was that, in 2013–2014, the HSE inspected 23,470 premises (HSE 2014a: 11), just 10% of the 1985 total. This era of reduced inspection has coincided with a new framing of risk-based practice as a matter of making explicitly normative value judgements about which areas of risk *ought* and *ought not* to be regulated via inspection. Dodds (2006) has referred to this as a shift from 'better regulation' to 'risk-tolerant deregulation', that is, to setting limits on the use of intrusive tools like inspection according to categorical criteria about whether intervention is 'rational', in cost–benefit terms, or whether the risk in question 'ought' to be tolerated for the wider economic benefit of society (2006: 534). Risk is cast as a positive feature of contemporary society, rather than an undesirable side effect of business or social activity, and regulatory intervention is "*an always regrettable means of correcting market failure*" (Prosser 2010: 1).

The 2005 Hampton Review laid the foundations for this approach, identifying the need to reduce "*unnecessary inspections*", which were costly, inflexible, inefficient, and a "*barrier to growth*" (2005: 25–27). While decision-making was couched in the technical language of risk-assessment, the weighting of these risks was increasingly undertaken via reference to political interests; as an example, the Rogers Review of LA enforcement recommended that enforcement be targeted according to the likelihood and severity of harm, plus consideration of public and political priority (2007: 36–43), leading to a decline of 95% in rates of LA inspection between 2009 and 2014 (HSE 2014b: 2). In the sphere of health and safety, policy reviews have recommended that 'low-hazard' workplaces (shops, offices, schools, public spaces: Young 2010: 28), self-employed workers, and areas of unquantified 'hazard', should be taken outside the scope of formal inspection (Löfstedt 2011: 43). The regulated population was sorted into risk categories ('high risk areas', 'areas of concern', and 'low-risk areas') for this purpose, according to assessments of sectoral hazard levels, with only the first of these (including the construction and waste industries) receiving inspection visits (DWP 2011: 9). This placed some sectors which appear relatively hazardous, such as quarries, agriculture, and health and social care, outside the scope of inspection, as it was "*unlikely to be effective*" (2011: 9) in those areas. These choices were made at a sector level and on the basis of aggregated, rather than individual, assessments of risk and outcome, thus mirroring the 'actuarial' decision-making found in areas like insurance (Ewald 2002). It also reflected the instrumental value of inspection; agriculture, although high-risk, remains a diffuse industry made up of small undertakings scattered across the countryside, where inspection was recognised, here by HSE's Deputy Chief Executive (2008–2015), as less likely to bear fruit:

> the role of Regulator is to...change behaviours so there is less likely to be market failure...It's difficult to get that into the agriculture sector, for example, which is much more dispersed...you can't get the 20 biggest farmers around the table and know you're speaking to people who are influencing 40% of the industry. (Kevin Myers Interview, para. 31)

On the one hand, this shows the current centrality of efficiency and value-for-money as considerations driving regulatory decisions; on the other, it also reflects a politicisation of these choices, which ensures that '*the duty-holder, and not the regulator, owns the risk*' (John Rimington Interview, para. 89; Kevin Myers Interview, para. 20; Robens 1972: para. 41). The settlement of interests around issues of risk is now undertaken via reference to political values of economic self-determinism and individual autonomy, or tightly defined measurements of cost and benefit; the range of contexts where it is legitimate for regulators to engage with duty-holders is narrowed (Almond 2015). So the use of licensing-type arrangements ('Fee For Intervention'[21]), which place the regulatory relationship onto a more commercial footing, the restriction of inspection to being an exceptional and voluntary measure used only in relation to certain 'deserving' industries, and the related reduction in regulatory accountability for sectors and workplaces that fall outside that scope (Rothstein et al. 2015: 221), all reflect the "*pressures [on regulators] to become...more like the organisations they regulate...because this is what is expected*" (Power 2007: 91). Difficult areas, and issues that challenge regulatory performance and hence, organisational legitimacy, can be circumvented in favour of a focus on 'core' areas of regulatory business (Black 2005: 515; Black and Baldwin 2012; Power 2007: 89–91). Being 'risk-based' is thus a means of safeguarding, and avoiding threats to, the legitimacy of the regulatory undertaking (Almond and Esbester 2018); it reflects a narrowing of the limits of what regulators are willing to bear responsibility for, and a redrawing of the limits of the functional legitimacy claims and challenges that are accepted within the modern regulatory state.

CONCLUSION

The changes that have occurred in the last 60 years around the operation of health and safety in practice have tended to involve reassessments of the balance of responsibilities between public regulators, on the one hand, and private interests and regulated stakeholders, on the other. Whether it is a matter of regulators accepting new responsibilities for public safety, or sharing their traditional responsibilities with new

---

[21] http://www.hse.gov.uk/fee-for-intervention/.

stakeholder groups and professions, or withdrawing from performing the symbolic enforcement functions that only a state regulator is meant to do, the core issue being addressed is one of functional legitimacy. This involves the consideration of some key questions; who should do this? How should they do it? And how do we assess whether this is done well, or well enough? Certain core values (of efficiency, expertise, and effectiveness: Baldwin 1995: 45; Black 2008: 146; Hood 1991: 6) have tended to be deployed to make these assessments, and those involved in the field of health and safety have taken proactive steps to demonstrate their fidelity to these values at different times. Values of good governance, risk-based decision-making, and responsiveness, have all been deployed to demonstrate that health and safety remains an area where it is legitimate for some coordinated control of public, social, and economic life to take place.

But some persistent and recurring challenges to this process can be identified across the four thematic areas explored in this chapter. First, as we have seen most clearly in relation to the question of 'public' safety, the boundaries between 'public' and 'private' regulation can be highly contested; this means that it is often unclear where the responsibilities of public, private, and individual regulatory actors might begin and end. While regulatory theory has moved to accommodate and recognise this 'polycentricity' (Black 2008), its effects in terms of broader social and political legitimacy formations are more contested and changeable than might be thought. For example, demands for regulators to take on additional responsibility become more pronounced at times of crisis or after disasters, and recede at other times, especially when the broader political climate aligns against regulation in general. And the interrelationship between safety regulators and safety professionals, or non-state agents, is so complex as to blur the boundaries of that distinction. The second theme is the increasing importance of this expanded infrastructure of regulation to the delivery of the core outcomes that the regulatory system aims to provide. As the 'core' infrastructure of state-led regulation has receded over the last 30 years, leading (as we have seen) to a reduction in the use of inspection and enforcement, so the reliance on arms-length delivery of health and safety interventions by a 'periphery' of other organisations has grown, to the point where health and safety could not now be delivered or secured otherwise. Finally, and overarching each of these considerations, the functional legitimacy of health and safety depends not only on how it does what it does, but also on how that is measured or valued by policy actors within the political context that surrounds the issue. Being adjudged effective or efficient is contingent not only on the actual performance of the actors or institutions in question, but also upon the criteria used to weight or assess those values, and this is reflective of political decisions, not merely bureaucratic ones.

REFERENCES

PRIMARY SOURCES

Barry, R. (1967). Accident Prevention in the Iron and Steel Industry. Talk to the Central Metropolitan Group, London Industrial Accident Prevention Committee, 13 March 1967, Noted in the Minutes of the Meeting, p. 3. MRC, MSS.292B/146/2/4.

Bell, C. (1977). *Interviewed on BBC TV Programme 'Red Alert', Aired on 5 July 1977.* Caversham: BBC Written Archives.

Better Regulation Executive (BRE). (2008). *Effective Inspection and Enforcement: Implementing the Hampton Vision in the Health and Safety Executive.* http://www.nao.org.uk/wp-content/uploads/2008/03/HSE_Hampton_report.pdf.

British Rail. (1977, August 16). *Fatal Accidents to B.R.B. Employees.* TNA, AN 156/936.

British Rail Board and HSE. (1982). *An Outline Report on OSH by the Accident Prevention Advisory Unit.* TNA, AN 16/157.

Brown, P. (1988, November 16). Balance Sheets and Shrouds. *Guardian*, 1.

Building Businesses…Not Barriers. (1986). Cmnd.9794. London: Crown.

Cook, S., Tirbutt, S., & Hearst, D. (1985, January 11). Flats Gas Explosion Enquiry Ordered. *Guardian*, 1.

Department of Employment. (1974). *Annual Report, HM Chief Inspector of Factories, 1973.* Cmnd. 5708. London: Crown.

Department of Employment. (1975). *The Flixborough Disaster. Report of the Court of Inquiry.* London: Crown.

Department of Employment and Productivity. (1968). *Annual Report of HM Chief Inspector of Factories, 1967.* Cmnd. 3745. London: HMSO.

Department of Employment and Productivity. (1970). *Annual Report of HM Chief Inspector of Factories, 1969.* Cmnd. 4461. London: Crown.

Department of Environment, Transport and the Regions (DETR). (2000). *Revitalising Health and Safety: Strategy Statement.* London: Crown.

Department of Work and Pensions. (2011). *Good Health and Safety, Good for Everyone.* http://www.dwp.gov.uk/docs/good-health-and-safety.pdf.

Employment Committee. (1988). *The Work of the Health and Safety Commission and Executive*, minutes of evidence. HC 267. London: Crown.

Employment Committee. (1993). *Health and Safety Matters, 1992/3, Minutes of Evidence.* HC755-I. London: Crown.

England, B. (c. 1979). Discussion. *Work Hazards, 21,* 18. Samuel Barr Collection, GCU DC 140/2/1/2.

Guardian. (1974, June 3) The holocaust at Flixborough. 10.

Hampton, P. (2005). *Reducing Administrative Burdens: Effective Inspection and Enforcement* (The Hampton Review). London: HM Treasury.

Home Office. (1933). *Annual Report, HM Chief Inspector of Factories, 1932.* Cmnd. 4377. London: HMSO.

House of Commons. (2004). *Committee of Public Accounts, Health and Safety Executive: Improving Health and Safety in the Construction Industry, Fifty-Second Report of Session 2003-4* (HC 627). London: HMSO.

House of Lords. (2000). *Science and Society: Third Report of the Select Committee on Science and Technology.* London: HMSO. http://www.publications.parliament.uk/pa/ld199900/ldselect/ldsctech/38/3801.htm.

HSE. (1980). Health and Safety, Manufacturing and Service Industries. Report for 1978. London.

HSE. (1981). Health and Safety, Manufacturing and Service Industries. Report for 1979. London.

HSE. (1988). *The Tolerability of Risks from Nuclear Power Stations.* London: HMSO.

HSE. (2014a). *The Health and Safety Executive Annual Report and Accounts 2013/14* (HC228). http://www.hse.gov.uk/aboutus/reports/1314/ar1314.pdf.

HSE. (2014b). *HELA Data Collection–Analysis of LAE1 2013/14 Data from Local Authorities.* [Paper H17/01]. http://www.hse.gov.uk/aboutus/meetings/committees/hela/310714/data-collection%E2%80%93analysis.pdf.

HSE. (2015). *Enforcement Policy Statement.* http://www.hse.gov.uk/pubns/hse41.pdf.

Kinnock, N. (1973, May 21). Workers' Safety and Health. HC Deb, Hansard Vol. 857, cc. 62–117.

Layfield, F. (1987). *Sizewell B Public Inquiry: Report by Sir Frank Layfield.* London: HMSO.

Lifting the Burden. (1985). London: HMSO.

Löfstedt, R. (2011). *Reclaiming Health and Safety for All: An Independent Review of Health and Safety Legislation* (The Löfstedt Review). London: Crown.

Ministry of Labour. (1962). *Annual Report of the Chief Inspector of Factories, 1961.* Cmnd. 1816. London: Crown.

Ministry of Labour. (1966). *Annual Report of H.M. Chief Inspector of Factories, 1965.* Cmnd. 3080. London: Crown.

Paterson, A. (1975, February 7). Faults in Safety Laws. *Morning Star,* 4.

Robens, L. (1972). *Safety and Health at Work: Report of the Committee 1970–72* (The Robens Report). London: HMSO.

Rogers, P. (2007). *National Enforcement Priorities for Local Authority Regulatory Services* (The Rogers Review). London: HMSO.

RoSPA. (1971, June 14). *Verbal Evidence on Behalf of RoSPA to the Robens Committee.* TNA, LAB 96/75.

RoSPA. (1971, September 22). *National Publicity Committee Minutes.* RoSPA Archives, D.266/2/4/3.

RoSPA. (1976). *The Role and Function of the Safety Officer.* RoSPA National Occupational Health and Safety Committee, 25 November 1976, Point 4.6. RoSPA Archives, D.266/2/24.

Thompson, M. (c. 2000). University of Aberdeen 'Lives in the Oil Industry' Project, Tape 2 Side B. British Library, London, C963/13.

Times. (1973, January 31). Accidents Caused by Unguarded Minds. 7.

Times. (1974, April 4). Powers to Deal with Accidents at Work. 16.

Times. (1985, January 11). Nine Die as Blast Rips Flats. 1.

Williams, R. (c. 1979). A Key Issue. *Work Hazards, 21*(c. 1979), 3. Samuel Barr Collection, GCU DC 140/2/1/2.

Young, L. (2010). *Common Sense, Common Safety* (The Young Review). London: Crown.

SECONDARY SOURCES

Aldrich, M. (1997). *Safety First. Technology, Labor and Business in the Building of American Work Safety, 1870–1939.* Baltimore: Johns Hopkins University Press.

Almond, P. (2006). An Inspector's-Eye View: The Prospective Enforcement of Work-Related Fatality Cases. *British Journal of Criminology, 46*(5), 893–916.

Almond, P. (2007). Regulation Crisis: Evaluating the Potential Legitimizing Effects of 'Corporate Manslaughter' Cases. *Law and Policy, 29*(3), 285–310.

Almond, P. (2009). The Dangers of Hanging Baskets: Regulatory Myths' and Media Representations of Health and Safety Regulation. *Journal of Law and Society, 36*(3), 352–375.

Almond, P. (2013). *Corporate Manslaughter and Regulatory Reform.* Basingstoke: Palgrave Macmillan.

Almond, P. (2015). Revolution Blues: The Reconstruction of Health and Safety Law as 'Common-Sense' Regulation. *Journal of Law and Society, 42*(2), 202–229.

Almond, P., & Colover, S. (2010). Mediating Punitiveness: Understanding Public Attitudes Towards Work-Related Fatality Cases. *European Journal of Criminology, 7*(5), 323–338.

Almond, P., & Colover, S. (2012). Communication and Social Regulation: The Criminalization of Work-Related Death. *British Journal of Criminology, 52*(5), 997–1016.

Almond, P., & Esbester, M. (2016). Il/Legitimate Risks? Public Opinion and Health and Safety Regulation in post-1960 Britain. In T. Crook & M. Esbester (Eds.), *Governing Risks: Danger, Safety and Accidents in Modern Britain, c. 1800–2000* (pp. 295–314). London: Palgrave Macmillan.

Almond, P., & Esbester, M. (2018). Regulatory Inspection and the Changing Legitimacy of Health and Safety. *Regulation & Governance, 12*(1), 46–63.

Almond, P., & Gray, G. C. (2017). Frontline Safety: Understanding the Workplace as a Site of Regulatory Engagement. *Law & Policy, 39*(1), 5–26.

Atherley, G. R. C., & Hale, A. R. (1975). Pre-requisites for a Profession in Occupational Safety and Hygiene. *Annals of Occupational Hygiene, 18*(4), 321–334.

Ayres, I., & Braithwaite, J. (1992). *Responsive Regulation: Transcending the Deregulation Debate.* Oxford: Oxford University Press.

Baldwin, R. (1995). *Rules and Government.* Oxford: Clarendon Press.

Baldwin, R. (1996). Regulatory Legitimacy in the European Context: The British Health and Safety Executive. In G. Majone (Ed.), *Regulating Europe* (pp. 83–105). London: Routledge.

Baldwin, R. (2004). The New Punitive Regulation. *Modern Law Review, 67*(3), 351–383.

Baldwin, R., & Black, J. (2016). Driving Priorities in Risk-Based Regulation: What's the Problem? *Journal of Law and Society, 43*(4), 565–595.

Ball, D., & Ball-King, L. (2013). Safety Management and Public Spaces: Restoring Balance. *Risk Analysis, 33*(5), 763–771.

Bandle, T. (2007). Tolerability of Risk: The Regulator's Story. In F. Bouder, D. Slavin, & R. Löfstedt (Eds.), *The Tolerability of Risk: A New Framework for Risk Management* (pp. 93–103). London: Earthscan.

Bardach, E., & Kagan, R. A. (1982). *Going by the Book: The Problem of Regulatory Unreasonableness.* New Brunswick: Transaction Press.

Bartrip, P., & Fenn, P. T. (1983). The Evolution of Regulatory Style in the Nineteenth Century British Factory Inspectorate. *Journal of Law and Society, 10*(2), 201–222.

Bartrip, P., & Hartwell, R. (1997). Profit and Virtue: Economic Theory and the Regulation of Occupational Health in Nineteenth and Early Twentieth Century Britain. In K. Hawkins (Ed.), *The Human Face of Law: Essays in Honour of Donald Harris* (pp. 45–84). Oxford: Oxford University Press.

Beck, U. (1992). *Risk Society: Towards a New Modernity* (M. Ritter, Trans.). London: Sage.

Black, J. (2000). Proceduralizing Regulation: Part I. *Oxford Journal of Legal Studies, 20*(4), 597–614.

Black, J. (2005). The Emergence of Risk-Based Regulation and the New Public Risk Management in the United Kingdom. *Public Law,* 512–549.

Black, J. (2008). Constructing and Contesting Legitimacy and Accountability in Polycentric Regulatory Regimes. *Regulation & Governance, 2*(2), 137–164.

Black, J., & Baldwin, R. (2010). Really Responsive Risk-Based Regulation. *Law & Policy, 32*(2), 181–213.

Black, J., & Baldwin, R. (2012). When Risk-Based Regulation Aims Low: Approaches and Challenges. *Regulation & Governance, 6*(1), 2–22.

Boudia, S., & Jas, N. (2007). Introduction: Risk and "Risk Society" in Historical Perspective. *History and Technology, 23*(4), 317–331.

Braithwaite, J. (2008). *Regulatory Capitalism: How It Works, Ideas for Making It Work Better*. Cheltenham: Edward Elgar.

Carson, W. G. (1970). White-Collar Crime and the Enforcement of Factory Legislation. *British Journal of Criminology, 10*(4), 383–398.

Carson, W. G. (1979). The Conventionalisation of Early Factory Crime. *International Journal of the Sociology of Law, 7*(1), 37–60.

Dalton, A. J. P. (1998). *Safety, Health and Environmental Hazards at the Workplace*. London: Cassell.

Dawson, S., Willman, P., Clinton, A., & Bamford, M. (1988). *Safety at Work: The Limits of Self-Regulation*. Cambridge: Cambridge University Press.

de Magalhães, C. (2010). Public Space and the Contracting-Out of publicness: A Framework for Analysis. *Journal of Urban Design, 15*(4), 559–574.

Dingwall, R. (1999). "Risk Society": The Cult of Theory and the Millennium? *Social Policy and Administration, 33*(4), 474–491.

Dingwall, R., & Fenn, P. (1987). "A Respectable Profession"? Sociological and Economic Perspectives on the Regulation of Professional Services. *International Review of Law and Economics, 7*(1), 51–64.

Dingwerth, K. (2007). *The New Transnationalism: Transnational Governance and Democratic Legitimacy*. Basingstoke: Palgrave Macmillan.

Dodds, A. (2006). The Core Executive's Approach to Regulation: From 'Better Regulation' to 'Risk-Tolerant Deregulation'. *Social Policy & Administration, 40*(5), 526–542.

Dunlop, C. (2014). *Health and Safety Myth-Busters Challenge Panel: Case Analysis* (University of Exeter Research Paper).

Esbester, M. (Forthcoming). *The Birth of Modern Safety: Preventing Worker Accidents on Britain's Railways, 1871–1948*. Abingdon: Taylor & Francis.

Ewald, F. (2002). The Return of Descartes's Malicious Demon: An Outline of a Philosophy of Precaution. In T. Baker & J. Simon (Eds.), *Embracing Risk: The Changing Culture of Insurance and Responsibility* (pp. 273–302). Chicago: University of Chicago Press.

Fay, E. (1968). Disaster Enquiries. *Medico-Legal Journal, 36*(4), 161–174.

Field, S. (1990). Without the Law? Professor Arthurs and the Early Factory Inspectorate. *Journal of Law and Society, 17*(4), 445–468.

Fressoz, J.-B. (2007). Beck Back in the 19th Century: Towards a Genealogy of Risk Society. *History and Technology, 23*(4), 333–350.

Furedi, F. (1997). *Culture of Fear: Risk-Taking and the Morality of Low Expectation*. London: Cassell.

Geertz, C. (1975). Common Sense as a Cultural System. *The Antioch Review, 33*(1), 5–26.

Gill, P. (2002). Policing and Regulation: What Is the Difference? *Social and Legal Studies, 11*(4), 523–546.

Gray, G. C. (2006). The Regulation of Corporate Violations: Punishment, Compliance, and the Blurring of Responsibility. *British Journal of Criminology,* *46*(6), 875–892.

Haines, F. (2011). *The Paradox of Regulation: What Regulation Can Achieve and What It Cannot.* Cheltenham: Edward Elgar.

Halliday, S., Ilan, J., & Scott, C. (2011). The Public Management of Liability Risks. *Oxford Journal of Legal Studies, 31*(3), 527–550.

Hawkins, K. (1984). *Environment and Enforcement: Regulation and the Social Definition of Pollution.* Oxford: Clarendon Press.

Hawkins, K. (2002). *Law as Last Resort: Prosecution Decision-Making in a Regulatory Agency.* Oxford: Oxford University Press.

Hawkins, K., & Hutter, B. (1993). The Response of Business to Social Regulation in England and Wales: An Enforcement Perspective. *Law & Policy, 15*(2), 199–217.

Hood, C. (1991). A Public Management for All Seasons? *Public Administration, 69*(1), 3–19.

Hood, C., Rothstein, H., & Baldwin, R. (2001). *The Government of Risk: Understanding Risk Regulation Regimes.* Oxford: Oxford University Press.

Huising, R., & Silbey, S. S. (2011). Governing the Gap: Forging Safe Science through Relational Regulation. *Regulation & Governance, 5*(1), 14–42.

Hutter, B. (1986). An Inspector Calls: The Importance of Proactive Enforcement in the Regulatory Context. *British Journal of Criminology, 26*(1), 114–128.

Hutter, B. (1989). Variations in Regulatory Enforcement Styles. *Law & Policy, 11*(2), 153–174.

Hutter, B. (1997). *Compliance: Regulation and Environment.* Oxford: Clarendon Press.

Hutter, B. (2005). The Attractions of Risk-Based Regulation: Accounting for the Emergence of Risk Ideas in Regulation. *CARR Discussion Paper No. 33.* http://www.lse.ac.uk/accounting/CARR/pdf/DPs/Disspaper33.pdf.

Hutter, B., & Lloyd-Bostock, S. (1990). The Power of Accidents: The Social and Psychological Impact of Accidents and the Enforcement of Safety Regulations. *British Journal of Criminology, 30*(4), 409–422.

Hutter, B., & Manning, P. (1990). The Contexts of Regulation: The Impacts Upon Health and Safety Inspectorates in Britain. *Law and Policy, 12*(2), 103–136.

Kagan, R. A. (1984). On Regulatory Inspectorates and Police. In K. Hawkins & J. M. Thomas (Eds.), *Enforcing Regulation* (pp. 37–64). Boston: Kluwer-Nijhoff.

Klijn, E. H., & Koppenjan, J. (2016). *Governance Networks in the Public Sector.* London: Routledge.

Kydd, S. ('Alfred'). (1857). *The History of the Factory Movement: From the Year 1802, to the Enactment of the Ten Hours' Bill in 1847* (2 Vols.). London: Simpkin, Marshall & Co.

Levi-Faur, D. (2013). The Odyssey of the Regulatory State: From a 'Thin' Monomorphic Concept to a 'Thick' and Polymorphic Concept. *Law & Policy, 35*(1), 29–50.

Löfstedt, R. (2005). *Risk Management in Post-Trust Societies*. London: Earthscan.

Mascini, P., & van Wijk, E. (2009). Responsive Regulation at the Dutch Food and Consumer Product Safety Authority: An Empirical Assessment of Assumptions Underlying the Theory. *Regulation & Governance, 3*(1), 27–47.

May, P., & Burby, R. (1998). Making Sense out of Regulatory Enforcement. *Law and Policy, 20*(2), 157–182.

McAllister, L. K. (2010). Dimensions of Enforcement Style: Factoring in Regulatory Autonomy and Capacity. *Law & Policy, 32*(1), 61–78.

McLean, I., & Johnes, M. (2000). *Aberfan: Government and Disasters*. Cardiff: Welsh Academic Press.

McQuaid, J. (2007). A Historical Perspective on Tolerability of Risk. In F. Bouder, D. Slavin, & R. Löfstedt (Eds.), *The Tolerability of Risk: A New Framework for Risk Management* (pp. 87–92). London: Earthscan.

Mendeloff, J., & Gray, W. B. (2005). Inside the Black Box: How Do OSHA Inspections Lead to Reductions in Workplace Injuries? *Law & Policy, 27*(2), 219–237.

Meyer, J. W., & Rowan, B. (1991). Institutionalized Organizations: Formal Structure and Myth and Ceremony. In W. W. Powell & P. J. DiMaggio (Eds.), *The New Institutionalism in Organizational Analysis* (pp. 41–62). Chicago: University of Chicago Press.

Moran, M. (2003). *The British Regulatory State: High Modernism and Hyper Innovation*. Oxford: Oxford University Press.

Németh, J., & Schmidt, S. (2011). The Privatization of Public Space: Modeling and Measuring Publicness. *Environment and Planning B: Planning and Design, 38*(1), 5–23.

Norrie, A. (2001). *Crime, Reason, and History* (2nd ed.). Cambridge: University Press.

Okun, A., Watkins, J., & Schulte, P. (2017). Trade Associations and Labor Organizations as Intermediaries for Disseminating Workplace Safety and Health Information. *American Journal of Industrial Medicine, 60*(9), 766–775.

Otter, C. (2016). Artificial Britain: Risk, Systems and Synthetics Since 1800. In T. Crook & M. Esbester (Eds.), *Governing Risks in Modern Britain: Danger, Safety and Accidents, c. 1800–2000* (pp. 79–104). London: Palgrave.

Pantti, M. K., & Wahl-Jorgensen, K. (2011). "Not an act of God": Anger and Citizenship in Press Coverage of British Man-Made Disasters. *Media, Culture and Society, 33*, 105–122.

Pearce, F., & Tombs, S. (1990). Ideology, Hegemony, and Empiricism. *British Journal of Criminology, 30*(4), 423–443.

Pidgeon, N. (1998). Risk Assessment, Risk Values and the Social Science Programme: Why We Do Need Risk Perception Research. *Reliability Engineering and System Safety, 59*(1), 5–15.

Pidgeon, N., Walls, J., Weyman, A., & Horlick-Jones, T. (2003). *Perceptions of and Trust in the Health and Safety Executive as a Risk Regulator [Research Report 100]*. Sudbury: HSE Books.

Pires, R. R. (2011). Beyond the Fear of Discretion: Flexibility, Performance, and Accountability in the Management of Regulatory Bureaucracies. *Regulation & Governance, 5*(1), 43–69.

Power, M. (1997). *The Audit Society: Rituals of Verification.* Oxford: Oxford University Press.

Power, M. (2004). *The Risk Management of Everything.* London: Demos.

Power, M. (2007). *Organized Uncertainty: Designing a World of Risk Management.* Oxford: Oxford University Press.

Prassl, J., & Risak, M. (2015). Uber, Taskrabbit, and Co.: Platforms as Employers-Rethinking the Legal Analysis of Crowdwork. *Comparative Labor Law & Policy Journal, 37*(3), 619–652.

Prosser, T. (2010). *The Regulatory Enterprise: Government, Regulation, and Legitimacy.* Oxford: Oxford University Press.

Radaelli, C. M. (1999). The Public Policy of the European Union: Whither Politics of Expertise? *Journal of European Public Policy, 6*(5), 757–774.

Rhodes, G. (1981). *Inspectorates in British Government. Law Enforcement and Standards of Efficiency.* London: Allen and Unwin.

Rhodes, R. (2001). *Understanding Governance: Policy Networks, Governance, Reflexivity and Accountability.* London: Open University Press.

Rothstein, H., Borraz, O., and Huber, M. (2013). Risk and the Limits of Governance: Exploring Varied Patterns of Risk-Based Governance across Europe. *Regulation & Governance, 7*(2), 215–235.

Rothstein, H., Beaussier, A., Borraz, O., Bouder, F., Demeritt, D., de Haan, M., et al. (2015). When 'Must' Means 'Maybe': Varieties of Risk Regulation and the Problem of Trade-Offs in Europe. *HowSAFE WP, 15*(1), 1–21.

Samuels, A. (1963). Offices, Shops and Railway Premises Act, 1963. *Modern Law Review, 26*(5), 539–542.

Scharpf, F. W. (1999). *Governing in Europe: Effective and Democratic?* Oxford: Oxford University Press.

Scott, C. (2004). Regulation in the Age of Governance: The Rise of the Post-Regulatory State. In J. Jordana & D. Levi-Faur (Eds.), *The Politics of Regulation: Institutions and Regulatory Reforms for the Age of Governance* (pp. 145–174). Cheltenham: Edward Elgar.

Sirrs, C. (2016). Risk, Responsibility and Robens: The Transformation of the British System of Occupational Health and Safety Regulation, 1961–1974. In T. Crook & M. Esbester (Eds.), *Governing Risks in Modern Britain: Danger, Safety and Accidents, c. 1800–2000* (pp. 249–276). London: Palgrave.

Sparrow, M. (2000). *The Regulatory Craft.* Washington, DC: Brookings Institution Press.

Suchman, M. (1995). Managing Legitimacy: Strategic and Institutional Approaches. *The Academy of Management Review, 20*(3), 571–610.

Swuste, P., Gulijk, C., & Zwaard, W. (2010). Safety Metaphors and Theories: A Review of the Occupational Safety Literature of the US, UK, and the

Netherlands, Till the First Part of the 20th Century. *Safety Science, 48*(8), 1000–1018.

Teubner, G. (1987). Juridification: Concepts, Aspects, Limits, Solutions. In R. Baldwin, C. Scott, & C. Hood (Eds.), *A Reader on Regulation* (pp. 389–440). Oxford: Oxford University Press.

Thomas, M. W. (1948). *The Early Factory Legislation: A Study in Legislative and Administrative Evolution*. Leigh-on-Sea: Thames Bank.

Thornton, D., Gunningham, N., & Kagan, R. A. (2005). General Deterrence and Corporate Environmental Behavior. *Law and Policy, 27*(2), 262–288.

Tombs, S., & Whyte, D. (2007). *Safety Crimes*. Cullompton: Willan.

Tombs, S., & Whyte, D. (2010). A Deadly Consensus: Worker Safety and Regulatory Degradation under New Labour. *British Journal of Criminology, 50*(1), 46–65.

Tombs, S., & Whyte, D. (2013). The Myths and Realities of Deterrence in Workplace Safety Regulation. *British Journal of Criminology, 53*(5), 746–763.

Tyler, T. R. (2011). The Psychology of Self-Regulation: Normative Motivations for Compliance. In C. Parker & V. L. Nielsen (Eds.), *Explaining Compliance: Business Responses to Regulation* (pp. 78–102). Cheltenham: Edward Elgar.

van der Heijden, J. (2017). Brighter and Darker Sides of Intermediation: Target-Oriented and Self-Interested Intermediaries in the Regulatory Governance of Buildings. *The ANNALS of the American Academy of Political and Social Science, 670*(1), 207–224.

Walls, J., Pidgeon, N., Weyman, A., & Horlick-Jones, T. (2004). Critical Trust: Understanding Lay Perceptions of Health and Safety Risk Regulation. *Health, Risk, and Society, 6*(2), 133–150.

Walters, D., Johnstone, R., Frick, K., Quinlan, M., Baril-Gingras, G., & Thébaud-Mony, A. (2011). *Regulating Workplace Risks: A Comparative Study of Inspection Regimes in Times of Change*. Cheltenham: Edward Elgar.

Warburton, R. M. (1980). The Factory Inspectorate: Its Changing Roles. *Employee Relations, 2*, 6–11.

Wells, C. (2001). *Corporations and Criminal Responsibility* (2nd ed.). Oxford: Oxford University Press.

Wilson, G. K. (1983). *The Politics of Safety and Health: Occupational Safety and Health in the United States and Britain*. Oxford: Clarendon Press.

Wilson, J. Q. (1980). The Politics of Regulation. New York, NY: Basic Books.

CHAPTER 7

Justice and Values:
Whose Interests Are Served?

INTRODUCTION

The previous chapter explored how the perceived legitimacy of health and safety has been influenced by 'functional' aspects: in essence, by the ways it has been implemented and applied. This was related to the 'inputs' that those involved in regulation brought to the changing field of health and safety: in particular, ideas of expertise and efficiency. This tied back to Chapter 5, which examined *who* had a role in making decisions about the health and safety priorities and the democratic mandate (or not) of action. However, these aspects remain only a part of the legitimacy equation. Following Beetham (1991), there are important normative elements which require consideration if we are to develop a rounded picture of the legitimacy of health and safety over time. These elements relate to Black's (2008) 'justice' claims—so, is what has been done in the best interests of society? And whose values were found within health and safety as it was enacted? These factors relate, therefore, to a broad consideration of perceptions of what health and safety has achieved. These are matters of *output legitimacy* (Scharpf 1999). According to Suchman (1995: 579) and Black (2008: 145), this involves exploring whether health and safety has been viewed by society as pursuing outcomes that are seen as 'positive' or as improving welfare (Meyer and Rowan 1991: 50). Understanding how far what has been achieved is perceived as aligning with the moral views and normative preferences of

© The Author(s) 2019

P. Almond and M. Esbester, *Health and Safety in Contemporary Britain*, https://doi.org/10.1007/978-3-030-03970-7_7

society is particularly significant when it comes to areas of potential contest such as health and safety regulation (Almond 2007).

This chapter accordingly examines the legitimacy of health and safety since 1960 from the perspective of justice and values, discussing whose interests might have been served. In particular, and in contrast to the primarily institutional focus of much of the preceding three chapters, which were concerned largely with the sources and outcomes of processes of control, and the contexts in which they operate, the question of competing interests moves our attention towards the individuals who are *subjects of* health and safety. Workers are vitally important participants around health and safety, and their interests are notionally most central to the protections that it introduces, but they are often overlooked within regulatory studies in favour of a broader interest in institutions, processes, and strategic outcomes (Almond and Gray 2017). At the same time, concerns about imbalances in terms of the values represented and the interests pursued by health and safety tend to relate to a long-standing sense that workers, and marginalised workers in particular, are the constituency who are most likely to miss out or not be represented within health and safety as it is enacted in practice. How far is this true? Has health and safety continued to evolve in the direction of enacting a welfarist, protectionist social agenda (which arguably motivated health and safety reforms in earlier eras), or has there been a shift in the core values that drive action in this area?

Answering such questions, and providing an appropriately rounded response to the overarching question of legitimacy that motivates this study, requires that proper attention be paid to the granular and lived realities of health and safety at the level of individuals. There are four key features to this. The first relates to a particular understanding of the interests seen as being served that has risen to prominence since the 1990s: the apparent commercialisation of health and safety. As demonstrated in Chapter 5, the economics of health and safety have been an underlying concern for many actors since the nineteenth century at least. However, the more recent focus on commercialisation reflected a change in direction, in terms of who was involved in overseeing or promoting health and safety, and what their vision or motives might be. This was intimately connected to the rolling back of the state under a neoliberal agenda (using 'New Public Management' models) and to the rise of private stakeholders to fill the gap—the safety profession and the insurance industry.

The second major aspect touches on an issue raised at various points in the book, about the 'harmony of interests' and supposed rise and decline of consensus in health and safety. This would presuppose some degree of alignment between the parties involved and an idea that all interests have been served. However, this chapter will demonstrate that the supposed mutual interest was rather more contested than imagined. Focusing particularly on the 1974 Health and Safety at Work Act, the emergence of the Health and Safety Executive (HSE) and Health and Safety Commission (HSC), it will show that increasingly from the early 1980s onwards, conflict about health and safety became more open and increasingly embedded in the sphere of high politics and public debate. Ideas about 'shared values' therefore require rather careful handling. The final two elements to the chapter are focused at a more 'every-day' level, looking at how much choice and autonomy workers had in health and safety 'on the ground.' This will be used to reflect upon the degree to which workers' interests, as opposed to those of others, such as employers, were being addressed. This will include some consideration of worker agency and autonomy, including the steps they were able to take to protect themselves. Finally, the chapter will consider how far wider societal and workplace cultures, shaped around notions of masculinity and femininity, had an impact on health and safety outcomes, and the degree to which these impacts might be seen as detrimental.

Commercialisation

Whilst the economics of health and safety regulation have been subject to long-standing debate (for example, Breyer 1982; Ogus 1994; Pouliakas and Theodossiou 2013; Veljanovski 1983; Viscusi 1983), this has traditionally focused on the financial implications of accidents or implementing safeguards, and was often couched in rather straightforward terms of monetary cost. However, the rather broader issue of commercialisation—selling health and safety services and advice—has prompted new questions about the impact it has on perceptions of those involved in the field. Since approximately the 1990s, this has become a fundamental tension, influencing perceptions of health and safety and decisions about what is legitimate in the area. This is particularly germane because social science research has tended to suggest that the normative values that regulation seeks to pursue (essentially welfarist, prosocial, altruistic character and public service ethos) are central to the legitimation of health and safety as an area of activity (Mascini et al. 2013; Pidgeon et al. 2003; Walls et al. 2004; Zwetsloot et al. 2013). The evidence explored in Chapter 2 also suggested that the enduring public support for health and safety was linked to its character as a matter of guaranteeing basic standards of 'social citizenship' (Prosser 2006), and that moves which compromised this normative positioning were problematic in terms of perceived legitimacy. In particular, the commercialisation of 'health and safety', as a field, prompted very strong expressions of disapproval from focus group respondents. This was seen in the mixed attitudes towards the emergent safety profession. The more it was seen as a 'job creation exercise', as self-serving, and as something that was paid for rather than undertaken because it was the right thing to do, the greater attitudes towards health and safety hardened. Of course, these perceptions are variable and reflect a range of different factors, but they pose a crucial question; what impact has the increasing commercialisation of the field of health and safety had on its legitimacy?

The connection of financial imperatives to issues of health and safety is not new, of course. The Workman's Compensation Act of 1897 provided for a minimum level of compensation for some injured workers (or their relatives) without the need to prove negligence on the part of the employer (Asher 2003; Bartrip and Burman 1983). This statutory scheme provided some (though crucially not all) workers with a means of redress which they previously lacked (Banks 2009), although it did lead to the distancing of these cases from the criminal law, categorising them as issues of private compensation rather than as public wrongs (Bartrip and Fenn 1983: 66). This created the foundations for a parallel system of non-state 'regulation' running alongside the statutory, state-led oversight of health and safety, one driven by the commercial imperatives of insurance, compensation, and private risk-amelioration (Hutter 1997: 183; Viscusi 1983). Once a system of compensation was in place, employers started to seek to manage their potential exposure via new schemes of Employers' Liability Insurance (Brown 1931), and this was subsequently made a compulsory feature of health and safety landscape by the Employers' Liability (Compulsory Insurance) Act 1969, which sought to ensure that employees should not be left at the mercy of employers who could not cover the cost of compensating for injury, and to ensure that breaches of any insurance contract by the employer did not rebound onto employees (Simpson 1972). This Act was passed under the same Government which instigated the Robens Committee in 1970 (though it was introduced as a Private Member's Bill), and formed part of an effort to rebalance the distribution of health and safety risk within the workplace.

Commercial imperatives have acted as drivers of health and safety provision via the competing monetary interests of workers and employers, and the perceived costs of compliance (which have long motivated resistance to regulation). Trades unions have, since the 1870s, sought to secure the interests of their members via the use of compensation claims and legal action following injury (Asher 2003), although this focus has at times been viewed negatively, as in 1970, when it was claimed that *"the Unions have made a negligible contribution to accident prevention, preferring to concentrate on the pursuit of compensation"*.[1] However, the unions' pursuit of compensation has, in more recent historiography, been seen as part of a 'mixed economy' of approaches to pushing employers to improve safety, initially, and later health (Long 2011; McIvor 2013: 165; Melling 2003). Health and safety has also been as an issue around which unions can 'organise' and recruit members, especially in newer sectors and areas of industry (Dan Shears Interview, para. 73). And, like any business cost, employing companies have sought to limit those associated with health and safety (both compliance and breach), through the use of the legal system, professional health and safety advice, and other products and services designed to reduce costs for a fee.

In four specific fields, the extent to which commercial considerations have impacted upon health and safety has increased significantly in recent years, prompting considerable disquiet in many quarters. The first key trend was analysed in Chapter 6: the emergence of a safety profession made up of individuals and organisations that profited from their work in this field. IOSH and its members, RoSPA, and the British Safety Council, among others, provide services to safety professionals and duty-holders (training, accreditation, materials, and professional development). At the same time, the self-regulatory, open-ended nature of the 1974 Act (discussed in Chapter 4) has meant that duty-holders have increasingly sought to source health and safety advice, compliance and management services from professionals. Since the compulsory introduction of written risk assessments via the COSHH Regulations (part of

[1] N. T. Freeman (BTP Ltd.), 'Safety in the Seventies', in Minutes of Teesside Industrial Accident Prevention Committee meeting, minute 3248, 25 November 1970. Royal Society for the Prevention of Accidents [RoSPA] Archives, D.266/2/18/11.

the 1992 'six pack' of regulations: John Rimington Interview, para. 81), this has arguably, in some areas, provided a degree of impetus towards a form of ritualism, with processes of paper audit and policy provision being undertaken as 'rituals of verification' in order to pacify legal or institutional needs rather than to address real problems in a meaningful way (Power 1997; Braithwaite 2008: 141). Typically this has been seen in those organisations with more limited past experience of health and safety regulation.

Some employers have therefore seen 'off the peg' health and safety services as a means of ticking a box for little cost, as Andrew Hale observed:

> every employer had to have qualified advice from a safety professional [...] a lot of companies simply signed contracts for a very small number of hours a year, and said, 'OK, that's safety and health solved, I've got my person on the books, they come in twice a year, we have a nice cup of tea, and they're gone before they can be a nuisance'. (Andrew Hale Interview, para. 65)

In some sectors, particularly those where there is a less well-established interest and engagement with health and safety management, these attitudes have endured, as one former HSE inspector recalled:

> you say, have you got any paperwork? [...] they'll open up this file and there'll be something that had been done by a consultant and everything was page 1 of 1. And their belief was, I've paid somebody a lot of money to do this and now I'm covered because I can get you a load of paperwork that says this has been done. The fact that it was completely generic and they just literally had taken another company's name off it. (Dan Shears Interview, para. 25)

Insurers, professionals, and other parties also recognised this tendency (Peter Kinselley Interview, para. 19; Huw Andrews Interview, para. 135; Young 2010). In practical terms, it is felt that 'skin-deep' legal compliance of this sort precludes the internalisation of safety norms, to the longer term detriment of the regulatory regime. In terms of the legitimacy implications, this is an issue with which the health and safety profession has yet to engage fully and requires further work if it is not to have a continuing negative impact in the future. Whether professional organisations such as IOSH and others should position themselves primarily as commercial providers, or as public-interest advocates, is a complex point of positional politics that has implications for the way the profession is viewed.

The second area where commercial imperatives have impacted upon health and safety practice is in the provision of insurance to employers, and the subsequent influence that insurers have had on the health and safety agenda. Compulsory Employers' Liability insurance was introduced in 1969 as a 'backstop' to general health and safety provision and management, to ensure that injury claims did not fall out of the system entirely (Leka et al. 2012: 10; Simpson 1972), and from the outset, insurers were empowered (under Regulation 2(2)) to recover the costs of any liability that they are exposed to as a result of the failure of an employer to comply with legal requirements. From the 1970s, then, insurers were able to begin exerting a degree of influence over insured employers in terms of their health and safety provision. This involved taking a more proactive approach to the assessment of workplace risks, developing greater industrial expertise to allow for more accurate risk assessment and premium-setting. In the 1970s and 1980s, the changing industrial profile of the United Kingdom, along with the emergence of new technologies of production, meant that insurers had to develop their knowledge base (Leka et al. 2012: 12), as one insurance industry professional recalled:

> the [insurance] industry had developed certain approaches to risk, but in my view it was never sufficiently structured to actually say, 'what are the real risk issues?' There has historically been a reliance on past experience as being a good guide to the future. Which it is, if the future's no different to the past, but you couldn't guarantee that. I think there was always a belief that there was an amount of regulation out there, it should be helpful, particularly if the regulator is well resourced, and is a real regulator, and is likely to turn up. (Jim Wilkes Interview, para. 20)

Insurers thus became more able and willing to intervene in the maintenance and regulation of workplace safety, certainly, and, arguably increasingly, health. As a source of decentred regulation, insurers took on a degree of involvement in setting out expectations around prevention, precaution and risk management around health and safety, and thus where the parameters of responsibility ought to lie (Black 2008; Ewald 2002; Scott 2004).

A shift in attitude and approach in this direction can be identified in the 1990s (Hawkins 2002: 256; Leka et al. 2012: 12; David Eves Interview, para. 79). At this time, there was perhaps a recognition that unscrutinised Employers' Liability insurance (sold as an add-on to wider insurance packages) had exposed insurers to excessive liabilities; Dalton (1998: 11) cites insurance industry figures from 1994, suggesting that for every £1 of Employers' Liability premium received, insurers were paying out £1.24 in claims and costs. The threat of increased premiums, and a desire to minimise these liabilities, changed the relationship between insurers and their clients, as the head of Casualty Practice for the UK's largest Employers' Liability insurer, observed:

> we're processing tens of thousands of claims every year, we're developing knowledge and insight about what 'good' looks like for their sector [...] there's an expectation from our customers that they will get something to differentiate the insurance proposition and that they see us very much as an advisor to their organisations, that we can improve their performance to minimise the cost of risk in their business. (Huw Andrews Interview, para. 102)

When combined with the message that the largest proportion of costs incurred following a workplace failure were not actually insured at all (Dalton 1998: 10), this role allowed insurers to reposition themselves as drivers of compliance. It is fair to say that this positive appreciation of insurers has not endured, being challenged and, to an extent, superseded, by an emergent public narrative (identified in Chapter 2) that casts insurers as engines of 'over-regulation', pushing for standards of compliance that are much higher than those laid down in law, and which require the employment of health and safety consultants (Young 2010). At the same time, there have been concerns that this has led to a reconstruction of some of the specificity and prescriptiveness which the 1974 Act was meant to have removed, as one former HSE policy actor observed:

> if you can get your insurance company to insure you by having a list of rules that everybody must obey and a list of rules that says whose fault it is if anything goes wrong, then for an awful lot of busy, small [firms...] It's tempting to have a ready-made code [...] the authorities are blamed for the fact that a lot of it seems nit-picky and disproportionately meddlesome. (Helen Leiser Interview, para. 69)

Overall, at the end of the 2010s, attitudes towards insurers remain highly controversial. Many interviewees voiced very critical attitudes towards insurers, particularly around their tendencies to *"spread risk rather than concentrate[d] on the areas where action needs to be taken"* (Jenny Bacon Interview, para. 60), to *"over-contest"* liability (John Armitt Interview, para. 30), and to take an amoral attitude to risk:

> this guy had been standing under a vat of acid and the vat opened [...] I raised that with an American insurance company [...] He said, 'it's not costing us money mate, if they pay the premiums then we don't mind [...] we're making our money and the rest of it's their fault'. (Frank Doran Interview, para. 28)

Such accusations were refuted by the insurers themselves, however, who saw themselves as service providers like any other, and a convenient target for others to blame for any shortcoming in the system:

> It's a myth. And we seek to challenge it in every way we can. We are a convenient target because historically we've never really stood up and argued our case. We also accept that very often it's legitimate for a company in trying to put through some risk control measure, which is unpopular in a business, to say, it's an insurance requirement. (Jim Wilkes Interview, para. 143)

It is interesting to note the extent to which regulatory myths (and thus media and political discourse) have echoed these criticisms (Almond 2009). After all, it is hard to conceive of a means of controlling risk which is *less* solidaristic and collectivist, and more compatible with an individualised and marketised approach to governance (Ewald 2002). Given that insurance has long been seen as something *"responsible businessmen normally do as a matter of prudence"*,[2] it would seem that insurance ought to be wholly compatible with the broader movements towards self-regulation and individualism. But 'actuarial' modes of thinking, which are increasingly pervasive in areas of public risk management, involve the division of populations into those who are more or less deserving of protection according to their levels of self-reliance

---

[2] Lord Drumalbyn, H. L. Deb., Hansard vol. 304, col. 1061, 23 July 1969.

and moral character (Ericson and Doyle 2003; Ewald 2002; O'Malley 2004); this can involve the exercise of very significant, and often seemingly unaccountable and arbitrary, power, meaning that the legitimacy of insurers, as a now-central component of the health and safety system, is deeply problematic (Daniels and Sabin 1997), and thus poses problems for other actors in the system as well.

The third key shift is one which is closely linked to that of insurance—litigation and the pursuit of compensation for injury and ill-health. At the end of the 1960s, it was estimated that no more than 20% of injured workers pursued civil liability claims for negligence or breach of duty (Simpson 1972: 68). But since approximately 2000, concern has grown over the emergence of a 'compensation culture', the perception that society has become increasingly litigious and blame-focused, that the numbers of personal injury tort cases and the amount of compensation awarded have increased markedly, and that the result has been a diminution of notions of personal responsibility and an increase in risk aversion (Morris 2007). Both the Young Review (2010) and Löfstedt Review (2011) engaged with this issue as a matter of course, prompted in no small part by a very pronounced political agenda of resistance to the 'burdens' of civil liability in this area (Almond 2015). While the empirical data on rates of claiming lends little support to assertions that this has been a major development in recent years, there is evidence that a *perception* of excessive claiming has had an influence over health and safety decision-makers (Wright et al. 2008). Those in the insurance industry are typical of many, in that they regarded lawyers and claims management firms as having driven much of the 'problem' of risk aversion:

> claimant lawyers, they are seeking to expand their businesses. Why wouldn't they? So they're always looking for areas that perhaps could develop claims [...] some of the advice we're able to give to people, it's not so much just health and safety, but it's to say, by the way, these are the areas that a claimant's lawyer may seek to attack you on. (Jim Wilkes Interview, para. 105)

In common with other groups (particularly the general public), those within the health and safety sector saw civil compensation as a new, pervasive problem, and were quite censorious in attitude towards those who make claims, and those who facilitate them in doing so, as this former HSC Commissioner illustrates:

there is an issue around blaming work, and that ethos, rightly or wrongly, has changed in the last 50 years, from workers putting up with all sorts of terrible conditions [to now...] looking for someone to blame. 'It's not my fault, somebody's done it to me' [...] that's led to the increase in litigation, and it's not that people are successful, it's just the [...] have-a-go culture. (Sayeed Kahn Interview, para. 74)

For much of the period since 1960, regulators tended to distance themselves from the issue of civil litigation, seeing it as something that did not properly concern them:

the Commission said, 'our job is to deal with the criminal side and not to get involved in the civil side'. We should just leave it, it wasn't our responsibility [...] the unions [...] see the role of compensation and the unions pursuing compensation claims as a key element of the health and safety system. A lot of the flak that was coming HSE's way [...] should be directed rightly or wrongly to those who were taking compensation cases. (Bill Callaghan Interview, para. 113)

Callaghan's comments suggest a helpful chronology on this issue: that for much of our period, the pursuit of compensation was something that was driven and facilitated by the trade unions as a means of protecting and advancing their members' interests. Suspicion about the benefits paid to injured employees, both in terms of encouraging 'malingering' (Leneman 1993) and as being overly generous, was not new, of course: to one meeting of accident prevention officers in the railway industry in 1967, it *"seemed significant that in certain respects, viz. industrial injury benefits and income tax rebate, it almost paid a man to be off from work due to an accident"*.[3] But, in contrast to previous eras, the resistance in the twenty-first century has been less one of scaling back worker's rights to compensation (although see the Löfstedt Review's recommendations around strict liability (2011: 91) and the subsequent Enterprise and Regulatory Reform Act 2013 s.69) and more framed in terms of preventing undeserving claims in the area of health and safety beyond the workplace. It is now rarely explicitly asserted that workers should not be

[3] British Rail, Meeting of Regional Accident Prevention Representatives (16 May 1967), minute 1. The National Archives of the UK, London [TNA], AN 208/218.

compensated for injury; rather, objections focus on the public as 'undeserving' participants, mirroring the focus group discussions explored in Chapter 2.

The effects of compensation claims were most clearly seen and felt in the field of perception and preventative activity, rather than actual litigation case-load. In general terms, it appears that, in recent years, companies and duty-holders have become more cautious and restrictive in their activities, and to be unduly led by a desire to preclude liability (for example, see Halliday et al. 2011). Public commentators and right-leaning politicians, in particular, have seen this as having made companies unduly risk-averse (Young 2010), and also resistant to the cooperative agenda of the health and safety regulators, who were now cast as potential prosecutors. This also led to a tendency to resist engagement with post-incident investigations (Roger Bibbings Interview, para. 65; Graeme Collinson Interview, para. 66; David Maidment Interview, paras. 94–95). An associated effect has been a tendency towards producing local, self-imposed sets of highly prescriptive and restrictive rules for duty-holders and employees to follow. Partly, this links to a legalistic approach to defending and contesting responsibility by reference to specific requirements and rules:

> I put it down more to the lawyers than the safety professionals, that this picture of gold plating and rules which were not necessary, because that fitted into the compensation culture. You had to have simple rules which you could prove people had broken, otherwise you wouldn't get compensation out of the company. So it was more in the interests of the lawyers to have the detailed regulations, rather than the safety people. (Andrew Hale Interview, para. 63)

This would constitute an example of a process of juridification, the process by which legal requirements become more precise, detailed, pervasive, and numerous over time (Teubner 1987). One of the potential negative effects of juridification of this sort is a reduction in the perceived legitimacy of the law, which became more self-referential, insular, and less understandable and accessible as a result. This was observed by the Deputy Director of the Engineering Employers' Federation in the early 1960s: "*the danger we have to guard against today is of passing the point of diminishing returns so far as legislation is concerned. Once we get past this point there is a risk of the law being brought into disrepute and this is a situation against which we must at all times endeavour to protect ourselves".*[4] All the elements of the commercialisation agenda set out here have arguably had similar effects; the degree to which perceptions and expectations have driven the agenda here demonstrate that this distancing might underpin at least some of the current concerns around health and safety.

The fourth element of commercialisation to have impacted upon the legitimacy of health and safety is the status of the regulators themselves, and the degree to which their activities appear to involve a degree of commercial service provision. Whilst it might be assumed that the inspectorates have, until relatively recently, been understood as 'just' enforcers who engage in responses to the efforts and actions of regulated duty-holders, actually, the pressures to develop commercial services are surprisingly long-standing. In 1967, for example, the National Union of Agricultural Workers was voicing concern over the *"bad psychology"* of Ministry of Agriculture inspectors holding conflicting advisory, enforcement, and production-oriented responsibilities.[5] The Robens Committee asked witnesses *"whether the [Factory] inspectorate can or should be turned into a consultancy as well as an enforcement agency"*. One of the RoSPA witnesses replied: *"the Inspectorate feel like the Police – they do not want to be restricted to enforcement. They want to be able to consult and to advise [...] they want to be thought of in a positive role instead of in a purely negative one"* before agreeing that a consultancy role would be appropriate.[6] At a lower level, newly introduced charges for HSE guidance notes met with complaint in 1979, with one concerned individual noting that requiring a deposit of ten pounds before orders would be accepted *"in*

---

[4]T. A. Swinden, paper presented to Joint Industrial Conference on the prevention of accidents, 13 November 1962, p. 10. Trades Union Congress [TUC] Library, HD 7273.

[5]National Union of Agricultural and Allied Workers, Legal Department, Memorandum to Executive Committee following meeting with Minister of Agriculture, 21 March 1967, p. 2. Museum of English Rural Life, University of Reading [MERL], SR 5NUAW/B/XXI/4.

[6]Verbal evidence of W. G. Alexander on behalf of RoSPA to Robens Committee, 14 June 1971, transcript, pp. 4–5. TNA, LAB 96/75.

effect stops any shop steward getting them".[7] The linking of income to inspectorate activities has therefore been an ongoing process, at least discussed for most of the past fifty years.

The depth of challenge that this poses to legitimacy has increased over approximately the last ten years, however, as HSE has been pushed towards an increasingly commercialised position, with elements of its work either charged for, or reorganised along commercial lines. Newly 'marketised' regulatory strategies, such as the 2012 introduction of 'Fee for Intervention' (FFI), whereby HSE undertakes cost recovery when it intervenes in response to breaches of the law,[8] have arguably shifted the regulator into a new relationship with regulated firms (Almond 2015: 226–227; Baldwin and Black 2016: 575; Temple 2014). On the one hand, in certain areas, a 'payment for approval' relationship has been a relatively longstanding feature of specialist, high-hazard industries (such as nuclear and offshore) where site licensing has been a standing prerequisite to operation. These measures reflected the high degree of specialisation required, and the particular needs of the regulatory relationship in those sectors, and constituted a particularly onerous arrangement, as a former HSE Director-General explained:

> [it was] an absolutely necessary feature of a very strict safety regime [...] It's a very severe thing in several ways because it means that you can take away someone's livelihood [...] In general HSE have never been in favour of licensing arrangements, because, in effect, they transfer the responsibilities and much of the liability for an accident on to the regulator whereas the central principle of the Robens arrangements is that the responsibility for safety must lie on the operator. (John Rimington Interview, para. 89)

It is clear that, over time, government has sought to redraw the limits of these relationships, seeking to pass the costs of regulating from the public to the private sectors wherever possible, and expanding the rationality of commercialised service-provision to a wider range of workplaces. Indeed, Select Committee hearings of the early 1990s asked whether HSE could be privatised entirely and its operations passed on to private organisations and voluntary bodies (Employment Committee 1992:

---

[7]E. C. Parker, letter to *Work Hazards* 21 (c. 1979), p. 17. Samuel Barr Collection, Glasgow Caledonian University [GCU] DC 140/2/1/2.

[8]See http://www.hse.gov.uk/fee-for-intervention/ and Temple (2014) for discussion.

paras. 41–59), something that senior officials at HSE at the time had significant reservations about:

> That raises very great difficulties. We do charge for certain activities. We charge wherever we issue a license [...] in principle, anyone can charge for anything, if he has a statute saying he could do it, but we could not charge fairly. (John Rimington Interview, paras. 47–48)

Later in the 1990s, individuals working within HSE were being pushed again to look at charging regimes. According to one offshore inspector, within HSE "*there was unease [...] because we felt that [...] if a company has to pay for coming along and having a meeting with us to talk about something where they're going to be proactive in terms of health and safety that that might actually discourage these sort of meetings, you know, and [...] the full interchange of information*".[9] Other senior managers within HSE were also sceptical about licensing as it creates a dynamic whereby the license-granter would always be residually responsible, on some level, for having 'approved' the actions of the license-holders ("*Always ask for a licence if you want to shift the responsibility*": Jim Hammer Interview, para. 21). Similarly, senior figures within the regulatory infrastructure were also sceptical of such arrangements, such as Sir Frank Davies, HSC Chair 1993–1999 and a very experienced businessman:

> I disapprove, very strongly, of charging. All the time I was [at HSC] I was under pressure from the Government and from the HSE to charge [...] my Commission resisted it and we did not let it in. Once I went charging started [...] it's made things very unpopular. I think it's a bit tough if you're charged, because we're going to prosecute and we're going to charge you while we collect the evidence. I can tell you this, the big companies I've associated with, don't like it. (Frank Davies Interview, para. 66)

Davies's concern was twofold; that the difficult and sometimes tense relationship between HSE's facilitative and investigative functions (Hawkins 2002: 286–293) would be damaged, and that the motives of regulatory actors would be seen as less legitimate, if they were tied to a commercial imperative to intervene. At the same time, according to

---

[9]Martin Thompson, University of Aberdeen 'Lives in the Oil Industry' project, tape 4 side a. British Library, C963/13.

those in HSE who implemented these changes, a government commitment to making big business bear the costs of operating was opposed by industry voices, but they were won round over time as the procedural safeguards (ensuring *"transparency, proportionality, and fairness"*) on the licensing system developed and took force (Neal Stone Interview, paras. 37–41).

By the 2010s, the move towards charging had re-emerged in the guise of FFI, a policy which proposed that all firms subject to regulatory inspection should be charged for the costs of processing and addressing any 'material breaches' of the law that were found as a result of HSE intervention or investigation. This was prompted by a pressure from government during the post-2008 'austerity' era for HSE to become more self-financing and to recover a greater proportion of its operating costs from its business activities rather than Treasury underwriting. The issue of fairness was invoked in relation to ensuring that regulatory burdens are placed upon those who 'deserve' to bear them, something that the current HSE Deputy Chief Executive was keen to reiterate:

> FFI is the health and safety equivalent of the polluter pays. We have got limited resources at our disposal and so therefore it is right, in my view that we ought to focus those limited resources on the people that are most in need of those resources […] There are risks in terms of managing that Chinese wall - you can't be a consultant one day and a regulator the next day. (Kevin Myers Interview, para. 61)

However, many stakeholders (such as business organisations)[10] were critical of the new arrangements, characterising this as a form of 'safety tax' on businesses, which would fatally undermine the impartiality and neutrality of HSE inspectors, who would now become revenue-collectors. The Temple Review, a triennial government review of HSE's operations, concluded that this perception ran a serious risk of undermining trust in the regulator, making them seem more like revenue-raisers than impartial regulators (Temple 2014).

---

[10]For example, the Institute of Directors: http://www.iod.com/influencing/policy-papers/regulation-and-employment/hse-proposal-for-extending-cost-recovery, and the CBI (in 2004): http://www.publications.parliament.uk/pa/cm200304/cmselect/cmworpen/456/456we42.htm (at para. 20).

These concerns about commercialisation were mirrored elsewhere in relation to perceptions of the legitimacy of regulators themselves, with some observers suggesting that the broader, ongoing commercialisation agenda would undermine the credentials and effectiveness of the regulator:

> HSE is being turned into a delivery agency with a significant element of self-funding [...] what is lost in that potentially is the ability of the civil servants who run it to think about improvement and think about how things can be done differently and better. And I think there's a great danger at the moment that HSE will not be able to [...] exercise the level of creativity required to take health and safety forward. (Roger Bibbings Interview, para. 35)

Commercialisation has been broached in terms of 'monetising' the expertise of HSL, the specialist HSE laboratories, as well as realising the value of assets and other forms of revenue-raising activity (publications, products, etc.). HSE had done this for many years, as the Director-General of HSE pointed out in 1992, "*we charge for about a hundred things we do [...] we sell a lot of publications [...] certificates [...] licenses*".[11] However, the pressure to generate income intensified as the austerity policies of the era after the 2008 financial crash took hold within government, something that regulatory insiders expressed concern about:

> I'm not in favour of the commercialisation agenda of HSE [...] which has been forced on them by Ministers. There are things they can sell, but they should be careful because once you go past a certain point they will both get conflicts of interest and opportunity costs. (Senior HSE source Interview, para. 168)

For the most part, stakeholders indicated that they saw commercialisation as a legitimate means of extending the reach of the regulatory system, and thus as a good thing. But this balance is difficult, because doing so does change the terms and conditions on which health and safety engagement occurs. Public attitudes suggest that the perceived moral

---

[11] *Employment Committee, The Work of the Health and Safety Commission and Executive* (1991/1992 HC 263), Minutes of Evidence 12 February 1992, para. 49.

stance of those working in health and safety is one of the key strengths that those bodies have; this is lost when there is a sense of undue commercialisation. The other aspect of commercialisation is the notion that it renders health and safety a service to be paid for, rather than a process to invest in—that companies outsource it, and pay lip service as a result. As we have already seen, these risks have been identified and have been realised at different points in time as well, with potentially toxic associations and implications for the legitimacy of health and safety and regulation.

The Limits of Consensus?

As Chapter 3 noted, the idea that a 'consensus of interest' between political parties and stakeholders existed during the middle part of the twentieth century is one of the foundational tenets of the field of health and safety. Most of the social science literature casts the formative period and the major reforms between 1960 and 1980 in this light (Baldwin 1987, 1995; Dawson et al. 1988; Hutter 1997; Sirrs 2016), and indeed, the idea of an 'identity of interest' between affected parties was core to the Robens philosophy (Robens 1972, para. 21). What is clear, however, is that this supposed consensus and agreement was more patchy, partial, and problematic than might be thought, with disagreement, variation, and tension existing just beneath the surface (McIvor 2013; Sirrs 2016). Sometimes this disagreement was more explicit, as in this summary of the Robens Report by Neil Kinnock MP during Parliamentary debate in 1973:

> the report has a grievous deficiency which makes it an inadequate basis for the kind of action required. Audacity, the quality that is most sorely needed in bringing about a meaningful change in this sphere of affairs, is completely and totally absent from the report [...] The report is afraid to give any support to the concept of giving workers any meaningful power in matters of health and safety [...] these major defects are the direct consequence of a fundamental confusion about the reason for the shortcomings in the present system.[12]

The sense of health and safety as an area of disagreement is also borne out via historical (Carson 1979; McIvor 2013) and critical literature

---

[12] 'Workers' Safety and Health', HC Deb 21 May 1973, Hansard vol. 857 cc62–117, 65.

(Beck and Woolfson 2000; Dalton 1998; Nichols 1997; Tombs and Whyte 2010, 2013; Tucker 1995) that demonstrates the roots of these conflicts in the enduring tension between capital and labour interests. As such, we have to exercise great caution before accepting that there was ever a true era of 'consensus' around health and safety (or indeed post-war politics more generally: Kerr 2005; Thane 2018). That said, there is evidence of some form of substantive change, in both the nature of debate, and in where the frontiers of conflict were drawn, over time, and it appears that this shift has been towards a fracturing of consensus. How far this is accurate is worth some consideration.

To begin with, it is worth reflecting on the degree of consensus that surrounded the major changes that were introduced in 1972–1974; there were points of conflict and disagreement, and many who felt that either Robens or the 1974 Act (or both) gave an undue advantage to the interests of those on the other side of the political spectrum. For many trade unionists and critical observers, the lack of a developed role for safety representatives, and the emphasis on 'too much law' and 'worker apathy' were unpalatable (Nichols and Armstrong 1973); for many conservatives, the extension of state oversight via such a wide-ranging Act ran contrary to their small-state instincts. The Labour government of 1974 sought, when introducing the Act, to position it as a socialist project, partly as a response to the fact that development of the legislation had occurred under a Conservative administration:

> The first author of the Bill, indeed the author of many good things, is my right hon. Friend the Secretary of State for Social Services [Barbara Castle]. It was her decision to invite Lord Robens to preside over the Committee on Safety and Health at Work [...] It would have been the one disadvantage of that happy outcome [Labour winning the 1974 Election] if in the process the Bill had been lost [...] We believe that it is a good idea that this Bill should have been brought forward.[13]

The general tenor of the debates around the Bill was positive, with opposition MPs speaking in its favour.[14] There was a broad acceptance of the need for reform, and of the direction being taken.

---

[13]Michael Foot, 'Health and Safety at Work Etc. Bill', HC Deb 3 April 1974, Hansard vol. 871 cc1286–1394, 1287.

[14]William Whitelaw, Conservative, ibid., 1302; Cyril Smith, Liberal, ibid., 1322.

This said, it was not the whole story. There were some areas where attitudes did diverge, one of which was the exclusion of agricultural workers from the scope of the Bill. The Agriculture Inspectorate was founded in 1956 via the Agriculture (Safety, Health and Welfare Provisions) Act 1956, in order to cope with the particular challenges that this sector threw up, including isolated workplaces, the employment of casual workers (often children or migrants), and the use of hazardous heavy machinery which intersected with the activities of the public (particularly children on farmland).[15] As the passing of the 1974 Act began under a Conservative government, the responses to the concerns of bodies such as the Country Landowners' Association about the impact of including agriculture within the scope of the 1974 Act betrayed a particular political leaning. The Ministry of Agriculture thus retained responsibility for agricultural safety per se, while HSE would enforce 'general safety', which might apply on farms as elsewhere. The concerns of the National Union of Agricultural and Allied Workers, and others,[16] meant that the minority Labour government which brought the Act into law following the 1974 election included agricultural workers within the Act, but withdrew them again at a late stage (in order to preserve cross-party consensus). As a result, "*agriculture was always different, it had its own ministry and problems*" (Jenny Bacon Interview, para. 14). There was clearly significant conflict around this issue, including around the difficulty of instituting adequate tripartite relations in the agricultural sector; negotiations over the implications of the new legislation and its general duties (which would apply to self-employed workers and those on farms in any case) continued for some time. The Robens-era 'consensus' around tripartism, the unification of inspectorates, the role of safety representatives, and the scope of the general duties was always shaky and limited; reform involved ongoing negotiation and dispute.

Despite this, the process of passing the new Act was relatively unproblematic because of the extensive Robens review process and the sheer volume of stakeholder input that went into the law's creation, as Jenny

[15] National Union of Agricultural and Allied Workers, Minutes of Health Safety and Welfare Committee, 12 October 1960, p. 1; 25 October 1961, p. 2. MERL, SR5NUAW BXXI/I.

[16] Such as Neil Kinnock, 'Workers' Safety and Health', HC Deb 21 May 1973, Hansard vol. 857 cc62–117, 65–66.

Bacon, who worked as a civil servant on the Bill during its passage (and who later became Director-General of HSE) recalled:

> I've had dealing with several bits of legislation during my civil service career and this was, by far, the easiest […] because a lot of the ground-work had been done by Robens and then by a good follow-up consultation process. Actually getting it through the House of Lords and the House of Commons and into the Act was far easier than one might have expected for a potentially controversial area. (Jenny Bacon Interview, para. 14)

Where the HSWA 1974 did prompt friction (the 'first-class Whitehall row' mentioned by Michael Foot during this debate: Hutter 1997: 27), it was around the problems of reshaping existing government infrastructure. The politics of implementation was more complex, as Jim Hammer, HSE's first Chief Inspector of Factories, recalled:

> the [1974] Bill was the most bipartisan operation in Westminster and the most divisive one in Whitehall […] Whitehall was up in arms. How can you steal from the Home Office the explosives inspectorate? How can you take the mines inspectorate from the Department of Power, the nuclear inspectorate from Power, the agricultural inspectorate from Ag and Fish […] they were very reluctant, it was very much resisted. (Jim Hammer Interview, para. 45)

Part of the political success of the 1974 Act, and the HSE, was the creation of the Commission, which served as a forum within which decisions about health and safety policy could be made at one step removed from the cut and thrust of partisan politics and the frontline of wider industrial relations. Unlike in other areas of policy, decisions were insulated from the emergence of new interest groups and from bodies, less politically moderate than the Confederation of British Industries (CBI) and Trades Union Congress (TUC), who were not part of the tripartite system; these actors and the issues they flagged, which engaged with the 'high politics' of social justice or industrial relations, were not functionally compatible with HSC/E's operational dynamics and focus on the 'low politics' of day-to-day decision-making (Moran 2003: 138).

As the overview of HSC confirms, it functioned as a successful 'steering mechanism' for many years, diverting potential political disagreement into a forum of manageable social interaction (Almond 2007;

Habermas 1976, 1981: 356–373). This 'consensus by committee' was to endure well into the twenty-first century, as a senior HSE figure recalled, explaining how certain voices were 'inside the regulatory tent' and others were not:

> industry weren't worried by HSE at all. They were perfectly happy [...] the fracture was between the more mainstream unions who, on the whole, still believed in tripartism [...] and didn't 'cause trouble', and people like UCATT and [the Centre for Corporate Accountability] who were peddling a particular view of the world and sought to regard HSE as the great enemy, these people got no political support [...] It was people who held that view speaking to people who held that view. (Senior HSE Source Interview, para. 114)

In general terms, the greatest feature of HSE and the health and safety system has been its enduring character and stability over time; even changes in the broader political culture have not always impacted on regulators or on the wider politics of the issue as directly or profoundly as might have been expected (Bill Callaghan Interview, para. 16). To a lesser or greater degree, there has been a broad acceptance of the regulation of health and safety as a general idea, and a tolerance of the mechanisms by which this is done. As the current Deputy Chief Executive of HSE points out, were it not so, things would have changed a long time ago: "*I actually think there is an underlying recognition of the legitimacy of health and safety regulation [...] if there wasn't would we have survived [...] the Thatcher administration?*" (Kevin Myers Interview, para. 59).

Most influential in this regard is the support given to health and safety regulation by employers and business actors, who have to some degree protected or insulated HSE via their engagement with the regulatory system, something recognised within HSE:

> the thing which affected Chris Grayling [Minister of State for Work and Pensions 2010-2012...], the reason why HSE is still there today is purely because, in my opinion, in political terms, ministers have never managed to get British industry to say they don't want it. It's as simple as that. Industry, CBI, IoD, whoever, do not say to Ministers get rid of HSE. If they did, HSE would be dead in the water. (Senior HSE Source Interview, para. 147)

Many interviewees spoke about the perceived role of a *broad* acceptance of health and safety on the part of major employers and large companies in shoring up political and social support for HSE and the British legislative system. This support can be understood in two ways. On the one hand, from an economic perspective, regulation has always involved the maintenance of the conditions governing access to markets (Braithwaite 2008; Carson 1979). It ensures a level playing field for agents within a marketplace, and acts to exclude others who may wish to enter that market but who cannot yet achieve the standards required (Braithwaite 2008: 17). This would certainly account for many of the trends encountered around the limits in practice of EU regulation (as actors from new entrant economies struggle to meet extant standards). On the other hand, business support can be understood in terms of the desire to ensure that businesses run as smoothly as possible, avoiding costly incidents and legal conflicts (Senior Government Source Interview, para. 54). This is an example of Suchman's concepts (1995) of cognitive and pragmatic legitimacy; regulation is both an inevitable and, for the most part, generally useful, fact of life. Acceptance on these grounds to some extent precludes or renders unnecessary any deeper consideration of moral or normative legitimacy, where value conflicts might arise.

That said, from the 1980s onwards, there is ample evidence of a breakdown of political consensus around the issue of health and safety, at least at the level of party politics, ideology, and media coverage. Many interviewees testified to the contested nature of policy in these areas, and the changing political fortunes of 'health and safety' at different times. Particular individuals and governments have been more or less open to health and safety at different times:

> When I was appointed [HSC Chair in 1993], at my interview I was questioned, what was the point of it and why was it necessary. There was no doubt there was some scepticism about it. (Frank Davies Interview, para. 146)

As well as the ideological and policy preferences of Ministers and governments, there is a perception that a broader political shift has occurred since c. 1980 towards the language and terminology of a neoliberal settlement (Almond 2015; Beck and Woolfson 2000; Gray 2009; Tombs and Whyte 2010), something that goes beyond immediate policy issues, and beyond health and safety itself, to encompass a new way of seeing the world. The most obvious influence of this shift has been towards a restriction of budgets and the 'desirable' role of the state in regulating business, and an antipathy towards 'burdens on business', as one HSE insider of the 1980s recalled:

> We were very used to promises of 'bonfires of regulations'. We used to be subject to investigative inquiry about every three years. Because the HSE was a nice compact unit of about 4000 people, it could be reviewed fairly quickly. Also it was generally not particularly well known, not very popular and could easily get turned over [...] these regular inquiries were also aimed at reducing the number of senior people. (Jim Hammer Interview, para. 57)

Over time, similar concerns would become entrenched across the political spectrum; New Labour governments would also become fixated on the reduction of regulatory 'burdens' and on a view of the public/private divide that envisaged a much less prominent role for state bureaucracies like HSE in overseeing the economic and business sectors (Beck and Woolfson 2000; Dodds 2006; Tombs and Whyte 2010).

Most recently, the post-2010 Conservative-led governments of David Cameron initially showed a much more pronounced resistance to the notion of health and safety regulation. Discussion around the rolling back of 'red tape' and state-led regulatory burdens intensified, and there were concerted efforts made to more narrowly target interventions and cut the financial cost of regulating in the area of health and safety (Almond 2015; James et al. 2013). One example of this trend was the publication of an influential policy report by a right-wing think-tank (called '*Health and Safety: Reducing the Burden*'; Policy Exchange 2010), which presents a fairly unvarnished small-state, socially and economically conservative argument for the reduction in the scope of health and safety regulation, and which decisively influenced the policy framing of the subsequent Young (2010) and Löfstedt (2011) reviews. In many ways, a consistent and enduring commitment to 'risk-based' regulatory approaches (Black 2005), and principles of accountability and review (the New Public Management: Hood 1991) can be discerned, which amounted to a continuation of previous Governments' approaches (as found, for example, in the Hampton Review of 2005). What arguably became more noticeable after 2000 was the degree to which the ideological climate hardened, particularly as reflected in government rhetoric, towards a broader rejection of a welfare-oriented outlook. This peaked at certain points (2010–2012) and is arguably still more pronounced than in the past, as Lord McKenzie, Labour's Health and Safety Minister from 2007 to 2010, argued:

> the rhetoric of this government has shown an antipathy to health and safety; the rhetoric at the highest levels. I'm sure the previous government would never have done that, simply it's not in their DNA to do it. Now I sense that probably some messages have got through to the highest levels, we haven't heard so much of it as we used to. (Lord McKenzie Interview, para. 51)

So, while established governmental tools of review and efficiency drive were deployed again after 2010, the thought processes that were guiding them were more explicitly *political* in nature, in that they sought to produce outcomes that mapped onto a specific worldview based around support for individualised (rather than mandated) decision-making, rationalised (rather than universal) intervention, and business-oriented (rather than welfarist) regulation (Almond 2015: 221). Underpinning this approach, which arguably moved some way away from some of the principles (state-mandates, universality, welfarism) of the 'Robens approach' and the 'consensus-era' approach to health and safety, was a much more populist, aggressive rejection of *the idea* of health and safety regulation, not just of the means by which it had traditionally been pursued. For some commentators, such as Lawrence Waterman, former Head of health and safety for the 2012 Olympic Delivery Authority, this reflected a return to a deep-seated antipathy towards solidaristic, welfare-oriented issues on the right of British politics:

> [health and safety] never really got on the Conservative Party agenda, and ever since they've been [...] 'God, we really didn't mean to join Europe and we'd like to leave'. 'We didn't really mean to support the Health and Safety at Work Act and we'd like to move backwards.' So why not express it in those terms? Because that would make them sound as though they're callously indifferent to injuries to workers, which they probably, in the main, are. (Lawrence Waterman Interview, para. 10)

Across the period of this book, health and safety has thus potentially moved from being situated first and foremost in the realm of 'low politics' of implementation and enactment, within a broadly consensual, closed field of known interests and considerations, to become a more prominently discussed feature of the 'high politics' of the day, and imbued with some of the wider ideological meanings that this might imply. Some of this has, of course, been muted since the 2016 referendum on membership of the EU and subsequent 'Brexit'-dominated politics. It remains to be seen how far health and safety might return to prominence in the political agenda, either as a result or corollary of these processes (Smismans 2017), or once the policy shock of Brexit has subsided.

The media interest in 'regulatory myth' stories since approximately 2000 and the desire of a wider public constituency for this kind of coverage and discussion have 'politicised' much of the debate around health and safety in a manner which has called the legitimacy of the issue into question. This has reflected to some extent a newly prominent, determinedly right-wing and 'anti-political correctness' thread within popular culture which has gained ground in the last decade; some central threads of this position (anti-European, anti-health and safety, 'common sense', the *Daily Mail*) combine in former Top Gear presenter Jeremy Clarkson's reference to HSE as "*having done more damage to British industry than the Luftwaffe*" (also Sayeed Kahn Interview, para. 123).[17] Ragnar Löfstedt, reflecting on the reception and impact of his Review in 2011, argued that politicians' self-interested willingness to pander to this constituency has proved to be relatively resistant to evidence-based input:

> there is still belief amongst certain politicians, and more on the right than the left, as we've seen post-Löfstedt review, they still say that health and safety is a problem [...in 2012] I was asked on the spot what I thought of the Prime Minister's discussion on health and safety that's published in the Evening Standard, and I said it was very unfortunate because it wasn't true [...] why are some politicians trying to win political brownie points with health and safety? (Ragnar Löfstedt Interview, para. 49)

That said, the evidence gathered about public attitudes, discussed in Chapter 2, does remind us of the need to separate these political currents from the general attitudes of the public; to this day, there remains a

---

[17] http://www.dailymail.co.uk/news/article-505788/Jeremy-Clarkson-Prime-Minister.html.

deep-rooted acceptance of the idea of health and safety, and recognition of the right to safety, although these preferences are embedded beneath a layer of negative public opinions, which obscure them. This is also, to some degree, true of the political sphere—health and safety is not really ever rejected *outright*, except by the most extreme sceptics. Health and safety retains a grounding of popular and political acceptance and support. The project is not as badly perceived as might be thought. There remain areas where conflicts arise, and these continue to cluster around pinch points where the needs of the job, and the tension between safety and profit, collide. While this has long been the case, this also opens doors for improvement where the business case for health and safety can be demonstrated.

Free Labour, Agency & Structural Factors

This idea of a business case for health and safety is, as discussed in Chapter 5, not new (Breyer 1982; Pouliakas and Theodossiou 2013; Viscusi 1983), though expressing outcomes in terms of monetary costs and benefits has sometimes been problematic (Nichols and Armstrong 1973): are deaths, injuries and ill-health reducible to hard figures? How are less tangible impacts accounted for? The business case model also suggests that there is a power dynamic in operation: there are some aspects which employers or other organisations can address but which trades unions or individual employees might not be able to change. When considering outcomes, it is therefore important to consider how individual agency and more structural factors might be involved, and the impact that these areas might have upon the 'justice' elements of legitimacy. Can health and safety outcomes be legitimate if those subject to risks do not fully understand the nature of those risks? Have the power imbalances between groups produced outcomes which serve one party's interests over another, or over those of wider society?

One enduring feature of health and safety as a social issue, then, is the status of risk assumption: the extent to which individuals fully understood the hazards to which they have been exposed and took them on either willingly or in exchange for some additional benefit. This remains a powerful idea affecting perceptions of issues around risk. Whilst the ability of individuals to determine their health and safety has been constrained in particular ways, today the idea that people can and should be responsible for the risks they apparently voluntarily assume, and a notion of victim-blaming, can still attach to public perceptions of health and safety; the idea that (other) people should take responsibility for their own safety. Framing many of these debates were long-standing historically grounded ideas about 'free labour' being able to negotiate (as equals) with employers about their contracts and terms of employment (Almond 2013: 94–97; Banks 2009; Bronstein 2008; Kostal 1994; Tucker 1995). According to this logic, workers knew about risks 'inherent' in the job and either took action to negotiate a higher wage (see below) or refused to take the job. Elements of this outlook remained far into the twentieth century, deeply ingrained in long-established heavy industries. Recalling the 1930s, Portsmouth dockyard worker Thomas Lee was open about the difficulty and dirt of the working environment—and also the acceptance that it was simply a part of the job.[18] In 1968, one Glaswegian shipbuilder observed that it "*is within the province of nightshift to protest against unsafe conditions and request alternative work*"[19]; therefore any who remained were assenting to the dangers of the role.

Something that developed (in the nineteenth century) alongside this view of the willing assumption of risk was the idea of the 'careless' worker, another concept which found long expression in the twentieth century. Health and safety were therefore often cast as personal problems, to be resolved by individual action (and hence, overwhelmingly, under the civil law: Bartrip and Fenn 1983). To some extent, this might have been a positive move, recognising some degree of worker agency and encouraging buy-in and greater views of the legitimacy of health and safety. Engaging with shopfloor workers, in particular, has long been seen as a challenge—often one that employers have claimed could be met through voluntary adoption of 'safety consciousness' or latterly, 'safety culture' (however problematic those concepts may be: Silbey 2009). One means employed in Britain since 1913 has been safety education (Esbester *forthcoming*): trying to persuade people to change their behaviour, often using media such as posters or films, intended to

---

[18] Thomas Lee, Portsmouth Dockyard worker c. 1933–1981, including as Safety Officer c. 1963–1976. Recording 1, c. 22:00. Portsmouth Royal Dockyard Historical Trust [PRDHT] Interview, accessed at Portsmouth History Centre [PHC], PD3/AD/185.

[19] Letter from Assistant Shipyard Manager to Shipyard Convenor of Shop Stewards (18 April 1968), p. 2. Samuel Barr Collection, GCU DC 140/1/1/5.

'stop carelessness'.[20] This drive towards safety education was based on the understanding that improving safety and health was a cooperative endeavour, in which, in the words of RoSPA's 1966 National Industrial Safety Campaign, 'we all shoulder a responsibility for safety.' So it was that in 1970 one senior British Rail Safety and Welfare Officer stated that the intention was "*to make every man his own safety officer*".[21] Yet, using education to spread messages about safe and unsafe work practices and to change behaviour has been, and remains, fraught with difficulties.

The voluntary approach ran the risk of placing the majority of responsibility for health and safety on individual workers (an early form of what would come to be termed a '*responsibilization*' strategy: Gray 2009), who were often not in a position to make fully-informed decisions or had insufficient power to change all the factors involved in producing unsafe and unhealthy work environments. Too often, material was produced by well-intentioned people, but who were lacking in expertise in the challenges posed by behaviour change. RoSPA recognised this in the 1960s, bringing in experienced advisors from the worlds of public relations, advertising, and design to help produce safety education.[22] How successful this was is impossible to judge—measuring the effectiveness of safety education is notoriously difficult—but it attempted to speak to the shopfloor worker, and to create individual experts who fitted within a system of self-regulation.

Education's focus on carelessness and personal choices also obscured other, more structural, factors that often lay outside the control of the individual. Some of these were formal: perhaps most significantly, the distribution of power in the workplace. The idea of employer and employee as workplace equals was belied by the reluctance (certainly in older industries, such as the railways) to involve unions or workers in decisions over health and safety. The defence of managerial prerogative was seen perhaps most visibly in the debates over the introduction of safety representatives in the 1970s.[23] This was manifest in the

[20] K. R. Allen, 'Training for Safety', talk given to Central Metropolitan Group of the London Industrial Committee (9 November 1963), p. 2. MRC, MSS.292B/146/2/3.

[21] Correspondence written by D. A. Verdon-Smith, 9 June 1970. TNA, AN 174/1522.

[22] RoSPA, 'RoSPA's experience with posters as an aid to accident prevention' (1968), p. 2. MRC, MSS.292B/146/17/2.

[23] See, for example, employer responses to the CBI's call for contributions on the HSC's consultation on the issue, at MRC, MSS.200/C/3/EMP/4/40.

oil industry, which was extremely weakly unionised before Piper Alpha in 1988; in the early 1990s, one platform's management was characterised as *"'old hands' who feared that [Safety Reps] would open the door to union-isation."*[24] This fear of loss of control was played out on the ground in various ways. One former HSE inspector noted of the 1970s and early 1980s that employers were unwilling to allow inspectors to talk to workers unaccompanied:

> I think it's a question of power relationships. The inspector is in a position of power, as is the employer [...] the employer's relationship with the inspector increases the power of the employer vis-a-vis his or her workers. The inspector talking to and getting information from the workers has the potential to undermine the employer's power so [...] that might explain it. (David Morris interview para. 87)

So, in some industries (particularly those involving unskilled labour), the ways in which management responded to health and safety issues could have an impact on how able workers were to prioritise health and safety, limiting choices and ideas of voluntary assumption of risks. It is likely that this had an impact on perceptions of legitimacy, though accessing testimony to interrogate this is problematic.

The notion of the voluntary assumption of risk is further complicated by the distribution of knowledge of dangers across workplaces. A consistent feature of worker testimony about the earlier part of this book's period is the lack of awareness of health risks, notably surrounding asbestos. Dock workers recalled working without any form of protective clothing and even playing with asbestos in the 1950s and 1960s: *"it was bags of asbestos and if it burst open the dockers enjoyed playing with it – sticking it up as a beard, side whiskers [...] no one knew it was dangerous"*.[25] Another docker went further about the dangers, which were *"[u]nknown tae us at that particular time but the employers have known since 1947 how dangerous it was. But omitted to let us know. Still not a*

---

[24] University of Aberdeen report to HSE, 'The effectiveness of offshore safety representatives and safety committees' (1993), pp. 70–71. TUC Library, HD 7269.

[25] Anonymous interviewee, interviewed by David Walker in 2009, about his work at Glasgow docks. Glasgow Dock Workers Project, Scottish Oral History Centre, University of Strathclyde Archives and Special Collections, GB249 SOHC18 (2009).

word from the employers".[26] In his mind, then, the risks were *not* assumed voluntarily. Power relations might also be gendered, with a particular impact upon occupational experiences of health and safety. In the mid-1980s, it was observed that "*old problems have emerged as new health and safety issues for women. Sexual harassment is one example [...] Not only is it humiliating, unpleasant and distressing to most of us, it can make us anxious, stressed and affects our work*".[27] In this articulation, then, there were particular challenges to safety and health experienced by women in the workplace which were not voluntarily assumed and could not be reasonably foreseen.

There were also cases in which workers were ordered or expected to continue practices potentially detrimental to their health and safety. At a 1966 meeting of British Rail's accident prevention officers, "*attention was focussed on malpractices which were tacitly permitted at local level. Regional Accident Prevention Officers must continue to press Management about the need for safe practices at all times*".[28] One Glaswegian recalled an incident in the late 1970s in which a former worker tried to warn about the dangers of asbestos: "*he wanted tae come in and talk tae the dockers but the foremen wouldn't let him come in just tae tell us about the stuff* "*Naw, naw, communist,*" *you know, things like that [...] daft*",[29] This is not solely an historical problem, either (Gray 2009). Paul Clyndes of the RMT was concerned about pressures that might be applied to rail workers in 2015, noting that if they refused to do something they judged dangerous the manager would: "*get someone more vulnerable to do the task. He's going to find some way of putting pressure on that individual*" (Paul Clyndes Interview, para. 65).

At the same time, some employers evidently felt compelled to act in ways that endangered worker (and sometimes public) lives. One key

[26] Thomas O'Connor, interviewed by David Walker in 2009. Glasgow Dock Workers Project, Scottish Oral History Centre, University of Strathclyde Archives and Special Collections, GB249 SOHC18 (2009).

[27] 'Danger: Women at Work', *GLC Women's Committee Bulletin* (23 March 1985), p. 48. TUC Library, HD 7268.

[28] British Rail, Minutes of meeting of Regional Accident Prevention Officers, 28 September 1966, minute 11. TNA, AN 208/218.

[29] James McGrath, interviewed by David Walker in 2009. Glasgow Dock Workers Project, Scottish Oral History Centre, University of Strathclyde Archives and Special Collections, GB249 SOHC18 (2009).

factor was economic, including the pressures of production and the profit motive. David Morris observed that health and safety "*was ignored where it was convenient to ignore it and far too often it was convenient to ignore it, particularly safety, health has always been another issue*" (David Morris Interview, para. 44). This was particularly acute in industries that were operating on the margins of profitability. One inspector in the 1990s saw this in the offshore industry, a volatile international market:

> the main operating costs are people, so the tendency is to slim the crews down to [...] as low as actually needed, and of course there's a fine dividing line that they must not cross because once you start removing people then [...] something has got to give [...] the regulator needs to keep on top of things to make sure they don't actually go beyond the bounds of this line of going from a safe, very efficient operation to a ultra-efficient operation.[30]

David Morris also saw this in the construction industry in the 1970s, with cost of work as the determinant of health and safety measures and a pressure to be competitive by 'cutting corners' (David Morris Interview, para. 42).

Economic considerations and the concomitant pressures that they place upon the voluntarism and choice of individual workers were capable of having a significant effect on the perceived legitimacy of health and safety, not least because they gave rise to the perception that these rules are not universal, and not focused on the type of universalistic, social solidarity-oriented outcomes that underpin justice-based legitimacy claims (Black 2008; Prosser 2006). Instead, there was a sense that the rules worked selectively, for, and to the benefit of, only one side of the workplace power nexus, thus undermining both commitment and compliance, as this reminiscence from one former Chair of the HSC, reflects: "*Unfortunately in those days [the 1960s] when you talked to the men about the folly of [choosing to break safety rules], they'd say, 'Boss's rules'. And they wouldn't take any notice*" (Sir Frank Davies Interview, para. 36).

Of course, economic factors did not just affect the actions of employers, but also had an impact upon employees. Workers exhibited a variety of sometimes contradictory responses to health and safety issues,

---

[30]Martin Thompson, interviewed as part of the University of Aberdeen's 'Lives in the Oil Industry' project. British Library, C963/13, tape 4, side a.

demonstrating elements of choice in their actions, as well as the impact of external constraints beyond the individual employer. The ways in which financial imperatives might encourage risky behaviours were clearly important. Piecework was one aspect of this, as was noted in 1972: "*There is little doubt that the method of payment tends to affect workers' attitudes to safety, particularly if the work itself is such that it can be expedited by failure to observe safety precautions or by using improvised equipment*".[31] There might also be financial implications that discouraged employees from reporting incidents, observed in relation to the recent past of the nuclear industry: "*What is the onus on the individual to actually report a [medical] condition that might invalidate the particular work that they actually need to do? This is probably a new area that we're facing*" (Chris Marchese Interview, para. 49). In a number of settings, and across the period since 1960, there is therefore evidence that economic factors have reduced the choices workers felt they had available.

The place of health and safety within bargaining has been a controversial one. A 1974 Trade Union Research Unit publication aimed "*to demonstrate how concern over safety and industrial accidents can be brought constructively and effectively into collective bargaining*".[32] This is certainly one area in which worker agency was demonstrated. According to one former ICI worker and TGWU shop steward in the 1970 and 1980s, "*if you can bring safety into an issue, no matter what it is, a wages issue or a conditions issue of God knows what, if safety is there and you are on firm ground it is a major part of your artillery*".[33] The management at the Upper Clyde Shipbuilders Yard appeared resentful of this approach in 1968: "*I feel that some undercurrent exists whereby they are prepared to make the most of all or any problems which may and will crop up despite our best endeavours*".[34] In the railway industry, it took the major crash at Clapham Junction in 1988 to change the prevailing culture:

---

[31] Written evidence of the Dock and Harbour Authorities' Association to the Robens Committee, n.d. (c. 1971), pp. 4–5. TNA, LAB 96/57.

[32] Trade Union Research Unit, 'Safety, accidents and collective bargaining', 1974, p. 1. TUC Library, HD 7262–7262.5.

[33] Doug May, interviewed by David Walker, 2005. Chemical Workers Project, Scottish Oral History Centre, University of Strathclyde Archives and Special Collections, GB249 SOHC7 (2004–2005).

[34] Letter from Assistant Shipyard Manager to Shipyard Convenor of Shop Stewards, 18 April 1968, p. 1. Samuel Barr Collection, GCU DC 140/1/1/5.

We said to the trade unions, 'we want to take safety out of the negotiating machinery and have safety reps who do nothing other than safety.' And the trade unions actually said, 'well, we've been telling you about the importance of staff safety for donkey's years.' So we had to eat a bit of humble pie but they agreed to take it out and have dedicated safety reps and that was a fundamental shift in about 1990. (David Maidment Interview, para. 33)

Just like other areas where perceptions of self-interest have clouded attitudes towards what is often felt to be an area of altruism and the 'greater good', there is some suggestion that the use of this issue as a bargaining tool has had some impact in terms of the legitimacy of health and safety, although in employment relationships, where economic rationality and financial motives might already predominate, the extent of this may be a little less pronounced in practice than elsewhere.

Officially, bargaining by using health and safety did not mean 'danger money', but instead using health and safety issues as a tool to extract improvements to working conditions. However, as late as 1979, the 'Coventry Health and Safety Movement' reminded workers that *"union journals should emphasise prevention rather than compensation and that negotiators should seek safety not danger money"*.[35] There is no doubt that dangerous or dirty jobs could be a welcome source of money for employees, though how far this form of incentivisation should be considered as voluntary assumption of risk is questionable. David Morris recalled that he:

was certainly conscious when inspecting the docks in London, donkey's years ago [in the 1970s], that the trade unions' principal interest often appeared to be getting danger money for their members rather than reducing or eliminating risk", something which reflected "the nature of society at the time. If you did a high risk job you got more pay for it and that was your compensation for the risk of injury. (David Morris Interview, paras. 28–30)

This was also played out in other ways: one Glaswegian docker recalled that even in the late 1970s workers who refused to unload asbestos cargoes were put to the *"back of the queue"* and would *"only be utilised in*

---

[35] 'CHASM', *Work Hazards* 21 (c. 1979), p. 19. Samuel Barr Collection, GCU DC 140/2/1/2.

another job if required.[36] Faced with the prospect of no work at all, it is little wonder that some opted for dangerous roles (Melling 2003). Employers were also known to plead that the costs of making changes would threaten the economic viability of a firm or plant: "*ICI being ICI said, 'well, if we've got tae dae all these things it's no going tae be viable so the plants will probably hae to shut doon.' And you've got young guys in there with mortgages and families [...] the argument was blown out of the water*".[37] Clearly, these are factors that influenced worker behaviour but which lay well outside the sphere of things that they could directly influence for themselves; whilst the evidence presented here has been largely historical in nature, it seems unlikely that these pressures and factors have disappeared in the twenty-first century. Instead, they are probably manifest in different ways, particularly in the newer sectors, including in relation to precarious working such as zero hours contracts and the gig economy (Fudge and Owens 2006; Quinlan et al. 2001).

Femininity, Masculinity, and Autonomy

Given there have been a wide range of factors in relation to health and safety over which workers had little or no control, when considering the justice elements of legitimacy and whose values and interests might have been served, it is important to consider what impact this had on ideas about agency and autonomy. Workers were not simply pawns, moved around at the will of those 'above' them, but were, in some circumstances, and with some limits, able to take steps to protect their health and safety. Before exploring this aspect, however, it is helpful to examine another area which shaped the responses workers were able to make: prevailing notions of gendered workplace roles. Needless to say, this area has seen an extremely complicated interaction with wider cultural and societal values, notably including expectations of 'appropriate' behaviours for men and women (Clarke and Wall 2009). There is considerable evidence to suggest that for most, if not all, of the post-1960

---

[36] Alfred McMillan, interviewed by David Walker 2009. Glasgow Dock Workers Project, Scottish Oral History Centre, University of Strathclyde Archives and Special Collections, GB249 SOHC18 (2009).

[37] KG, interviewed by David Walker, 2005, Chemical Workers Project, Scottish Oral History Centre, University of Strathclyde Archives and Special Collections, GB249 SOHC7 (2004–2005).

period, many workers in the largely male-dominated traditional heavy industries perceived some aspects of health and safety as unmanly and therefore as illegitimate (Bradley 2012: 108–154; Johnston and McIvor 2004, 2007; McIvor and Johnston 2002). Prevention, protection, and precaution were thus often negatively associated and so limited in their reach and capacity. Female workers were also subject to pressures to act in particular ways, including legal restrictions on what tasks they could undertake at work, as well as informal cultural expectations of roles and actions. There has been potential conflict, then, between the health and safety interests of workers and gendered expectations of behaviour and employment.

These wider social and cultural attitudes might also have had an impact on the legitimacy of health and safety. In male-dominated heavy industries, there was an acceptance of the normality of occupational casualty and ill-health, and potential resistance to unwanted welfare reforms (Hayes 2002), at least before the HSWA. So, in 1959, the Barrow Docks Safety First Committee was prepared to write off 16 injuries as "*a normal pattern on a dock undertaking*",[38] and in 1970, a shipyard council meeting on Clydeside was told by a management representative that he "*personally felt that the hazards were being exaggerated and were no worse than those encountered in normal shipbuilding*".[39] Given such attitudes, it is unsurprising that many workers appear to have paid limited attention to the possibilities of improving their working environment. According to one former HSE inspector:

> it was understood that, particularly manual labourers, would almost invariably be damaged by the work they did and their retirement, if they had one, would be fairly brief. That sounds horribly harsh but it was just the nature of society at the time. (David Morris Interview, para. 26; also David Eves Interview, para. 15)

This testimony was reflected in archival comments such as that from 1968 which claimed that "*no one is obliged to work on nightshift and anyone who feels he cannot face up to the normal hazards of this form of*

---

[38] Minutes of Barrow Docks Safety First Committee, 1 October 1959, p. 1. TNA, BK 9/48.

[39] Yard Council Meeting (Upper Clyde Shipbuilders Yard), 14 April 1970, p. 2. Samuel Barr Collection, GCU DC 140/1/1/5.

occupation can request a change".[40] It is questionable how many men would have admitted that they could not cope with what was portrayed as a normal working environment—this might have been taken as a challenge to masculinity in what was still a macho workplace. The social and cultural expectations of maleness were significant in shaping men's behaviour for much, if not all, of the book's period (H. Young 2010).

Such notions of masculinity were reflected across the more traditional industries, including construction,[41] offshore,[42] and the railways.[43] In the steel industry in 1963, it was publicly reported that "*Steel men say safety hats make them look ridiculous. Safety helmets were not worn by some workmen because they felt that they looked like 'guys'*".[44] In his report for the same year, the Chief Inspector of Factories argued that "*[w]here protective clothing is provided, he should wear it [...] His example is of vital importance to the young who are extremely sensitive to ridicule and to being thought 'cissy' by older workers*" (Ministry of Labour 1964: 50). Acceptance and buy-in to health and safety regulation was thus heavily patterned by wider social attitudes towards masculinity. At the same time, on rare occasions, masculinity could also help 'sell' precaution. British Rail tried to introduce high visibility clothing in the 1960s and 1970s, for workers who maintained the tracks—they were particularly vulnerable, as they worked in amongst moving trains. Experiments to make track workers more obvious to train crews with hi-vis clothing were made from 1964, but it was noted that "*[s]ome men were reluctant to wear H.V. garments. This appeared to be related to a psychological feeling of being over-conspicuous and as though some element of deliberate disregard for safety demonstrated a brave approach to their work*".[45]

These attitudes were not, of course, the 'fault' of the individual worker: they operated within a complex set of social norms that were difficult to challenge. Nevertheless, it did present a problem. Compulsory

---

[40] Letter from Assistant Shipyard Manager to Shipyard Convenor of Shop Stewards, 18 April 1968, p. 2. Samuel Barr Collection, GCU DC 140/1/1/5.

[41] John Armitt Interview, para. 52.

[42] Martin Thompson, interviewed as part of the University of Aberdeen's 'Lives in the Oil Industry' project. British Library, C963/13, tape 4, side b.

[43] David Maidment Interview, paras. 17–19.

[44] 'Steel Men Say Safety Hats Make Them Look Ridiculous', *Scarborough Evening News*, 3 May 1963.

[45] 'High Visibility Clothing' memorandum, 9 May 1967, p. 1. TNA, AN 171/683.

use of hi-vis clothing was considered, but rejected for fear that resistance would mean *"the garment will be instinctively associated with a dictatorial managerial attitude. Men will devise ways of wearing it when under observation and discard it immediately after"*.[46] Great care was taken by BR in trials of hi-vis clothing and in consultation with the workers and unions, with a gradual roll-out of hi-vis garments from the late 1960s. On occasion, dominant notions of masculinity were used to 'sell' the new clothing: a 1967 promotional poster featured a woman wearing a hi-vis bikini, for example. By 1971, when further use of hi-vis was proposed, it was noted that

> [h]igh visibility clothing was at one time very unpopular with the staff due to its anti-social appearance. However, this has now been largely overcome by building and other workers who are frequently issued with similar clothing and it is considered that the staff reaction will be favourable.[47]

When faced with societal norms of behaviour (masculine or feminine) which threatened individual health and safety, one key challenge was how to effect change both within a workforce *and* wider society.

Such examples were atypical, however; much more common was the way in which men were expected to be strong, silent and stoical. This was evident in one case when a more proactive approach was taken with regard to the health risks from vinyl chloride in the 1970s:

> [the] workers knew they were at risk and the trade unions were actually being very supportive of them, but this group, the one thing they hated was having to discuss it with their wives, families, children and so on, because it was answering impossible questions. I remember one guy [...] got sent to Coventry by all his mates for about a month because he'd appeared on the local radio station talking about the risks. (Tim Carter Interview, para. 40).

Cultural assumptions about appropriate roles also had an impact on the health and safety of women workers. Indeed, for much of the period, the typical worker was assumed by most parties to be a man. For example, in the 1980s, when Threshold Limit Values were set for exposure

---

[46] Ibid., p. 2.

[47] 'High Visibility Clothing' memorandum, c. May 1971, pp. 1–2. TNA, AN 171/683.

to chemicals and other causes of ill-health, only one value was given, based on a man, ignoring the physiological differences between men and women (Denning 1985: 13). As late as 1999, it was possible for the TUC to note that *"health and safety often concentrates on the risks men face, because of the way the labour market used to be. Work has changed, and women's health and safety should be addressed just as much as men's has been"*.[48] For a considerable period, there were legal restrictions in place, limiting women's employment at particular times or in particular roles. These were increasingly contested from the 1960s onwards; at the end of the 1970s, the Equal Opportunities Commission looked at women's status in relation to health and safety. The CBI's response was that *"all the present restrictions on the hours of women are archaic and irrelevant in present day conditions and are certainly an impediment in many situations on equal opportunities of employment for women"*.[49] This was in a context of trying to increase women's participation in the labour force:

> there weren't enough people to fill all the jobs, so they were trying to get women into employment. So there were views around about the kinds of job that it was safe and appropriate for women to do, from a health, safety and welfare point of view. (Jenny Bacon Interview, para. 6)

Much of the debate continued to focus on health risks to women's reproductive role[50]:

> It is interesting that we have neglected women's general health in favour of their reproductive health [...w]hile we must care for women workers and their unborn children, we must also move beyond the popular myth that reproductive hazards are specific to women workers. (Denning 1985: 14–15).

So, as well as masculinity, societal constructions of femininity had an impact on exposure to risks. Despite this, workers retained some degree of autonomy, and were able to take action to reduce some

[48]TUC, 'Restoring the Balance. Women's Health and Safety at Work', n.d. (c. 1999), inside pages. TUC Library, HD 7268.

[49]Equal Opportunities Commission, 'Health and Safety Legislation: Should We Distinguish Between Men and Women?' (March 1979), p. 80. MRC, MSS.222/AP/5/102.

[50]Council of Service Unions, 'Health of Women at Work', 2 March 1982, pp. 1–3. MRC, MSS.381/W/1/4/10.

risks, sometimes through formal routes (such as the post-1977 Safety Representatives and Committees, discussed in Chapter 5, or through collective bargaining) and sometimes through more informal means. Underlying much of this was the belief that workers had the responsibility to safeguard themselves, as seen in the 1963 talk which claimed that the aim of training was that "*[e]very person must be made personally aware that accidents can happen to him and will unless her [sic] personally does something to stop them happening*".[51] Section 7 of the HSWA 1974 also placed a duty upon employees to "*take reasonable care for the health and safety of himself and of other persons*", thus further formalising and legalising this notion of self-reliance ('*responsibilization*': Gray 2009). In terms of formal methods, workers might make concerns known to management committees, but these channels of communication were dependent upon the largesse of employers.

Alongside formal mechanisms, there existed a number of less formal techniques that workers might use to address health and safety issues. Some workers improvised solutions to dangers as best they could, showing a reluctance to accept risks; Portsmouth Dockyard worker Tom Smith recalled tying a rag over his nose to combat the polluted environment in the 1950s.[52] Some more outspoken workers might refuse dangerous work, as in the case of one Scottish chemical worker in the 1970s: "*If I thought something was unsafe I wouldnae do it and I had many rows with managers and supervisors o'er this, 'you want tae do it, you do it'*".[53] One dock worker who refused to work with asbestos in the early 1970s simply stated that "*I thought my health was more important than money*".[54] Gradually, and as a result of concerted worker action, asbestos was phased out of Royal Dockyards like Portsmouth as well as some commercial docks. However, this had an unintended consequence: hazards were pushed further down the contracting and supply chain, onto

---

[51] K. R. Allen, 'Training for Safety', talk given to Central Metropolitan Group of the London Industrial Committee (9 November 1963), p. 1. MRC, MSS.292B/146/2/3.

[52] Tom Smith, Portsmouth Dockyard worker c. 1950, transcript, p. 15. PRDHT, PHC, PD3/LF/3.

[53] KG, interviewed by David Walker, 2005. Chemical Workers Project, Scottish Oral History Centre, University of Strathclyde Archives and Special Collections, GB249 SOHC7 (2004–2005).

[54] Owen McIntyre, interviewed by David Walker, 2009. Glasgow Dock Workers Project, Scottish Oral History Centre, University of Strathclyde Archives and Special Collections, GB249 SOHC18 (2009).

sub-contractors (Bartrip 2014),[55] constituting a form of 'risk-shifting' that exposes workers to enhanced pressures while also rendering safety *"fungible"* (in that contractors negotiate to undertake it in exchange for money: McDermott and Hayes 2018). Nevertheless, this demonstrates some of the ways in which workers might assert control over their working environment and recognise the legitimacy and importance of health and safety.

The regulations introduced under the 1992 Management of Health and Safety at Work Act also conferred some powers to employees, enhancing their ability to refuse to under dangerous tasks through 'work safe' policies. According to Paul Clyndes of the RMT, these policies are widely used in the contemporary railway industry: *"workers will say 'no, I am not going to do that, I am not going to undertake that task' and they have a right not to be dealt with by the employer in a disciplinary way under those procedures"* (Paul Clyndes Interview, para. 64). Although sometimes limited in practice, and also subject very much to countervailing economic and management pressures (Gray 2009), these mechanisms provide a route through which workers could attempt to push back against workplace dangers. However, worker autonomy (howsoever circumscribed) brought with it a significant trade-off—the ability to choose also meant that it was possible to select unsafe work practices. In some heavy industries, there appears to have been a dislike of being told what to do: one Portsmouth Dockyard safety officer recalled trying to protect workers, but that there was a perception amongst some that this might be encroaching upon their liberty or to keep tabs on staff.[56] The Barrow Docks Accident Prevention Committee was, in 1978, reminded that use of safety helmets *"was primarily aimed at securing the safety of the individual and it was incorrect to view any such regulations as an infringement of personal liberty"*.[57] Similarly, safety interventions encountered resistance when they rubbed up against competing worker preferences; for instance, wearing noise protection equipment prevented

---

[55] Thomas O'Connor, interviewed by David Walker in 2009. Glasgow Dock Workers Project, Scottish Oral History Centre, University of Strathclyde Archives and Special Collections, GB249 SOHC18 (2009); Tom Smith, transcript p. 16. PRDHT, PHC, PD3/LF/3.

[56] Brian Stubbs, Portsmouth Dockyard worker c. 1955–1980s, including as Safety Officer. Comments at c. 1 hour 2 minutes. PRDHT, PHC, PD3/AD/123.

[57] Minutes of Workington Safety Committee, 7 November 1978, minute 5. TNA, BK 9/14.

factory workers from conversing or listening to the radio (Frank Davies Interview, paras. 32–34). Whether or not workers objected on the grounds that they were being told what to do by management (via what workers dismissed as "*boss's rules*" in Frank Davies's experience: para. 36) is unknown, but it is clear that for a variety of reasons, workers sometimes acted in ways contrary to those advocated by management.

While it may be the case that historically macho attitudes have been associated largely with more masculinised, manual, industrial settings, and the idea of '*just getting the job done*', this is not to say that similar issues have not existed in other sectors. In a 1985 article, healthcare workers were one prime example of employees who were sacrificing their well-being in order to deliver the best care to their patients (Denning 1985: 12–13). These trends have also not disappeared as employment contexts have changed. For example, in healthcare and education, this voluntary assumption of risk has become increasingly prominent and recognised as an issue of concern: professionals on the frontline sacrifice their own health to service the needs of those whose best interests they are committed to serving. The job remains more important than the person; increasingly, however, this manifests itself in more 'feminised', caring settings:

> in the NHS, there is a culture of looking after the wellbeing of patients but not necessarily the wellbeing of ourselves […] it's an odd imbalance, that you would have somebody who will put themselves out for a patient, do everything they can to get them better, but not look after themselves. (Ruth Warden Interview, para. 72)

Evidently, the acceptance of health and safety as an issue is intimately linked to the social context and values of the time, around gender, individual choice, and self-identity. While this has always been the case, and while many of these values have changed, the debates that they reflect have not gone away; instead, they have taken on new forms around new sectors and issues.

Conclusion

This chapter has explored some key sites at which it is possible to see whose values have been played out in health and safety in Britain since 1960. Central to all has been a sense of the contested nature of health and safety, and the ways in which values have been dependent upon a variety of contingent factors. This contest and contingency has been visible from the domain of high politics through to the actions of individual workers, albeit in rather different forms, via four themes which reoccurred across the period of this book. Firstly, the notion of the commercialisation of health and safety: that is, going beyond the moral case for health and safety and selling services, products and advice. This has existed in a relatively limited form since well before the 1980s, but the particular coalescence of factors from this point onwards heightened the tensions surrounding the legitimacy of commercial endeavours, risking the perception of moving from health and safety as morally legitimate to something carried out for profit. This was related to the rise of the health and safety profession (discussed in Chapter 6) and the role of the insurance industry as potentially 'gold-plating' health and safety, as well as the perception (however inaccurate) of a compensation culture driving increasing numbers of public claims for health and safety issues. It is suggested that a crucial driver in allowing these factors to gain traction has been the increasing economic pressure on the regulator since the 1980s, producing a sense of slow withdrawal from the field and a need for the HSE itself to charge in order to raise income. These factors have had a potentially significant impact upon the perception of health and safety as shifting from the morally correct thing to do to something whose legitimacy is at least questionable.

Secondly, the putative idea that from approximately 1970 until relatively recently health and safety was developed in a broadly consensual framework has also been questioned. Contest was always a part of health and safety, even around what was a defining moment in regulatory and legitimatory terms, the Robens Report and the 1974 Health and Safety at Work Act, and it has often provided a driver for many of the most significant changes to have occurred since then. That said, the mechanisms built into the post-1974 system—notably tripartism through the HSC—at least had the effect of diffusing some of the more potentially serious disagreements over health and safety, and the system has remained relatively stable, if not conflict-free, for over forty years. Nevertheless, the increasing resistance to the idea of health and safety in the twenty-first century has at times offered serious challenge to the legitimacy of the area, sometimes to the extent of moving debate into the arena of high politics. Whilst this has eased off in light of Brexit, it is unlikely that the contested areas have been resolved.

The third and fourth elements are related, as they touch upon experience and practice on the ground. Debates around agency and power in the workplace are significant, in that they might affect whose values and interests were represented in health and safety regulation and as it happened on the shopfloor. Ideas about worker carelessness and the voluntary assumption of risk fail to acknowledge fully the relative power imbalances that affect how much some workers knew about the risks to which they were exposed. There were countervailing factors which might push both employers and employees into producing or accepting unsafe or unhealthy conditions, often tied to economics and the (lack of) availability of other forms of employment. Poor workplace health and safety might have been understood as a problem and as illegitimate, but there might have been little apparent alternative, particularly in declining industries and de-industrialising regions from the 1970s onwards. There were also more cultural aspects which were beyond the control of individuals or even employers: gendered patterns of expectation and behaviour which pressured men and women to act in ways that appeared to indicate health and safety was illegitimate, notably in the case of men by exposing the body to dangers. At the same time, since the 1960s, there appears to have been an increasing drive to observe health and safety precautions, with formal and informal routes available to workers, allowing them to protect their bodies and minds. Agency has always been and remains a significant factor. Running throughout all of these themes, then, is a sense of the contingency involved in health and safety. Whose values and interests have been served, and how, have varied over time, with the result that the perceived legitimacy of health and safety has waxed and waned.

References

Primary Sources

Alexander, W. G. (1971, June 14). Verbal Evidence on Behalf of RoSPA to Robens Committee. The National Archives of the UK, London [TNA], LAB 96/75.

Allen, K. R. (1963, November 9). 'Training for Safety', Talk Given to Central Metropolitan Group of the London Industrial Committee. Modern Records Centre, University of Warwick [MRC], MSS.292B/146/2/3.

Anonymous Interviewee. (2009). Interviewed by David Walker. Glasgow Dock Workers Project, Scottish Oral History Centre, University of Strathclyde Archives and Special Collections, GB249 SOHC18 (2009).

Barrow Docks Safety First Committee. (1959, October 1). Minutes. TNA, BK 9/48.

British Rail. (1966, September 28). Minutes of meeting of Regional Accident Prevention Officers. TNA, AN 208/218.

British Rail. (1967, May 9). 'High Visibility Clothing' memorandum. TNA, AN 171/683.

British Rail. (1967, May 16). Meeting of Regional Accident Prevention Representatives. TNA, AN 208/218.

British Rail. (n.d., c. 1971). 'High Visibility Clothing' memorandum, (n.d., c. May 1971). TNA, AN 171/683.

Brown, J. H. (1980, March). Letter to *The Land Worker*. Museum of English Rural Life, University of Reading [MERL].

CBI. (1976). File of Employer Responses to the HSC's Consultation on Safety Representatives. MRC, MSS.200/C/3/EMP/4/40.

Council of Service Unions. (1982, March 2). Health of Women at Work. MRC, MSS.381/W/1/4/10.

Coventry Health and Safety Movement. (n.d., c. 1979). *Work Hazards* 21: 19. Samuel Barr Collection, Glasgow Caledonian University [GCU] DC 140/2/1/2.

Denning, J. (1985, January 17). The Hazards of Women's Work. *New Scientist*, pp. 12–15.

Dock and Harbour Authorities' Association. (n.d., c. 1971). Written Evidence to the Robens Committee. TNA, LAB 96/57.

Employment Committee. (1992). The Work of the Health and Safety Commission and Executive, Minutes of Evidence 12 February 1992 (HC 263; London: Crown).

Equal Opportunities Commission. (1979, March). Health and Safety Legislation: Should We Distinguish Between Men and Women? MRC, MSS.222/AP/5/102.

Esbester, M. (forthcoming). *The Birth of Modern Safety: Preventing Worker Accidents on Britain's Railways, 1871–1948*. Abingdon: Taylor & Francis.

Foot, M. (1974, April 3). 'Health and Safety at Work Etc. Bill', HC Deb, Hansard Vol. 871 cc1286–1394, 1287.

Freeman, N. T. (1970, November 25). 'Safety in the Seventies', in Minutes of Teesside Industrial Accident Prevention Committee meeting. Royal Society for the Prevention of Accidents [RoSPA] Archives, D.266/2/18/11.

Greater London Council. (1985, March 23). Bulletin: 'Danger: Women at Work'. Trades Union Congress [TUC] Library, HD 7268.

Hampton, P. (2005). *Reducing Administrative Burdens: Effective Inspection and Enforcement* (The Hampton Review). London: HM Treasury.

KG. (2005). Interviewed by David Walker. Chemical Workers Project, Scottish Oral History Centre, University of Strathclyde Archives and Special Collections, GB249 SOHC7 (2004–2005).

Kinnock, N. (1973, May 21). 'Workers' Safety and Health', HC Deb, Hansard vol. 857 cc62–117, 65.

Lee, T. (n.d.). Portsmouth Dockyard Worker c.1933–81, Including as Safety Officer c.1963–76. Portsmouth Royal Dockyard Historical Trust (PRDHT) Interview, Accessed at Portsmouth History Centre (PHC), PD3/AD/185.

Löfstedt, R. (2011). *Reclaiming Health and Safety for All: An Independent Review of Health and Safety Legislation* (The Löfstedt Review). London: Crown.

Lord Drumalbyn. (1969, July 23). H. L. Deb., Hansard vol. 304, col. 1061.

Lord Robens. (1972). *Safety and Health at Work: Report of the Committee 1970–72* (The Robens Report). London: HMSO.

Lord Young. (2010). *Common Sense, Common Safety* [The Young Review]. London: Crown.

May, D. (2005). Interviewed by David Walker. Chemical Workers Project, Scottish Oral History Centre, University of Strathclyde Archives and Special Collections, GB249 SOHC7 (2004–2005).

McGrath, J. (2009). Interviewed by David Walker. Glasgow Dock Workers Project, Scottish Oral History Centre, University of Strathclyde Archives and Special Collections, GB249 SOHC18 (2009).

McIntyre, O. (2009). Interviewed by David Walker. Glasgow Dock Workers Project, Scottish Oral History Centre, University of Strathclyde Archives and Special Collections, GB249 SOHC18 (2009).

McMillan, A. (2009). Interviewed by David Walker. Glasgow Dock Workers Project, Scottish Oral History Centre, University of Strathclyde Archives and Special Collections, GB249 SOHC18 (2009).

Ministry of Labour. (1964). *Annual Report of the Chief Inspector of Factories, 1963* (Cmnd. 2450; London: Crown).

National Union of Agricultural and Allied Workers. (1960–1961). Minutes of Health Safety and Welfare Committee, 12 October 1960, p. 1; 25 October 1961, p. 2. MERL, SR5NUAW BXXI/I.

National Union of Agricultural and Allied Workers. (1967, March 21). Legal Department Memorandum to Executive Committee MERL, SR 5NUAW/B/XXI/4.

O'Connor, T. (2009). Interviewed by David Walker. Glasgow Dock Workers Project, Scottish Oral History Centre, University of Strathclyde Archives and Special Collections, GB249 SOHC18 (2009).

Parker, E. C. (n.d., c. 1979). Letter to *Work Hazards* 21, p. 17. Samuel Barr Collection, GCU DC 140/2/1/2.

Policy Exchange. (2010). *Health and Safety: Reducing the Burden*. London: Policy Exchange.

RoSPA. (1968). RoSPA's Experience with Posters as an Aid to Accident Prevention. MRC, MSS.292B/146/17/2.

Scarborough Evening News. (1963, May 3). Steel Men Say Safety Hats Make Them Look Ridiculous.

Smith, C. (1974, April 3). Health and Safety at Work Etc. Bill', HC Deb, Hansard vol. 871, c1322.

Smith, T. (n.d.). Interview; Portsmouth Dockyard Worker c.1950s. PRDHT, PHC, PD3/LF/3.

Stubbs, B. (n.d.). Interview; Portsmouth Dockyard worker c.1955–1980s, Including as Safety Officer. PRDHT, PHC PD3/AD/123.

Swinden, T. A. (1962, November 12). Joint Industrial Conference on the Prevention of Accidents. TUC Library, HD 7273.

Temple, M. (2014). *Triennial Review Report: Health and Safety Executive* [The Temple Review]. https://www.gov.uk/government/publications/trienni-al-review-report-health-and-safety-executive-2014. London: HMSO.

Thompson, M. (c. 2000). Interviewed as Part of University of Aberdeen 'Lives in the Oil Industry' Project. British Library, London, C963/13.

Trade Union Research Unit. (1974). Safety, Accidents and Collective Bargaining. TUC Library, HD 7262–7262.5.

TUC. (n.d., c. 1999). Restoring the Balance. Women's Health and Safety at Work. TUC Library, HD 7268.

University of Aberdeen. (1993). The Effectiveness of Offshore Safety Representatives and Safety Committees, Report to HSE. TUC Library, HD 7269.

Upper Clyde Shipbuilders Yard. (1968, April 18). Letter from Assistant Shipyard Manager to Shipyard Convenor of Shop Stewards. Samuel Barr Collection, GCU DC 140/1/1/5.

Upper Clyde Shipbuilders Yard. (1970, April 14). Yard Council Meeting. Samuel Barr Collection, GCU DC 140/1/1/5.

Verdon-Smith, D. A. (1970, June 9). Correspondence. TNA, AN 174/1522.

Whitelaw, W. (1974, April 3). 'Health and Safety at Work Etc. Bill', HC Deb, Hansard vol. 871 c1302.

Williams, R. (n.d., c. 1979). 'A Key Issue', *Work Hazards* 21 (c.1979). Samuel Barr Collection, GCU DC 140/2/1/2.

Workington Safety Committee. (1978, November 7). Minutes. TNA, BK 9/14.

SECONDARY SOURCES

Almond, P. (2007). Regulation Crisis: Evaluating the Potential Legitimizing Effects of 'Corporate Manslaughter' Cases. *Law and Policy, 29*(3), 285–310.

Almond, P. (2009). The Dangers of Hanging Baskets: Regulatory Myths' and Media Representations of Health and Safety Regulation. *Journal of Law and Society, 36*(3), 352–375.

Almond, P. (2013). *Corporate Manslaughter and Regulatory Reform*. Basingstoke: Palgrave Macmillan.

Almond, P. (2015). Revolution Blues: The Reconstruction of Health and Safety Law as 'Common-Sense' Regulation. *Journal of Law and Society, 42*(2), 202–229.

Almond, P., & Gray, G. C. (2017). Frontline Safety: Understanding the Workplace as a Site of Regulatory Engagement. *Law & Policy, 39*(1), 5–26.

Asher, R. (2003). Experience Counts: British Workers, Accident Prevention and Compensation, and the Origins of the Welfare State. *Journal of Policy History, 15*(4), 359–388.

Baldwin, R. (1987). Health and Safety at Work: Consensus and Self-Regulation. In R. Baldwin & C. McCrudden (Eds.), *Regulation and Public Law* (pp. 132–158). London: Weidenfeld and Nicholson.

Baldwin, R. (1995). *Rules and Government*. Oxford: Clarendon Press.

Baldwin, R., & Black, J. (2016). Driving Priorities in Risk-based Regulation: What's the Problem? *Journal of Law and Society, 43*(4), 565–595.

Banks, S. (2009). Woodley v Metropolitan District Railway Company (1877). In C. Mitchell & P. Mitchell (Eds.), *Landmark Cases in the Law of Tort* (pp. 127–152). Oxford: Hart.

Bartrip, P. (2014). "Enveloped in Fog": The Asbestos Problem in Britain's Royal Naval Dockyards, 1949–1999. *International Journal of Maritime History, 26*(4), 685–701.

Bartrip, P., & Burman, S. (1983). *The Wounded Soldiers of Industry: Industrial Compensation Policy, 1833–1897*. New York: Clarendon Press.

Bartrip, P., & Fenn, P. T. (1983). The Evolution of Regulatory Style in the Nineteenth Century British Factory Inspectorate. *Journal of Law and Society, 10*(2), 201–222.

Beck, M., & Woolfson, C. (2000). The Regulation of Health and Safety in Britain: From Old Labour to New Labour. *Industrial Relations Journal, 31*(1), 35–49.

Beetham, D. (1991). *The Legitimacy of Power*. London: Macmillan.

Black, J. (2005). The Emergence of Risk-based Regulation and the New Public Risk Management in the United Kingdom. *Public Law*, 512–549.

Black, J. (2008). Constructing and Contesting Legitimacy and Accountability in Polycentric Regulatory Regimes. *Regulation & Governance, 2*(2), 137–164.

Bradley, D. (2012). *Occupational Health and Safety in the Scottish Steel Industry, c.1930–1988: The Road to 'Its Own Wee Empire'* (Unpublished PhD thesis, Glasgow Caledonian University).

Braithwaite, J. (2008). *Regulatory Capitalism: How It Works, Ideas for Making It Work Better*. Cheltenham: Edward Elgar.

Breyer, S. (1982). *Regulation and Its Reform*. Cambridge, MA: Harvard University Press.

Bronstein, J. (2008). *Caught in the Machinery: Workplace Accidents and Injured Workers in Nineteenth-Century Britain*. Stanford: Stanford University Press.

Brown, H. W. (1931). Employers' Liability Insurance. *Transactions of the Faculty of Actuaries, 13*(1), 1–66.

Carson, W. G. (1979). The Conventionalisation of Early Factory Crime. *International Journal of the Sociology of Law, 7*(1), 37–60.

Clarke, L., & Wall, C. (2009). "A Woman's Place is Where She Wants to Work": Barriers to the Entry and Retention of Women into the Skilled Building Trades. *Scottish Labour History, 44*, 16–39.

Dalton, A. J. P. (1998). *Safety, Health and Environmental Hazards at the Workplace*. London: Cassell.

Daniels, N., & Sabin, J. (1997). Limits to Health Care: Fair Procedures, Democratic Deliberation, and the Legitimacy Problem for Insurers. *Philosophy & Public Affairs, 26*(4), 303–350.

Dawson, S., Willman, P., Clinton, A., & Bamford, M. (1988). *Safety at Work: The Limits of Self-Regulation*. Cambridge: Cambridge University Press.

Dodds, A. (2006). The Core Executive's Approach to Regulation: From 'Better Regulation' to 'Risk-Tolerant Deregulation'. *Social Policy & Administration, 40*(5), 526–542.

Ericson, R. V., & Doyle, A. (2003). *Risk and Morality [eds.]*. Toronto: University of Toronto Press.

Ewald, F. (2002). The Return of Descartes's Malicious Demon: An Outline of a Philosophy of Precaution. In T. Baker & J. Simon (Eds.), *Embracing Risk: The Changing Culture of Insurance and Responsibility* (pp. 273–302). Chicago: University of Chicago Press.

Fudge, J., & Owens, R. (Eds.). (2006). *Precarious Work, Women, and the New Economy: The Challenge to Legal Norms*. London: Bloomsbury.

Gray, G. C. (2009). The Responsibilization Strategy of Health and Safety: Neoliberalism and the Reconfiguration of Individual Responsibility for Risk. *British Journal of Criminology, 49*(3), 326–342.

Habermas, J. (1976). *Communication and the Evolution of Society* (F. G. Lawrence, Trans.). Boston: Beacon Press.

Habermas, J. (1981). *The Theory of Communicative Action: Book Two: The Critique of Functionalist Reason* (T. McCarthy, Trans.). Cambridge: Polity Press.

Halliday, S., Ilan, J., & Scott, C. (2011). The Public Management of Liability Risks. *Oxford Journal of Legal Studies, 31*(3), 527–550.

Hawkins, K. (2002). *Law as Last Resort: Prosecution Decision-Making in a Regulatory Agency*. Oxford: Oxford University Press.

Hayes, N. (2002). Did Manual Workers Want Industrial Welfare? Canteens, Latrines and Masculinity on British Building Sites 1918–1970. *Journal of Social History, 35*(3), 637–658.

Hood, C. (1991). A Public Management for All Seasons? *Public Administration, 69*(1), 3–19.

Hutter, B. (1997). *Compliance: Regulation and Environment*. Oxford: Clarendon Press.

James, P., Tombs, S., & Whyte, D. (2013). An Independent Review of British Health and Safety Regulation? From Common Sense to Non-sense. *Policy Studies, 34*(1), 36–52.

Johnston, R., & McIvor, A. (2004). Dangerous Work, Hard Men and Broken Bodies: Masculinity in the Clydeside Heavy Industries, c.1930s–1970s. *Labour History Review, 69*(2), 135–152.

Johnston, R., & McIvor, A. (2007). *Miners' Lung. A History of Dust Disease in British Coal Mining*. Aldershot: Ashgate.

Kerr, P. (2005). *Postwar British Politics: From Conflict to Consensus*. London: Routledge.

Kostal, R. W. (1994). *Law and English Railway Capitalism 1825–1875*. Oxford: Oxford University Press.

Leka, S., Jain, A., Hollis, D., Andreou, N., & Zwetsloot, G. (2012). *The Changing Landscape of OSH Regulation in the UK: A Review*. Leicester: IOSH.

Leneman, L. (1993). Workmen's Compensation at the Wemyss Coal Company 1906–1924. *Scottish Economic & Social History, 13*(1), 43–55.

Long, V. (2011). *The Rise and Fall of the Healthy Factory: The Politics of Industrial Health in Britain, 1914–60*. Basingstoke: Palgrave Macmillan.

Mascini, P., Achterberg, P., & Houtman, D. (2013). Neoliberalism and Work-Related Risks: Individual or Collective Responsibilization? *Journal of Risk Research, 16*(10), 1209–1224.

McDermott, V., & Hayes, J. (2018). Risk Shifting and Disorganization in Multi-Tier Contracting Chains: The Implications for Public Safety. *Safety Science, 106*, 263–272.

McIvor, A. (2013). *Working Lives: Work in Britain Since 1945*. Basingstoke: Palgrave Macmillan.

McIvor, A., & Johnston, R. (2002). Voices from the Pits: Health and Safety in Scottish Coal Mining Since 1945. *Scottish Economic and Social History, 22*(2), 111–133.

Melling, J. (2003). *The Risks of Working and the Risks of Not Working: Trade Unions, Employers and Responses to the Risk of Occupational Illness in British Industry, c.1890–1940s* (LSE Discussion Paper 12).

Meyer, J. W., & Rowan, B. (1991). Institutionalized Organizations: Formal Structure and Myth and Ceremony. In W. W. Powell & P. J. DiMaggio (Eds.), *The New Institutionalism in Organizational Analysis* (pp. 41–62). Chicago: University of Chicago Press.

Moran, M. (2003). *The British Regulatory State: High Modernism and Hyper Innovation*. Oxford: Oxford University Press.

Morris, A. (2007). Spiralling or Stabilising? The Compensation Culture and Our Propensity to Claim Damages for Personal Injury. *Modern Law Review, 70*(3), 349–378.

Nichols, T. (1997). *The Sociology of Industrial Injury*. London: Mansell.

Nichols, T., & Armstrong, P. (1973). *Safety or Profit?: Industrial Accidents and the Conventional Wisdom*. London: Falling Wall.

O'Malley, P. (2004). *Risk, Uncertainty and Government*. London: Routledge-Cavendish.

Ogus, A. (1994). *Regulation: Legal Forms and Economic Theory*. Oxford: Clarendon Press.

Pidgeon, N., Walls, J., Weyman, A., & Horlick-Jones, T. (2003). *Perceptions of and Trust in the Health and Safety Executive as a Risk Regulator* [Research Report 100]. Sudbury: HSE Books.

Pouliakas, K., & Theodossiou, I. (2013). The Economics of Health and Safety at Work: An Interdiciplinary Review of the Theory and Policy. *Journal of Economic Surveys, 27*(1), 167–208.

Power, M. (1997). *The Audit Society: Rituals of Verification*. Oxford: Oxford University Press.

Prosser, T. (2006). Regulation and Social Solidarity. *Journal of Law and Society, 33*(3), 364–387.

Quinlan, M., Mayhew, C., & Bohle, P. (2001). The Global Expansion of Precarious Employment, Work Disorganization, and Consequences for Occupational Health: A Review of Recent Research. *International Journal of Health Services, 31*(2), 335–414.

Scharpf, F. W. (1999). *Governing in Europe: Effective and Democratic?* Oxford: Oxford University Press.

Scott, C. (2004). Regulation in the Age of Governance: the Rise of the Post-Regulatory State. In J. Jordana & D. Levi-Faur (Eds.), *The Politics of Regulation: Institutions and Regulatory Reforms for the Age of Governance* (pp. 145–174). Cheltenham: Edward Elgar.

Silbey, S. S. (2009). Taming Prometheus: Talk About Safety Culture. *Annual Review of Sociology, 35*, 341–369.

Simpson, R. C. (1972). Employers' Liability (Compulsory Insurance) Act 1969. *Modern Law Review, 35*(1), 63–68.

Sirrs, C. (2016). Risk, Responsibility and Robens: The Transformation of the British System of Occupational Health and Safety Regulation, 1961–1974. In T. Crook & M. Esbester (Eds.), *Governing Risks in Modern Britain: Danger, Safety and Accidents, c.1800–2000* (pp. 249–276). London: Palgrave Macmillan.

Smismans, S. (2017). Risk Regulation at Risk: Brexit, Trump It, Risk It. *European Journal of Risk Regulation, 8*(1), 33–42.

Suchman, M. (1995). Managing Legitimacy: Strategic and Institutional Approaches. *The Academy of Management Review, 20*(3), 571–610.

Teubner, G. (1987). Juridification: Concepts, Aspects, Limits, Solutions. In R. Baldwin, C. Scott, & C. Hood (Eds.), *A Reader on Regulation* (pp. 389–440). Oxford: Oxford University Press.

Thane, P. (2018). *Divided Kingdom: A History of Britain, 1900 to the Present*. Cambridge: Cambridge University Press.

Tombs, S., & Whyte, D. (2010). A Deadly Consensus: Worker Safety and Regulatory Degradation under New Labour. *British Journal of Criminology, 50*(1), 46–65.

Tombs, S., & Whyte, D. (2013). The Myths and Realities of Deterrence in Workplace Safety Regulation. *British Journal of Criminology, 53*(5), 746–763.

Tucker, E. (1995). And Defeat Goes On: An Assessment of Third-Wave Health and Safety Regulation. In F. Pearce & L. Snider (Eds.), *Corporate Crime: Contemporary Debates* (pp. 245–267). Toronto: University of Toronto Press.

Veljanovski, C. G. (1983). Regulatory Enforcement: An Economic Study of the British Factory Inspectorate. *Law & Policy, 5*(1), 75–96.

Viscusi, W. K. (1983). *Risk by Choice: Regulating Health and Safety in the Workplace*. Cambridge: Harvard University Press.

Walls, J., Pidgeon, N., Weyman, A., & Horlick-Jones, T. (2004). Critical Trust: Understanding Lay Perceptions of Health and Safety Risk Regulation. *Health, Risk, and Society, 6*(2), 133–150.

Wright, M., Beardwell, C., Pennie, D., Smith, R., Norton Doyle, J., & Dimopoulos, E. (2008). *Evidence-based Evaluation of the Scale of Disproportionate Decisions on Risk Assessment and Management* (HSE Research Report 536). London: HSE Books.

Young, H. (2010). Being a Man: Everyday Masculinities. In L. Abrams & C. G. Brown (Eds.), *A History of Everyday Life in Scotland: Twentieth-Century Scotland* (pp. 131–152). Edinburgh: Edinburgh University Press.

Zwetsloot, G., van Scheppingen, A. R., Bos, E. H., Dijkman, A., & Starren, A. (2013). The Core Values that Support Health, Safety, and Well-Being at Work. *Safety and Health at Work, 4*(4), 187–196.

CHAPTER 8

Conclusions

INTRODUCTION

Health and safety has long been a critical issue of public policy in Britain, and remains so today, being embedded and interacted with in workplaces, organisations, and public spaces across the country. But, as discussed in Chapter 1, there is a sense that, especially in the years since 2010, the legitimacy of the *idea* of health and safety, and public acceptance of the laws, policies, regulations, and practices that enact and pursue it, has significantly and dramatically declined. Numerous policy documents and discussion pieces have sought to make sense of this shift, and the negative consequences it might bring, and most have argued that it has been the result of some combination of over-regulation and over-zealousness, a lack of relevance to modern life, institutional 'mission creep', a tendency towards counter-productive precaution, or a public rejection of the core ideas of protective welfarism and social solidarity that health and safety represents (Policy Exchange 2010; Young 2010; Löfstedt 2011; 2011; Brown and Hanlon 2014). Much of this discussion tended to regard health and safety as being 'in crisis', or as facing an existential challenge that is both unprecedented and fundamental, which was set to fatally undermine the whole field of health and safety, and which thus required radical reform to address and resolve. Such concerns were subsequently interwoven with a broader reform agenda that did arguably challenge or undermine the scope, reach, and status of health and safety (Almond 2015; James et al. 2013), thus giving a sense

© The Author(s) 2019   291
P. Almond and M. Esbester, *Health and Safety in Contemporary Britain*, https://doi.org/10.1007/978-3-030-03970-7_8

that the concerns over the damage that a loss of legitimacy might have been borne out in practice after all. As such, the contextual background that prompted the writing of this book was one of great concern over the past, present, and future, of health and safety.

This book has sought to understand and make sense of the changes and tendencies that have been seen in this area, but to do so in a rather different way. By rooting itself in a closely evidenced discussion of public attitudes and perceptions (see Chapter 2), it has remained circumspect and realistic about the nature and extent of the trends that might exist in this area. By taking an historical viewpoint, and seeking to place these recent trends and developments within a much longer time frame (as overviewed in Chapter 3), it has been able to contextualise, problematise, and better understand these changes, and indeed see the broader continuities that underpin them. By balancing consideration of factors that span the institutional, public, individual, and contextual spheres (discussed within Chapters 4–7), it has been able to offer explanations and identify a broader range of relevant issues. And by approaching the question via the analytical lens of the concept of 'legitimacy', it has understood the tendencies identified elsewhere as part of a broader set of socio-political processes, and has broadened its focus beyond the commonly cited trends and explanations identified elsewhere. The result is not, alas, a simple and easily digested explanation of change over time that points to a single causal factor as the source of the recent woes faced by 'health and safety'; instead, we have been able to offer a detailed overview of the diverse, nuanced, and complex realities that attach to the legitimation of a social institution like health and safety. Rather than attempting to present a simple set of observations and take-away points at this stage, then, this concluding chapter will seek to briefly summarise the findings of the preceding chapters, and offer some core observations about the legitimacy of health and safety. It will then draw out and develop some cross-cutting themes that underpin, and help to make sense of, these observations. Finally, it will attempt to identify some ways in which issues of legitimacy can be addressed in the future.

Observations on the Legitimacy of Health and Safety

It is very difficult to talk with precision about the legitimacy of 'health and safety', because the subject matter under discussion is so diverse and varied that the focus of attention has to be both broad, and differentiated between a range of different features. On the one hand, health and safety is a system of laws and regulations, laid down by Parliament and disseminated for duty-holders to implement and observe. It is also a regulatory infrastructure, an umbrella term for a complex set of interrelated institutional actors (at the level of the state, of regulated firms, and of other stakeholder bodies) who give effect to those laws, and who implement processes of management and control that are intended to advance and fulfil them. Each of these is important, but 'health and safety' is much broader than this, also incorporating the front-line actors whose work interacts with those rules, policies, and processes on a day-to-day level; it also encompasses a much wider range of stakeholders, ideas, discussions, and decisions that permeate through wider society, the media, and the consciousness of the general public. Health and safety is a social institution and an idea, not simply a regulatory system, and this breadth and open-endedness has in part rendered this idea more vulnerable to legitimacy challenges than other areas of public or regulatory policy, which are perhaps more shielded by virtue of their distance and alienation from the end-users and diffuse stakeholders whose lives they affect. Health and safety has indeed become a feature of "*every part of our life*", but rather than an "*infection*", as David Cameron saw it,[1] this may be better characterised as a process of 'permeation', a gradual factoring-in or coming to relevance, rather than an undermining or a take over.

In a sense, then, the legitimacy of health and safety is vulnerable *because* of the success that it has had in advancing its agenda of protecting workers and others from risk; were it to be less visible, present, or 'everyday', it would not give rise to the critical and irreverent opinions that public audiences express towards the issue, and would not function as a lightning-rod for political and media opposition. Indeed, the defining feature of the public attitudes that were explored in Chapter 2 was a balancing and leavening of these critical sentiments with a broader, underlying trust in, and acceptance of, the ideas that underpin the pursuit of health and safety, along with a recognition that, while the implementation may not always be perfect, the wider outcome has been that things are generally better now than in the past, and so are less meriting

---

[1] Speech to the Conservative Party Conference, 1/10/2008; reported by *The Guardian*, at http://www.guardian.co.uk/politics/2008/oct/01/davidcameron.toryconference1.

of serious opposition than the alternatives might have been. Similarly, the historical narrative set out in Chapter 3 is, in broad terms, a story of development, expansion, and inclusion: from offering specific protections to general and universal ones; from protecting 'core' workers to including more marginalised populations and the general public; from focusing on key safety risks to incorporating broader issues of health and welfare; and from being a low-profile issue of local concern, to a central tenet of industrial relations policy, and a recognised feature of most modern workplaces. This has not been a straightforward or linear process of progress, and there have been exceptions, failures, and retrenchments along the way. But overall, health and safety has gone from a marginal consideration to a much more central feature of policy and society, and, like any other area of social endeavour, the more it matters to people, the more contested or complex its legitimation becomes.

The constitutional legitimacy of health and safety, relating to the exercise of power by public authorities, has been subject to multiple challenges, as discussed in Chapter 4. The bases of that authority have changed over time, as new institutions and influences have emerged, and new demands for accountability and due process have been voiced. Perhaps a core unifying feature here has been the challenge posed by the development of a regulatory system that is much less state-led, and much more decentred, operating at arm's-length from direct Government oversight and via policy networks and a broader range of empowered actors. These governance trends are relatively well-recognised and established within UK regulation (Moran 2003), but it is instructive to see how they have opened a field like health and safety to contest, as local empowerment gives rise to perceptions of unaccountability, uncertainty, and inconsistency, and as new regulators seek to standardise their approaches, while also seeming to become more managerialised. Modernisation and the norms of 'new' governance have also eroded the certainties of 'old' styles of government, and may not yet have established their own legitimacy so as to counterbalance this.

The significant role played by long-term and wide-ranging processes of change over time was demonstrated in Chapter 5, which focused on ideas of democratic legitimacy, including who has been involved in health and safety and why it has been enacted in particular ways since 1960. Broadly, the changes to the British economy and occupational structure—the decline of older, heavier industries, the rise of newer, office- and service-based roles (sometimes working in ways and locations that confound older notions of the 'workplace'), the inclusion of a more diverse population in the workforce, and the decline in the prestige and influence of trade unions—have posed significant challenges to the legitimacy of health and safety. How people might engage with, and be engaged, by the field has remained a difficult question, yet if people feel that health and safety is something done *to* them rather than *by* them, then 'buy-in' may be problematic. At the same time, the place of health within health and safety has altered significantly—remaining persistently difficult to address but with increasing attention having been focused on the area since approximately the 1980s. This has brought with it a greater scope to involve and affect more people, with both benefits and challenges to its legitimacy. Incorporating more people and a wider remit across British society has, not unproblematically, resulted in a greater acceptance of health and safety—indeed, in some cases, it has contributed to the perception of over-extension.

The functional legitimacy of health and safety, relating to questions of how it is done, and how effective it is, was discussed in Chapter 6. There, it was shown that the redrawing of lines of responsibility around the non-occupational public sphere, around the professionalisation of health and safety, and around the regulatory functions of inspection and enforcement, have opened assessments of functional value to a new degree of scrutiny and contest, something that has involved the internalisation of external policy norms (like those of the New Public Management: Hood 1991). When the efficiency and effectiveness of health and safety are assessed in this way, a range of political value judgements are brought to bear, for the simple reason that the outcomes being sought are not universally agreed on; what constitutes a desirable level of intervention via inspection, or a suitable form of health and safety provision in public spaces? What is an appropriate level of protection to aim for? As elsewhere, these contestations, while manifested within specific areas of change, stem from a broader set of political tensions, and the legitimacy challenges that arise here also mirror those conflicts.

This sense of conflict was also picked up on in Chapter 7, in relation to normative aspects of legitimacy, in particular in relation to ideas of justice and whose interests have been served by health and safety. Notions of a 'harmony of interests' between all parties involved (at least in the workplace), and a perceived high-point of consensual politics during the 1970s, were shown to be more fragile than the dominant public narrative of the time suggested. There is no doubt that this was a period in which the sentiments of cooperation and practice of tripartism were important—but it also masked existing divisions and the seeds of what would become, from the 1980s, an increasingly politicised and contested area. This was suggested to be the result of several factors, including those which are economic in nature or origin, in particular, the rise of more commercial pressures and drivers acting on private organisations and the state regulators. The motives of the safety profession, the insurance industry and the alleged growth of a compensation culture have all cast the 'rightness' of health and safety in doubt. The neo-liberal and New Public Management settlements of the 1980s and 1990s also contributed to moving health and safety more from the realm of enactment in the workplace to the sphere of high politics and public debate, something only increased in the twenty-first century. At the same time, the ability of those on the ground, most directly affected by health and safety breaches, to shape their environments and exercise autonomy was also questioned. Partly this was about power to make changes to improve health and safety outcomes, and partly it was about wider social pressures which shaped opportunities available, particularly on gendered lines. The element of contingency in justice in health and safety has remained strong in contemporary Britain, with a consequent variability in legitimacy on an individual as well as a collective level.

Overall, then, there is no single reason why the legitimacy of health and safety has declined since 1960; indeed, it is quite problematic to try and claim that it has actually lost legitimacy in any consistent or definitive way, as public attitudes remain relatively robust, and the institutions of 'health and safety' continue to endure. Indeed, in many ways, the current position of health and safety is relatively strong, or at least no worse, than it has been in the past. But a convergence of very visible indicators of dissatisfaction (as expressed in political debate, in the media, and around particular commercial approaches to health and safety) has attracted attention to questions of its legitimacy, and further investigation has identified the complexity and nuance of these considerations. Rather than concluding that the legitimacy of health and safety has indeed suffered a fundamental breakdown, it might be more accurate to say that society, regulation, public attitudes, and have all developed in such a way as to create *conditions under which* that legitimacy can be, and has been, challenged. This is a much less certain or definitive conclusion to draw, and predicts a little less than a simple answer might, but it more accurately captures the reality of the situation. Health and safety is contested, and will remain so, but such contest need not be a fundamental challenge to its future, just as it has not necessarily proved fundamental in the past.

A Reflection on Some Core Themes

A number of core themes can be identified running throughout the discussions of the different aspects of legitimacy set out in this book. Each cuts across the different legitimacy headings and substantive areas we have discussed, but together, they help to illustrate some of the drivers and tendencies that have shaped the different legitimacy challenges identified.

A: The Centrality of Participation

It is clear that for health and safety to be seen as legitimate—and if it is to have positive outcomes, in terms of preventing injury, death and ill-health—it is vital that those who might be affected should feel engaged and involved in the area. How that participation has been encouraged—or even if it has been sought at all—has varied significantly over time. This involves both democratic and normative legitimacy aspects. In broad terms, on the current understanding that individuals might bear a right to a safe and healthy life, the British experience of health and safety has broadened out (if incompletely) from the 1960s when traditional occupational issues were supreme and workers were unlikely to be heavily involved in decision-making, to a more extended public health and safety arena, encompassing more people (including those positioned beyond the workplace) and with at least some increase in opportunities for involvement. In broad terms, we might look across the coverage of this book and conclude that the areas where the legitimacy challenges to health and safety seem most pronounced and entrenched (such as around the safety profession, and rule-making within the UK and EU) are all marked by relatively limited levels of participation and engagement. The same is true for worker participation within firms (especially those which are non-unionised or outside traditional industrial sectors), and engagement with previously marginalised populations. This suggests that greater efforts are needed to involve more people, at more levels of society, in health and safety.

What format that involvement now takes remains to be seen—it might be through public consultation or (via recognised unions in the workplace) formal representation. Certainly in this latter area, formal roles are a key means of increasing legitimacy and effecting positive outcomes (James and Walters 2002; Walters 2006). However, as this book has suggested, it is important to consider (and has been so since at least the early 1960s) who has not been represented: capturing the 'public' (a misleading wide-ranging term), minority or traditionally underrepresented groups (like women or minority ethnic groups) or even the vast majority of workers who labour in sectors with less well-established trade unions, has been difficult and continues to pose challenges to the legitimacy of health and safety. In addition, how capable people feel they are of judging health and safety might have changed as it has become subject to professionalisation, and this may have created barriers to participation, legitimacy and acceptance.

B: The Difference Between Public Opinion and Public Attitudes

One of the most important findings to come out of our investigation into public perceptions of health and safety (see Chapter 2), apart from the sense that those perceptions were less negative, on the whole, than might be expected, was the clear division that exists between public *opinions* towards health and safety, on the one hand, and public *attitudes*, on the other. This distinction has been observed in relation to public attitudes more generally, such as towards questions of crime and punishment (Green 2006), and which seem to reflect much deeper cognitive processes (Tversky 2011). This distinction also helps to understand the depth of some of the public legitimacy challenges to health and safety that have been seen in recent years. Opinions are surface-level, reactive, and intuitive in nature; in relation to health and safety, they centred around the idea of 'jobsworths', 'red tape', and the familiar news stories of 'health and safety gone mad'. Public attitudes are more reflective, nuanced, and based on deeper engagement with the issue at hand; in relation to health and safety, they are based on critical trust, and assessment of the expertise and motives of those who 'do' health and safety. What became clear throughout the research we reviewed was that these two tiers of perception are capable of coexisting at the same time, and they perform seemingly distinct functions. Opinions are social, and form part of the shared consciousness of 'common sense' experience (Geertz 1975) which provides a focus for discussion and humour. Attitudes are cognitive, and relate to the underlying preferences that respondents would, on reflection, prefer to commit to as a basis for future action.

While there are many people whose opinions and attitudes align, both being either critical or acceptingly, for many people it was perfectly feasible to complain about health and safety in loose or abstract terms and then espouse its value in more concrete or applied terms. This distinction provides an important insight for future studies of policy-making and the politics of health and safety, and places past trends and events in a new light. It seems relatively clear that much political and policy dialogue around health and safety has taken the 'noise' of public opinion, as reported in the news and reproduced on social media, as reflective of the sentiments of society as a whole, while the more complex 'signal' of public attitudes, which is much less amenable to being reduced to a meme or soundbite, has been relatively overlooked. If we recognise that the public are indeed more fundamentally supportive of health and safety than they are often portrayed as being, then we are led to question not only the policy choices that are often made in their name, but also the more negative prognoses that are held out for the future of health and safety. Indeed, sobering events such as the 2017 fire at Grenfell Tower serve to remind us of the true attitudes that are held towards health and safety—reflecting seriously on the issue uncovers a much more nuanced vision of the world we actually want to live in, and of health and safety's role in bringing it about.

C: The Importance of Understanding Historical Continuities

Looked at only in relation to the recent past, perceptions of the legitimacy of health and safety in 2018 might appear bleak, particularly if we are not sensitive to tuning and fail to discriminate the 'signal' noted above from the 'noise' of public opinions. However, looking in greater depth—which involves an appreciation of the longer term past—provides us with a richer and more revealing perspective. This perspective is not always comfortable, as reflection F, below, further demonstrates: contest and argument have long been a facet of negotiating health and safety. Yet, recognising the challenging parts of the past equips us better to understand the challenging aspects of the present. In that sense, this book has demonstrated some consistent challenges, particularly surrounding the role of the state, who should be involved in setting the health and safety agenda (and how), and the autonomy of those involved and subjected to conditions which might be detrimental to individual health, safety, and well-being. One key continuity worth stressing would be the notion of voluntary solutions to occupational health and safety issues. This had a long pedigree, dating back into the nineteenth century, and in some form has continued to underlie much of the thinking since 1960—articulated (albeit in slightly different terms) in Robens' influential conclusions about self-regulation. Central to this narrative is the understanding of individuals as possessing sufficient rationality and autonomy to be able to safeguard their lives and health. This may have been, and may remain, debateable in practice, but it is possible to see it underlying key aspects of health and safety.

The challenges identified here might have varied in intensity and focus—often around moments of crisis—but they can be seen as deep-seated and ongoing. At the same time, the principle that protecting people from death, injury, and ill-health is a good thing has rarely been seriously contested in public, even if the form of that protection has been debated and great injustices have undoubtedly been perpetrated in the past. This principle of protection from harm (whether enabled by formal state protections or pursued via individual choices) might be seen as a reminder of the essential stability—certainly within the 40 years since the introduction of the Health and Safety at Work Act—of the health and safety system. There has been variation in state policy: as political parties have changed, so too have policies towards health and safety changed—just as they have done for the preceding 150 years or more. Here, a key example would be the movement towards a more consensual policy in the 1970s contrasted with a shift to a much more fractious relationship in the 1980s. In this sense, the period since 1960 has followed a much older pattern, visible since at least the nineteenth century. Understood in this way, we might perhaps have greater confidence in the robustness of the system and the key actors—diffuse as they once again now are—in their ability to allow space for necessary debate about the legitimacy of health and safety.

D: Change as a Local and Contingent Process

Whilst there are some 'higher order' continuities we have identified, one further stable factor is rather more mundane and therefore easily over-looked: the contingency involved in framing health and safety. Often most readily felt at the local or individual level, the ways in which the immediate circumstances surrounding an issue have pushed action or policy in particular directions are important. The wide-ranging coverage of the book has showed how 'pockets' of practice have existed which vary: so, for example, the rise of the offshore industry at the same time as the decline of other heavy industries; the role of the nuclear and rail industries as developers of cutting-edge, localised approaches to health and safety; and the peculiarities of health and safety regulation and practice in oil and gas that produced its own culture (and arguably contributed to the Piper Alpha disaster). Indeed, Chapter 4 identified the uncertainties produced by the open-ended duties of care introduced by the HSWA 1974 as producing questions about legitimacy on the basis of the differential application of ideas and standards in different settings, industries, and places.

Throughout, we have noted the role of economic factors as potentially constraining options available to those 'on the ground', especially in sectors that were existing at the margins of profitability, and it is also clear that the people and personalities involved in health and safety, at all levels, from the grassroots to the Prime Minister, have played a hugely significant role in shaping perceptions, practices, and outcomes. One lesson to learn from this investigation, therefore, is that the direction of development in relation to health and safety, and the progress made towards legitimising it, is very variable and often down to a convergence of factors that are neither intended, nor expected, to have the impact that they do. Steps that are intended to liberalise the system have had the effect of prompting increased anxiety and prescription; institutions like the Health and Safety Commission (HSC) that protect health and safety from political pressure at one time, become political liabilities at another; and solutions that work in one setting have markedly different impacts in another. Health and safety is not a 'one size fits all' topic, and the same is true of debates around its legitimacy.

E: The Dual-Edged Nature of Regulatory Development and Change

One pronounced feature of the legitimacy of health and safety is the emergence of challenges that are actually the result of more positive and constructive steps towards the legitimation of this issue. This relates to the issue of contingency (above) and unintended consequences; sometimes the solution to a problem brings problems of its own. So, as an example, the emergence of a self-regulatory legal system, which allowed duty-holders to determine the direction they would be take in relation to health and safety (as discussed in Chapter 4), proved to be a liberating change for many, but also created an area of uncertainty around the question of compliance (as the steps to be taken were not prescribed). As such, prescription and anxiety around regulation have arguably increased, and have certainly shifted from the level of the state regulator to that of the 'periphery' of safety professionals and regulated firms. The expertise of health and safety professionals, and others, has been painstakingly and effectively built up over time, and provides a powerful reason to invest trust in decision-makers and outcomes; but it also insulates and distances those who possess it from other decision-makers, workers, and the public.

Other examples of this trend abound throughout our time period. Professional and risk-based approaches to regulatory engagement help to ensure efficient implementation, but can be interpreted as signs of self-interest, and even disinterest, on the part of those involved. The public dislike *'jobsworths with clipboards'* running health and safety, but respect serious, focused, expert people taking charge of the issue. Regulating health and safety beyond the workplace provides an effective response to public disasters and key issues of significant public concern (such as school safety), but it also makes 'health and safety' vulnerable to charges of over-reaching and interference. And an increasing focus on health, and health-related issues, has broadened the scope of health and safety, ensuring that it keeps pace with new and emergent risks and trends (from the use of asbestos to the prevalence of workplace stress and computer usage); but it also requires intervention in an ever-wider range of areas, some of which do not 'seem' like legitimate areas of health and safety. Such double-edged outcomes seem unavoidable, and indeed, should arguably not be avoided; tackling these tensions head-on, and finding ways to resolve them openly and directly, is central to what those involve in health and safety should be aiming to do in the future.

F: The Centrality of Contest and Argument to the Past and Future of Health and Safety

As already identified, one of the most important continuities around health and safety has been its long-standing contested nature. The issues, whether or not to act, who should act and how have all been subject to extensive and sometimes bitter debate since the rapid industrialisation of the nineteenth century. Balances of power, particularly between labour and capital, between unions and state, and between public and private, have waxed and waned, but rarely has there been a point at which it was possible to say everyone agreed. Even at the supposed high point of consensual formation around health and safety—the 1970s and the creation of the HSC/E as a means of allowing debate in a contained fashion—disagreements remained and were publicly aired (notably over the initial lack of formal role for safety representatives, and the scope of the Health and Safety Executive's (HSE) coverage around agriculture, rail and offshore). Contest was never far from the surface, and arguably, the efforts to use the tripartism of the HSC as a means of diffusing tension ultimately helped to fuel the renewed politicisation of health and safety in the 1980s and the subsequent spill-over into the post-2000 era of anti-health and safety media coverage. For all of the downsides that argument has had in terms of legitimacy, it has been a significant and necessary form of expression in order to produce negotiated outcomes that paid at least lip-service to ideas of democratic involvement. On these grounds, then, it is likely that contest will remain just as important to health and safety in the future: if nothing else, demonstrating that those who argue are invested in the area. Questioning received wisdoms about health and safety is essential to engagement, so provided the tenor of discussion can be civil, debate could act as a very useful means of building legitimacy in the future.

WHERE SHOULD WE GO FROM HERE?

As was established in Chapter 1, the idea of engagement between the past, present and future, including concepts such as the 'useable past' (Divall 2010), is one which has been gaining traction amongst historians and others over the last decade. Health and safety is an area of particular possibility in this move, something demonstrated by the research project underlying this book's research, which was funded by the Institution of Occupational Safety and Health (IOSH), a professional and practice-focused body. Continuing to explore the potential insights offered by academic research for a broad base of stakeholders would appear to be an important methodological way forward, as well as the precise implications and potential for improving the legitimacy of health and safety as an area of practice. It is hoped that different actors and stakeholders might gain insights from the relevant sections of this book that relate to their areas of activity and engagement, as well as from the picture generated by the whole. Holding in mind the interconnected, complex, and contingent nature of health and safety, as well as the possibilities of unintended consequences that it presents, is useful for all involved in generating policy ideas and approaches; simple answers are rarely simple, and resonate further and deeper than might be expected.

How might legitimacy be addressed in the future? Clearly, the contingent nature of debate around health and safety makes overly prescriptive suggestion unhelpful. Instead, benefitting from the insight from the past suggested by the underlying research, it might be more useful to pose a number of questions which make explicit some of the issues encountered but often not clearly articulated. This might help to structure thinking and encourage all with an interest in this area to develop useful approaches which contribute to an improved profile for health and safety. Chief among these would be questions about who should be involved and how? The evidence from the past suggests that responses to these questions will be predicated by pre-existing assumptions about the merits and points of the project of health and safety, including whether a solidaristic or individualistic belief is paramount. How can values, ideas and practices be communicated? This might be as much about the day-to-day aspects as the high politics, so the ability to reach and engage in a dialogue with a broad range of constituents is significant. What constitutes a constructive or useful discussion about health and safety, and how far can disagreement be tolerated? Open debate, and efforts to get a message across to others, ought to be viewed as constructive opportunities rather than threats; the contested nature of health and safety is an engine for change as well as a source of difficulty, and so we might look to emphasise the value and potential of open, inclusive, rational debate about the role and purpose of health and safety, as a means of positively influencing its legitimacy in future.

Finally, perhaps the single biggest political issue at the point of writing: what impact will 'Brexit' have on health and safety? We provide some limited discussion of this issue in Chapter 4, but at the current time, this has to be treated as a prospective, rather than a manifested, issue, so many questions remain unanswerable, for now. In general terms, this has been the subject of occasional fevered speculation about the effect on workers' rights and welfare reforms, particularly those introduced via EU law and adopted in UK statute. More usually, however, concern has been voiced in abstract soundbites about 'taking back control', and sustained attention has not been directed at health and safety. Will that last? If not, how will the current legal framework fare? What will replace it, and how will the deep-seated ideological differences, and legitimacy debates, that relate to Brexit as a whole, filter down to the issue of health and safety? As Smismans (2017) has perceptively argued, many of the core concerns that have animated the Brexit process (mistrust of experts, concerns over accountability and excessive interference, the influence of 'fake news') have also been present within the recent history of health and safety. As such, we should expect that the legitimacy debates explored in this book will be reopened as the search for a post-Brexit settlement goes on. Those involved in health and safety should be ready to engage with this process, and to do so on the basis of a clear understanding of the complexity and nuance of their field. There are no simple answers, and this is territory where the past offers little consolation or useful direction. The best that can be said is that these remain, then, uncertain times—as has been the case throughout the past 60 years.

References

Primary Sources

Cameron, D. (2008, October 1). Quoted in *Guardian*. Online. http://www.guardian.co.uk/politics/2008/oct/01/davidcameron.toryconference1.

Department of Work and Pensions (DWP). (2011). *Good Health and Safety, Good for Everyone*. Online. http://www.dwp.gov.uk/docs/good-health-and-safety.pdf.

Löfstedt, R. (2011). *Reclaiming Health and Safety for All: An Independent Review of Health and Safety Legislation* (The Löfstedt Review). London: Crown.

Policy Exchange. (2010). *Health and Safety: Reducing the Burden*. London: Policy Exchange.

Young, L. (2010). *Common Sense, Common Safety* (The Young Review). London: Crown.

Secondary Sources

Almond, P. (2015). Revolution Blues: The Reconstruction of Health and Safety Law as 'Common-Sense' Regulation. *Journal of Law and Society, 42*(2), 202–229.

Brown, T., & Hanlon, M. (2014). *In the Interests of Safety: The Absurd Rules that Blight Our Lives and How We Can Change Them*. London: Sphere.

Divall, C. (2010). Mobilizing the History of Technology. *Technology and Culture, 51*(4), 938–960.

Geertz, C. (1975). Common Sense as a Cultural System. *The Antioch Review, 33*(1), 5–26.

Green, D. (2006). Public Opinion Versus Public Judgement About Crime: Correcting the 'Comedy of Errors'. *British Journal of Criminology, 46*(1), 131–155.

Hood, C. (1991). A Public Management for All Seasons? *Public Administration, 69*(1), 3–19.

James, P., & Walters, D. (2002). Worker Representation in Health and Safety: Options for Regulatory Reform. *Industrial Relations Journal, 33*(2), 141–156.

James, P., Tombs, S., & Whyte, D. (2013). An Independent Review of British Health and Safety Regulation? From Common Sense to Non-sense. *Policy Studies, 34*(1), 36–52.

Moran, M. (2003). *The British Regulatory State: High Modernism and Hyper Innovation*. Oxford: Oxford University Press.

Smismans, S. (2017). Risk Regulation at Risk: Brexit, Trump It, Risk It. *European Journal of Risk Regulation, 8*(1), 33–42.

Tversky, A. (2011). *Thinking, Fast and Slow*. London: Penguin.

Walters, D. (2006). One Step Forward, Two Steps Back: Worker Representation and Health and Safety in the United Kingdom. *International Journal of Health Services, 36*(1), 87–111.

APPENDIX: LIST OF PROJECT INTERVIEWEES AND SELECTED KEY AFFILIATIONS

REGULATORS

1. Bill Callaghan (TUC Chief Economist 1979–1997; HSC Chair, 1999–2007).
2. John Rimington (Department of Employment/Civil Service 1970–1977; Manpower Services Commission 1977–1981; HSE 1981–1984, Director-General, 1984–1995).
3. Jim Hammer (HM Factory Inspectorate 1953–1974; HSE, Chief Inspector of Factories 1975–1984, Deputy Director-General 1984–1989; NEBOSH, Chair 1992–1995).
4. David Eves (HM Factory Inspectorate 1964–1974; HSE 1975–1985, Chief Inspector of Factories 1985–1988/1992–2002, Deputy Director-General 1989–2002).
5. Tim Carter (British Petroleum, Medical Officer/Chief Medical Officer 1974–1983; HSE, Director of Medical Services 1983–1992, Director of Field Operations Division 1992–1996; Maritime and Coastguard Agency, Chief Medical Advisor 1999–2014).
6. Jenny Bacon (Ministry of Labour/Department of Employment 1967–1992; HSE Deputy Director-General 1992–1995, Director-General 1995–2000).
7. Senior HSE Source [Anonymous] (Regulatory roles; HSE role).

© The Editor(s) (if applicable) and The Author(s) 2019
P. Almond and M. Esbester, *Health and Safety in Contemporary Britain*, https://doi.org/10.1007/978-3-030-03970-7

8. Kevin Myers (HSE 1976–2000, Chief Inspector of Construction 2000–2005, Chief Inspector, Hazardous Installations 2005–2008, Deputy Chief Executive 2008–present/Acting Chief Executive 2013–2014).

9. Helen Leiser (TUC Economics Department 1968–1973; Department of Employment 1974–1983; Cabinet Office 1983–1985; HSE 1986–1993; Department of Trade and Industry, 1995–2003).

10. David Morris (HM Factory Inspectorate 1971–1974; HSE 1975–2002; HM Railway Inspectorate/Office of the Rail Regulator, Deputy Chief Inspector of Railways 2002–2010).

11. Steve Sumner (Local Government Association; HSE Local Authority Unit).

12. Sir Frank Davies (Alcan Aluminium (UK) Ltd. 1967–1977, Director 1977–1983; HSC Chairman, 1993–1999).

SAFETY PROFESSIONS/REPRESENTATIVE BODIES

13. Neal Stone (HSE 1992–2005, Policy advisor to HSC Chair 2005–2007; British Safety Council, Head of Policy 2008–2014, Acting Chief Executive 2015–present).

14. Roger Bibbings (Amalgamated Union of Electrical Workers 1972–1977; TUC, Healthand Safety advisor 1977–1994; RoSPA, Occupational Health and Safety advisor 1994–2014).

15. Richard Jones (Rolls-Royce 1998–2000; IOSH 2000–2011, Head of Policy and Public Affairs 2011–present).

16. Lawrence Waterman (Olympic Delivery Authority, Head of Health and Safety 2008–2013; Battersea Power station development, Director of Health and Safety 2013–present).

17. Sayeed Khan (Large manufacturer 1992–2002; Engineering Employers Federation, Chief Medical Advisor 2002–present; HSC Commissioner/HSE Board member 2005–2011).

18. Stan Barnes (IOSH President 1978–1979).

19. Peter Kinselley (Abbey National 2003–2005; Linklaters, Health and Safety Manager 2005–2013; Mondelez, European Safety Manager, 2014–present).

20. Sheila Pantry (United Steel Companies Ltd.; National Coal Board; HSE, Head of Information Services 1977–1993; Sheila Pantry Associates, Founder and Managing Director 1993–present).

TRADE UNIONS

21. Sarah Lyons (National Union of Teachers, Pay, Conditions and Bargaining Department, Health and Safety Policy Lead).
22. Kim Sunley (GMB; Royal College of Nursing, Senior Employment Relations Adviser, UK Policy Lead on Health and Safety 2007–present).
23. Paul Clyndes (ASLEF 1999–2004; RMT Health and Safety Officer 2004–present).
24. Dan Shears (HSE 2000–2008; GMB National Health and Safety Officer 2008–present).
25. Ian Tasker (Clydesdale Bank 1977–2001; Scottish TUC, Assistant Secretary 2001–present).
26. Peter Jacques (TUC Social Insurance and Industrial Welfare Department, 1968–1993; HSC Commissioner, 1974–1995).

BUSINESSES/EMPLOYERS

27. Janet Asherson (HSE 1976–1986; CBI, Head of Environment, Health, and Safety 1989–2008; International Organisation of Employers 2008–present).
28. Graeme Collinson (ICI 1978–1992; AstraZeneca 1992–2003, Head of Health and Safety 2003–2008; Centrica, Group Director—Health, Safety, Environment and Security 2008–2013; HM Government Office of the Chief Scientific Advisor 2013–2014; Siemens 2015–present).
29. Rex Symons (ICI 1958–1961; British Drug Houses Ltd. 1961–1981, Managing Director 1981–1989; Merck Deputy Chairman 1989–1991; CBI Health and Safety Group Member 1991–2002; HSC Commissioner 1989–2002; Better Regulation Taskforce member 2002–2006).
30. John Armitt (John Laing Construction 1966–1988, Joint Managing Director/Chairman 1988–1993; Union Railways, Chief Executive 1993–1997; Costain Group, Chief Executive 1997–2001; Railtrack Chief Executive 2001–2002; Network Rail Chief Executive 2002–2007; Berkeley Group, Director 2007–present; Olympic Delivery Authority, Chairman 2007–2014; National Express, Director 2013–present).

31. Chris Marchese (Central Electricity Generating Board/Nuclear Electric Plc. 1971–1993; Magnox Group 1994–2000, Head of Operations 2000–2004; British Nuclear Group Ltd., Chief Operating Officer, 2004–2007).

32. David Maidment (British Rail 1960–1990, Head of Safety Policy 1990–1996).

33. Rear Admiral Paul Thomas (Royal Navy 1963–1990, Director, Nuclear Propulsion 1990–1994, MoD Chief Strategic Systems Executive 1995–1998; BNFL, Director, Environment, Health, Safety and Quality 2001–2008; Rail Safety Standards Board, Chairman 2008–present).

34. Ruth Warden (NHS 1992–2010; NHS Employers, Assistant Director, Employment Services 2010–present).

35. Jim Wilkes (Zurich Insurance, Senior Underwriter) and Huw Andrews (Liberty Mutual Insurance 1993–1998; Engineering Employers Federation 1998–2000; Citex 2000–2003; Zurich Insurance, Head of Casualty Practice 2003–present).

POLICY-MAKERS AND OTHERS

36. Ragnar Löfstedt (University of Surrey 1994–2002; King's College London, Professor of Risk Management 2002–present).

37. Andrew Hale (Institute of Industrial Psychology 1966–1972; University of Aston 1972–1984; Delft University, Professor of Safety Science 1984–2009, Emeritus 2009–present).

38. Baron McKenzie of Luton (PriceWaterhouse 1980–1998; Luton Borough Council, Leader 1999–2003; Lord in Waiting (Government Whip) 2005–2007; Parliamentary Under-Secretary of State, Department of Work and Pensions 2007–2010).

39. Frank Doran (MP, Lab., Aberdeen S. 1985–1992, Aberdeen Central 1997–2005, Aberdeen N. 2005–2015; Shadow Energy Minister 1988–1992; Minster, Department of Energy 1997–2000).

40. Senior Government Source [Anonymous] (Government Department role, Parliamentary role).

INDEX

© The Editor(s) (if applicable) and The Author(s) 2019
P. Almond and M. Esbester, *Health and Safety in Contemporary Britain*, https://doi.org/10.1007/978-3-030-03970-7

The manufacturer's authorised representative in the EU is Springer
Nature Customer Service Centre GmbH, Europaplatz 3, 69115 Heidelberg,
Germany. If you have any concerns regarding our products, please
contact ProductSafety@springernature.com

Printed and bound by CPI Group (UK) Ltd, Croydon, CR0 4YY

23/04/2026

02095601-0009